OTHER A TO Z GUI
THE SCARECROW I

162. *The A to Z of Existentialism* by Stephen Michelman, 2010.
163. *The A to Z of Hegelian Philosophy* by John W. Burbidge, 2010.
164. *The A to Z of the Holiness Movement* by William Kostlevy, 2010.
165. *The A to Z of Hume's Philosophy* by Kenneth R. Merrill, 2010.
166. *The A to Z of Husserl's Philosophy* by John J. Drummond, 2010.
167. *The A to Z of Kant and Kantianism* by Helmut Holzhey and Vilem Mudroch, 2010.
168. *The A to Z of Leibniz's Philosophy* by Stuart Brown and N. J. Fox, 2010.
169. *The A to Z of Logic* by Harry J. Gensler, 2010.
170. *The A to Z of Medieval Philosophy and Theology* by Stephen F. Brown and Juan Carlos Flores, 2010.
171. *The A to Z of Nietzscheanism* by Carol Diethe, 2010.
172. *The A to Z of the Non-Aligned Movement and Third World* by Guy Arnold, 2010.
173. *The A to Z of Shamanism* by Graham Harvey and Robert J. Wallis, 2010.
174. *The A to Z of Organized Labor* by James C. Docherty, 2010.
175. *The A to Z of the Orthodox Church* by Michael Prokurat, Michael D. Peterson, and Alexander Golitzin, 2010.
176. *The A to Z of Prophets in Islam and Judaism* by Scott B. Noegel and Brannon M. Wheeler, 2010.
177. *The A to Z of Schopenhauer's Philosophy* by David E. Cartwright, 2010.
178. *The A to Z of Wittgenstein's Philosophy* by Duncan Richter, 2010.
179. *The A to Z of Hong Kong Cinema* by Lisa Odham Stokes, 2010.
180. *The A to Z of Japanese Traditional Theatre* by Samuel L. Leiter, 2010.
181. *The A to Z of Lesbian Literature* by Meredith Miller, 2010.
182. *The A to Z of Chinese Theater* by Tan Ye, 2010.
183. *The A to Z of German Cinema* by Robert C. Reimer and Carol J. Reimer, 2010.
184. *The A to Z of German Theater* by William Grange, 2010.
185. *The A to Z of Irish Cinema* by Roderick Flynn and Patrick Brereton, 2010.
186. *The A to Z of Modern Chinese Literature* by Li-hua Ying, 2010.
187. *The A to Z of Modern Japanese Literature and Theater* by J. Scott Miller, 2010.
188. *The A to Z of Old-Time Radio* by Robert C. Reinehr and Jon D. Swartz, 2010.
189. *The A to Z of Polish Cinema* by Marek Haltof, 2010.
190. *The A to Z of Postwar German Literature* by William Grange, 2010.
191. *The A to Z of Russian and Soviet Cinema* by Peter Rollberg, 2010.
192. *The A to Z of Russian Theater* by Laurence Senelick, 2010.
193. *The A to Z of Sacred Music* by Joseph P. Swain, 2010.
194. *The A to Z of Animation and Cartoons* by Nichola Dobson, 2010.
195. *The A to Z of Afghan Wars, Revolutions, and Insurgencies* by Ludwig W. Adamec, 2010.
196. *The A to Z of Ancient Egyptian Warfare* by Robert G. Morkot, 2010.
197. *The A to Z of the British and Irish Civil Wars 1637–1660* by Martyn Bennett, 2010.
198. *The A to Z of the Chinese Civil War* by Edwin Pak-wah Leung, 2010.
199. *The A to Z of Ancient Greek Warfare* by Iain Spence, 2010.
200. *The A to Z of the Anglo–Boer War* by Fransjohan Pretorius, 2010.
201. *The A to Z of the Crimean War* by Guy Arnold, 2010.
202. *The A to Z of the Zulu Wars* by John Laband, 2010.
203. *The A to Z of the Wars of the French Revolution* by Steven T. Ross, 2010.
204. *The A to Z of the Hong Kong SAR and the Macao SAR* by Ming K. Chan and Shiu-hing Lo, 2010.

205. *The A to Z of Australia* by James C. Docherty, 2010.
206. *The A to Z of Burma (Myanmar)* by Donald M. Seekins, 2010.
207. *The A to Z of the Gulf Arab States* by Malcolm C. Peck, 2010.
208. *The A to Z of India* by Surjit Mansingh, 2010.
209. *The A to Z of Iran* by John H. Lorentz, 2010.
210. *The A to Z of Israel* by Bernard Reich and David H. Goldberg, 2010.
211. *The A to Z of Laos* by Martin Stuart-Fox, 2010.
212. *The A to Z of Malaysia* by Ooi Keat Gin, 2010.
213. *The A to Z of Modern China (1800–1949)* by James Z. Gao, 2010.
214. *The A to Z of the Philippines* by Artemio R. Guillermo and May Kyi Win, 2010.
215. *The A to Z of Taiwan (Republic of China)* by John F. Copper, 2010.
216. *The A to Z of the People's Republic of China* by Lawrence R. Sullivan, 2010.
217. *The A to Z of Vietnam* by Bruce M. Lockhart and William J. Duiker, 2010.
218. *The A to Z of Bosnia and Herzegovina* by Ante Cuvalo, 2010.
219. *The A to Z of Modern Greece* by Dimitris Keridis, 2010.
220. *The A to Z of Austria* by Paula Sutter Fichtner, 2010.
221. *The A to Z of Belarus* by Vitali Silitski and Jan Zaprudnik, 2010.
222. *The A to Z of Belgium* by Robert Stallaerts, 2010.
223. *The A to Z of Bulgaria* by Raymond Detrez, 2010.
224. *The A to Z of Contemporary Germany* by Derek Lewis with Ulrike Zitzlsperger, 2010.
225. *The A to Z of the Contemporary United Kingdom* by Kenneth J. Panton and Keith A. Cowlard, 2010.
226. *The A to Z of Denmark* by Alastair H. Thomas, 2010.
227. *The A to Z of France* by Gino Raymond, 2010.
228. *The A to Z of Georgia* by Alexander Mikaberidze, 2010.
229. *The A to Z of Iceland* by Gudmundur Halfdanarson, 2010.
230. *The A to Z of Latvia* by Andrejs Plakans, 2010.
231. *The A to Z of Modern Italy* by Mark F. Gilbert and K. Robert Nilsson, 2010.
232. *The A to Z of Moldova* by Andrei Brezianu and Vlad Spânu, 2010.
233. *The A to Z of the Netherlands* by Joop W. Koopmans and Arend H. Huussen Jr., 2010.
234. *The A to Z of Norway* by Jan Sjåvik, 2010.
235. *The A to Z of the Republic of Macedonia* by Dimitar Bechev, 2010.
236. *The A to Z of Slovakia* by Stanislav J. Kirschbaum, 2010.
237. *The A to Z of Slovenia* by Leopoldina Plut-Pregelj and Carole Rogel, 2010.
238. *The A to Z of Spain* by Angel Smith, 2010.
239. *The A to Z of Sweden* by Irene Scobbie, 2010.
240. *The A to Z of Turkey* by Metin Heper and Nur Bilge Criss, 2010.
241. *The A to Z of Ukraine* by Zenon E. Kohut, Bohdan Y. Nebesio, and Myroslav Yurkevich, 2010.
242. *The A to Z of Mexico* by Marvin Alisky, 2010.
243. *The A to Z of U.S. Diplomacy from World War I through World War II* by Martin Folly and Niall Palmer, 2010.
244. *The A to Z of Spanish Cinema* by Alberto Mira, 2010.
245. *The A to Z of the Reformation and Counter-Reformation* by Michael Mullett, 2010.

The A to Z of Logic

Harry J. Gensler

The A to Z Guide Series, No. 169

THE SCARECROW PRESS, INC.
Lanham • Toronto • Plymouth, UK
2010

Published by Scarecrow Press, Inc.
A wholly owned subsidary of
The Rowman & Littlefield Publishing Group, Inc.
4501 Forbes Boulevard, Suite 200, Lanham, Maryland 20706
http://www.scarecrowpress.com

Estover Road, Plymouth PL6 7PY, United Kingdom

British Library Cataloguing in Publication Information Available

Library of Congress Cataloging-in-Publication Data

The hardback version of this book was cataloged by the Library of Congress as
follows:

Gensler, Harry J., 1945–
   Historical dictionary of logic / Harry J. Gensler.
      p. cm. — (Historical dictionaries of religions, philosophies, and
movements ; no. 65)
   Includes bibliographical references.
   1. Logic—Dictionaries. 2. Logic—History—Dictionaries. I. Title.
II. Series.
BC9.G36   2006
160.9—dc22   2005030716

ISBN 978-0-8108-7596-8 (pbk. : alk. paper)

♾™ The paper used in this publication meets the minimum requirements of
American National Standard for Information Sciences—Permanence of Paper
for Printed Library Materials, ANSI/NISO Z39.48-1992.
Printed in the United States of America

Contents

Editor's Foreword

It would seem logical that logic should be sufficiently clear and straight-forward that it would not take a book of 350 pages to sum it up, let alone thousands of books, written by thousands of philosophers over more than two thousand years, to clarify it. But it quickly appears that logic is indeed very complex, given the many different contexts in which it must apply and the many different tasks it must solve, not the least of which is discarding fallacies that constantly crop up. Moreover, despite its long and venerable history, logic is still a very lively field with rapidly multiplying aspects and specializations, all of which have been constantly evolving since Aristotle's time, with the most dramatic changes and innovations coming in the more recent past. Indeed, as traditional Aristotelian logic was challenged by classical symbolic logic, the latter in turn is now being challenged by various non-classical logics.

So, *Historical Dictionary of Logic* has an awful lot to do, and 352 entries in the dictionary section is barely enough for the task. These entries must present or explain, among many other things, the most important logicians from Aristotle to the present day, the most significant theories and issues arising out of Western philosophy and other traditions, the more common fallacies and argument forms, many of the basic concepts and terms and techniques, as well as logic's application to many different sectors, not only the rest of philosophy but also science, and more specifically mathematics and computers. To help put things in their place, the discipline itself is first anchored in a brief introduction, and the initially gradual and then more hectic evolution is followed in the chronology, while the note on notation helps us follow the symbols used in the book. Then the bibliography, carefully structured and organized, refers as broadly as possible to some of the most relevant literature on different aspects of logic as well as literature on or by different philosophers.

This volume was written by Harry J. Gensler, S.J., one of whose princi-pal interests is logic, which he has taught for over three decades, most recently at John Carroll University, where he is professor of philosophy. He has written many articles and several books on the topic, including an

Introduction to Logic. In addition, he has written related software, namely, the *LogiCola* computer program for learning logic. With this sort of background, it is not surprising that Dr. Gensler goes out of his way to make logic accessible to broader circles. This is not done by simplifying things that are inherently complex but by presenting them in a clear and straightforward manner, by providing additional explanations for more complicated aspects, and above all by adding useful examples and injecting a bit of comic relief for those, especially students, who may find the subject dry or intimidating. The book does not produce instant insight but does serve as an excellent tool both for beginners and those more advanced.

John Woronoff
Series Editor

Preface

"All men are mortal; Socrates is a man; therefore Socrates is mortal." This is an argument—and a *valid* one, since the conclusion follows from the premises. Logic is about arguments. The task of logic is to analyze arguments and to help us see which ones are valid. Like most areas of study, logic has grown more complicated over the years.

This book is an encyclopedia of logic. It introduces the central concepts of the field in a series of brief, nontechnical "dictionary entry" articles. These deal with topics like logic's history, its various branches, its specialized vocabulary, its controversies, and its relationships to other disciplines. While the book emphasizes deductive logic, it also has entries on areas like inductive logic, fallacies, and definitions—and on key concepts from epistemology, mathematics, and set theory that are apt to arise in discussions about logic. Following the series guidelines, *Historical Dictionary of Logic* tries to be useful for specialists (especially logicians in areas outside their subspecialties) but understandable to students and other beginners; so I avoid topics or explanations that are so technical that only math majors would understand.

The major part of this book is the dictionary section, with 352 entries. While these are arranged alphabetically, there is also an organization based on content. Four very general entries start with "**logic:**" and serve mainly to point to more specific entries (like "**propositional logic**"); these in turn often point to related topics (like "**negation**," "**conditionals**," "**truth tables**," and "**proofs**"). So we have here a hierarchy of topics. Here are the four "**logic:**" entries:

- **logic: deductive systems** points to entries like **propositional logic, modal logic, deontic logic, temporal logic, set theory, many-valued logic, mereology,** and **paraconsistent logic.**
- **logic: history of** is about historical periods and figures and includes entries like **medieval logic, Buddhist logic, twentieth-century logic, Aristotle, Ockham, Boole, Frege,** and **Quine.**
- **logic: and other areas** relates logic in an interdisciplinary way to

other areas and includes entries like **biology, computers, ethics, gender, God,** and **psychology.**

- **logic: miscellaneous** is about everything else (including technical terms) and includes entries like **abstract entities, algorithm, *ad hominem*, inductive logic, informal/formal logic, liar paradox, metalogic, philosophy of logic,** and **software for learning logic.**

The entries vary in length from a sentence or two to several pages.

The front of the book has three important parts:

- A short notation section gives the main logical symbols that I use in the book, along with alternative symbols that others sometimes use.
- A chronology lists some of the main events in the history of logic.
- An introduction tries to give an overall view of logic, the big picture, in order to give a broader context for the dictionary entries.

The back of the book has a substantial bibliography on related readings.

Writing a brief *Summa Logica* like this is difficult, because logic is so vast and complicated. It is humanly impossible to succeed completely; I ask your forgiveness for omissions or inaccuracies. I hope that you, the reader, find this book useful—either to get a quick fix on some particular topic or to explore in a broader way the fascinating field of logic.

To sum up: logic studies valid reasoning—and this book tries to provide a clear, basic introduction to a very broad range of logical topics.

Notation

The first symbolism is what is used in this dictionary; the other symbolisms are some of the more common ones used by other logicians. There are also variations among logicians about dropping parentheses (most tend to drop outer parentheses) and about what sorts of letters to use (for example, some use just "p" to "s" for statements). The articles on **Boole**, **Carroll**, **De Morgan**, **Frege**, **Peano**, **Polish notation**, **scope ambiguity**, and **set theory** say more about notation.

ENGLISH	SYMBOLS
Not-A	$\sim$A, -A, $\neg$A, Na
Both A and B	(A $\cdot$ B), (A & B), (A $\wedge$ B), Kab
Either A or B	(A $\vee$ B), Aab
If A then B	(A $\supset$ B), (A $\rightarrow$ B), Cab
A if and only if B	(A $\equiv$ B), (A $\leftrightarrow$ B), Eab
For all x, x is F	(x)Fx, $\forall$xFx, ΠxFx
For some x, x is F	($\exists$x)Fx, $\exists$xFx, ΣxFx
It is necessary that A	$\Box$A, LA
It is possible that A	$\Diamond$A, MA

Chronology

1,000,000 BC Fred Flintstone is the first to put a valid argument into words: "If I hide, then the saber-tooth tigers will not eat me. But I will hide. So they will not eat me." Logical reasoning is born!

600–400 BC Ancient Greek interest in reasoning grows: Pythagoras and others prove geometric theorems, Heraclites and Parmenides dispute the law of non-contradiction, Zeno and others reason about philosophical paradoxes, and Sophists debate about politics.

470–347 BC Plato and Socrates reason about philosophical questions and theorize about first principles, the *a priori*, and the forms.

384–322 BC Aristotle presents us with the first known correct formulation of a logical principle, the first use of logical form, and the first axiomatic system. He does syllogistic logic (about "all," "no," and "some") and modal logic (about "necessary" and "possible"); he also defends the laws of non-contradiction and excluded middle. Formal logic is born!

371–286 BC Theophrastus, a disciple of Aristotle, refines his teacher's ideas about syllogisms and modal logic; he also develops a logic of conditionals, patterning these after his teacher's syllogisms.

315–285 BC Diodorus Cronus, who admires Socrates more than Aristotle, studies conditionals, modal logic, and the liar paradox. Tradition has it that he died of shame when he could not solve a logical problem.

c. 300 BC Euclid writes his geometry classic, *The Elements*, which becomes a model of logical reasoning.

c. 300 BC In India, Tibet, and China, an Eastern logic tradition begins that is parallel to and independent of the Western tradition. This Eastern tradition includes mostly Buddhists, but also some Hindus and others. It studies patterns of inference, grammar, debate rules, fallacies, reference, contradictions, and universals (with realists debating nominalists).

c. 300 BC Philo of Megara proposes that "If A then B" is true, if and only if it is not the case that A is true and B is false; others dispute this material-implication analysis.

279–206 BC The Stoic Chrysippus systematizes the logic of "and," "or," and "if-then." At first, the Stoic propositional logic is seen as a rival to Aristotle's syllogistic logic. Later the two merge into the "traditional logic" that dominates in higher education until the 20th century.

c. 50 AD The Nyāya (logical) school of Hindu philosophy begins in India. It teaches four paths to knowledge: experience, inference, analogy, and testimony.

129–216 AD Galen, a medical doctor, writes the *Institutio Logica*, which is the oldest surviving logic textbook of the ancient world. Other logicians of his time include Lucius Apuleius and Alexander of Aphrodisias.

480–524 AD Boethius writes works on logic (including commentaries) and translates Aristotle's logic into Latin. Most of these translated works are lost until the 12th century, except for *Categories* and *On Interpretation*, which are the main sources of logic for the next few centuries; the tradition based on these was later called the *logica vetus* (old logic). Except for Boethius, European Christians did little creative work on logic until the 11th century.

c. 800–1200 The Arab world dominates in logic; some Arab logicians are Christian, but most are Moslem. Baghdad and Moorish Spain are centers of logic studies. Figures include Al-Farabi, Al-Tayyib, Avicenna, and Averroës; they work on topics like modal logic, conditionals, universals, predication, existence, and categorical statements.

1033–1142 Anselm and Peter Abelard start a revival of logic in Christian Europe. Later, four more of Aristotle's logical works become available: *Prior Analytics*, *Posterior Analytics*, *Topics*, and *Sophistical Refutations*; the tradition based on these is called the *logica nova* (new logic).

1215–1277 Peter of Spain and William of Sherwood develop the new logic. Syllogisms are systematized further; the *Barbara, Celarent* verse becomes a tool for teaching logic at the new medieval universities. Peter's *Summulae Logicales* is a popular logic textbook for several centuries. There is much interest in universals and in the different ways a term can signify. Many of the Latin logic terms are still used today, like *modus ponens, a priori*, and *de re/de dicto*.

1235–1315   Ramón Llull tries to reduce knowledge to its 54 simplest concepts, from which other ideas can be constructed using a special logical notation. Based on these ideas, and in order to demonstrate beliefs about God, he built a "logic machine"—perhaps the first ever built.

c. 1285–1349   William of Ockham writes an influential *Summa Logicae* ("Summary of Logic"). He gives nominalist analyses of terms like "man" and "humanity," supports the "Ockham's razor" simplicity criterion, and develops principles of modal logic.

1300–1358   Jean Buridan formulates the standard rules for valid syllogisms. He also raises questions about choice, claiming that a dog placed at an equal distance between two bowls of food would choose one randomly. His disciple Albert of Saxony (c. 1316–1390) carries on his work, especially through his popular *Perutilis Logica* textbook.

1400–1800   Various logicians write textbooks in the Aristotelian tradition: Paul of Venice, Philipp Melanchthon, Peter Ramus, the Port-Royal Jansenists Antoine Arnauld and Pierre Nicole, and the Jesuit Gerolamo Saccheri (who first investigates non-Euclidian geometry). Immanuel Kant states that Aristotle not only was the first to conceive of logic but had also substantially brought the subject to its completion.

1646–1716   Gottfried Wilhelm Leibniz hints at changes to come. He proposes the idea of an artificial language that would reduce reasoning to arithmetic calculation. He creates a logical notation like that of George Boole; but this was not published until 1903. He also invents an "arithmetic machine" that can multiply and divide—and he coinvents calculus.

1700–1900   Various logicians contribute to syllogistic logic. Leonhard Euler diagrams "all A is B" by putting an A-circle inside a larger B-circle. William Hamilton adds statements like "all A is all B" and "all A is some B": the first affirms while the second denies that all B is A. Lewis Carroll entertains us with his silly syllogisms and his points about logic in *Alice in Wonderland*. John Venn gives us diagrams for testing syllogisms. And Christian Ladd-Franklin introduces antilogisms.

1716–1790   Gottfried Ploucquet and Johann Heinrich Lambert experiment with ways to symbolize logical notions.

1770–1883   Georg Wilhelm Friedrich Hegel proposes that a better logic would accept contradictions in nature as keys to understanding how thought historically evolves; one view provokes its opposite, and then the two come together in a higher synthesis. Karl Marx follows Hegel in seeing contradictions in the world as real; he applies this idea to political struggles. Most logicians complain that this "dialectical logic" confuses conflicting properties in the world (like hot/cold, or capitalist/proletariat) with logical self-contradictions (like the same object being both white and, in the same sense and time and respect, also non-white).

1829   Nikolay Lobachevsky creates non-Euclidean geometry. Geometry is increasingly looked at as a logical derivation from axioms and not as a description of the properties of physical space.

1832   Charles Babbage designs his steam-driven Analytical Engine, a computer that would (if actually built) accept punch-card programs. British scientists in 1991 built a version of Babbage's earlier Difference Engine, and it was accurate to 31 digits.

1843   Augusta Ada King, the Countess of Lovelace and the first computer programmer, writes the definitive essay explaining Babbage's Analytical Engine and showing how it relates matter to abstract logical processes.

1843   John Stuart Mill's *System of Logic* discusses deductive and inductive reasoning and informal fallacies. Mill sees logical and arithmetic principles as empirical; they seem necessary only because of our weak minds' inability to imagine their falsity. He emphasizes inductive logic and develops "Mill's methods" for arriving at causal explanations.

1847   Augustus De Morgan's *Formal Logic* introduces new (but clumsy) ways to symbolize logical relationships. The "De Morgan laws" are named after him.

1847 George Boole's *Mathematical Analysis of Logic* mathematically analyzes the logic of sets and propositions in his "Boolean algebra." Something akin to mathematical calculation can be used to check the correctness of inferences; further refinements are added by William Stanley Jevons, Ernst Schröder, and Charles Sanders Peirce. Mathematical logic is born!

1874 Georg Cantor's *Transfinite Numbers* creates modern set theory and the theory of transfinite numbers. Cantor proves that the set of real numbers is larger than the set of positive integers; his *continuum hypothesis* proposes that no sets are intermediate in size between these two.

1879 Gottlob Frege's *Begriffsschrift* ("Concept Writing") presents the first modern quantifiers and the first complete formalization of propositional and quantificational logic. Classical symbolic logic is born!

1884 Gottlob Frege's *Foundations of Arithmetic* argues for the logistic thesis that arithmetic can be reduced to logic.

1880s Charles Sanders Peirce, independently of Frege but later, invents quantifiers, using "$\prod x$" for the universal and "$\sum x$" for the existential. He also works out a logic of relations and invents logic gates, which simulate truth-table functions electrically.

1889 Giuseppe Peano's *Principles of Arithmetic* proposes the five Dedekind axioms as the basis for arithmetic and develops an improved logical notation; he also introduces the term "mathematical logic."

1899 David Hilbert's *Foundations of Geometry* proves that Euclidian geometry is consistent if arithmetic is. Later he proves that Nikolay Lobachevsky's non-Euclidian geometry is consistent if arithmetic is.

1900 At a mathematics conference in Paris, David Hilbert proposes 23 research problems for mathematicians and logicians of the 20th century —including proving the consistency of arithmetic.

1900 At a philosophy conference in Paris, the young Bertrand Russell, in an event he calls the turning point of his intellectual life, meets Peano.

1900 Edmund Husserl's *Logical Investigations* rejects the then-popular view that logic describes the psychology of human thinking. He defended this psychologism earlier but gives it up after a critique by Frege.

1900 Vacuum tubes are invented, which several decades later will be used in computers to execute logical functions.

1902 Bertrand Russell on 16 June sends a letter to Frege that uses "Russell's paradox" to show the inconsistency of Frege's system. Frege, who is crushed, on 22 June writes back accepting the criticism.

1903 Gottlob Frege's *Basic Laws of Arithmetic*, volume 2, continues the task of showing how to reduce arithmetic to logic. It explains Russell's criticisms and attempts to patch things up; but now several of the central proofs no longer work and the contradiction is still derivable.

1905 Bertrand Russell's "On Denoting" gives an analysis of definite descriptions, like "the king of France."

1908 Ernst Zermelo gives a reconstruction of set theory that avoids Russell's paradox and is adequate for mathematics. Later refinements are added by Thoralf Albert Skolem and Abraham Adolf Fraenkel.

1910–1913 Bertrand Russell and Alfred North Whitehead's three-volume *Principia Mathematica* tries to show in detail how arithmetic can be reduced to logic. Using an improved notation adapted from Peano, and a theory of types designed to avoid Russell's paradox, it brings together ideas from Frege, Peano, Cantor, Boole, and Schröder. *Principia Mathematica* becomes the definitive formulation of classical symbolic logic.

1912 Luitzen Brouwer rejects the reduction of arithmetic to logic; he takes an "intuitionist" approach to arithmetic, seeing it as arising from a pure Kantian intuition of time. He also proposes changing propositional logic to block some proofs by Cantor and others about infinite sets; he rejects the classic law of excluded middle, "A or not-A," since we may not be in a position to show the truth of either alternative.

1913 Henry Sheffer shows that the "Sheffer stroke" connective "(P | Q)," which means "not both A and B," suffices to define "and," "or," "if-then," "not," and all the other truth-functional connectives.

1915–1920   The Löwenheim-Skolem theorem shows that any consistent set of first-order quantificational formulas all come out true under some interpretation in the realm of natural numbers. So first-order quantificational logic cannot characterize non-denumerable infinities.

1917   Jan Łukasiewicz suggests a three-valued logic, shows the redundancy of one of the *Principia* axioms, formalizes Aristotle's syllogistic, and introduces Polish notation.

1918   Bertrand Russell's *Philosophy of Logical Atomism* uses the new logic to construct a metaphysics.

1918   Clarence Irving Lewis's *Survey of Symbolic Logic* treats the new logic from an historical perspective.

1921   Ludwig Wittgenstein invents truth tables. His *Tractatus Logico-Philosophicus* sees truth-functional logic as giving the logical structure of language and reality and as providing the basis for all *a priori* thinking.

1921   Emil Post (independently of Wittgenstein) invents truth tables, using "+" and "–" for "true" and "false"; he uses them to show the consistency, soundness, and completeness of the axiomatization of propositional logic in *Principia Mathematica*. He also (independently of Łukasiewicz) invents a three-valued logic.

1927–1939   Stanisław Leśniewski presents his logic in three related systems: protothetic, ontology, and mereology. The unconventional nature of his approach tended to discourage its wider influence.

1928   David Hilbert and Wilhelm Ackermann's *Principles of Mathematical Logic* proves the consistency of first-order quantificational logic.

1928   Rudolph Carnap's *Logical Structure of the World* uses logic to try to construct the world from sense experiences.

1930   Kurt Gödel proves the completeness of first-order quantificational logic.

1931   Kurt Gödel shows ("Gödel's theorem") that arithmetic cannot be completely formalized in any axiomatic system and that any proof of the consistency of arithmetic requires principles stronger (and thus less certain) than arithmetic itself.

1931   Alonzo Church shows ("Church's theorem") that the problem of determining validity in quantificational logic cannot be reduced to an algorithm (a finite mechanical procedure). His related "Church's thesis" proposes that the intuitive idea of an algorithm matches a more precise mathematical definition in terms of "recursive functions."

1932   Clarence Irving Lewis and Cooper Harold Langford's *Symbolic Logic* introduces various systems (S1 to S9) of modal logic.

1933   Alfred Tarski's "Concept of Truth in Formalized Languages" develops a semantic theory to capture the notion of *truth* in formal systems. Semantic methods in logic become increasingly popular.

1934   Gerhard Gentzen develops an *inferential* proof system, which he calls "natural deduction." The inferential approach to proofs is much easier to use than the earlier *axiomatic* approach.

1936   *The Journal of Symbolic Logic* begins publication, with Alonzo Church and Cooper Harold Langford as coeditors.

1937   Alan Turing's "On Computable Numbers" shows that Church's thesis is equivalent to proposing that the intuitive idea of an algorithm (a finite mechanical procedure) matches what could be done by a simple kind of computer called a "Turing machine."

1938   Claude Shannon, independently of Peirce's work in the 1880s, suggests simulating logical truth-functions electrically. This idea leads to the modern electronic computer.

1938   Kurt Gödel proves that the axioms of set theory are consistent with Cantor's continuum hypothesis.

1945   Logicians John von Neumann and Arthur Burks collaborate with mathematicians and engineers to create the ENIAC—the first large-scale electronic computer. Its purpose was to compute missile trajectories.

1946   Logicians John von Neumann, Arthur Burks, and Herman Goldstine write a paper outlining the "von Neumann" computer architecture, which puts programs and data in the same memory space. Most computers today use this architecture; this was an improvement over the ENIAC, which had to be rewired physically to change programs.

1946   Ruth Barcan Marcus develops a system of quantified modal logic. She debates with Willard Van Orman Quine, who opposes modal logic.

1947   The transistor is invented, which is crucial for modern computers and their logic devices.

1949   Leon Henkin presents a simpler proof for the completeness of first-order quantificational logic.

1950   Peter Strawson's "On Referring" criticizes Russell's analysis of definite descriptions.

1950   Alan Turing proposes the "Turing test," which reformulates "Can computers think?" into "Can computers be developed whose typed answers to questions cannot be distinguished from those of humans?"

1951   Georg von Wright introduces a system of deontic logic, modeled after standard modal logic; "ought" mirrors "necessary" and "all right" mirrors "possible."

1951   Willard Van Orman Quine's "Two Dogmas of Empiricism" attacks the analytic/synthetic distinction and the reduction of empirical statements to sense data.

1952   John Rosser and Atwell Turquette's *Many-Valued Logics* defend the idea of having more truth values than just "true" and "false."

1953   Ludwig Wittgenstein's *Philosophical Investigations* promotes a less logical way of doing philosophy.

1953   Willard Van Orman Quine's *From a Logical Point of View* promotes a more logical way of doing philosophy. He suggests "To be is to be the value of a bound variable" as a way to characterize a person's ontology. He also suggests that we avoid Russell's paradox by having the abstraction axiom "$(\exists x)(y)(y \in x \equiv \ldots y \ldots)$" require stratification.

1954   Nelson Goodman's *Fact, Fiction, and Forecast* raises problems about our ability to formulate adequate principles of inductive logic.

1955   The first computer programming languages appear, modeled after formal systems of logic and mathematics.

1955   Alan Newell, J. C. Shaw, and H. A. Simon write an artificial intelligence program, LOGIC THEORIST, to prove theorems in propositional logic using *Principia* axioms. Over the next few decades, practical instructional software would emerge to help students learn logic.

1956   Paul Grice and Peter Strawson's "In Defense of a Dogma" defends the analytic/synthetic distinction from Quine's criticisms.

1957   Arthur Prior's *Time and Modality* gives a system of temporal logic.

1958   Integrated circuits are invented, making possible small-cheap-fast logic devices for computers.

1959   Hao Wang writes a computer program that generates proofs for all first-order theorems of Russell and Whitehead's *Principia Mathematica*.

1959   Saul Kripke gives a possible-worlds semantics for modal logic. The 19-year-old Kripke bases this essay on earlier work that he did as a high school student. Modal logic and possible worlds grow in popularity.

1961   Joseph Bocheński writes *A History of Formal Logic*.

1962   William and Martha Kneale's *The Development of Logic* describes how logic developed from the ancient Greeks to the present.

1962   John Austin's *How to Do Things with Words* gives an "ordinary language" approach to philosophy of language.

1962   Jaakko Hintikka's *Knowledge and Belief* applies the techniques of formal logic to knowing and believing.

1963   Paul Cohen shows that the axioms of set theory are consistent with the denial of Cantor's continuum hypothesis.

1964 Jean Piaget and Bärbel Inhelder's *The Early Growth of Logic in the Child* uses psychological tools to study how logical thinking develops.

1967 Hilary Putnam's "Mathematics without Foundations" suggests that we need not worry about the foundations of mathematics.

1967 Peter Heath's article on "Nothing" in the *Encyclopedia of Philosophy* breaks new ground for philosophical humor.

1970 Willard Van Orman Quine's *Philosophy of Logic* gives a concise introduction to his philosophical views.

1972 Saul Kripke's *Naming and Necessity* gives a more mature statement of his views on modal logic, semantics, and metaphysics. This work is sometimes said to have made metaphysics respectable again.

1972 Richard and Val Routley work out a paraconsistent semantics in which "A, not-A ∴ B" fails, because sometimes "A" and "not-A" are both true while "B" is false.

1973 David Lewis's *Counterfactuals* investigates conditionals like "If A had been true, then B would have been true."

1973 Michael Dummett's *Frege: Philosophy of Language* begins a series of important studies on Frege, who is increasingly recognized as the central figure in the development of modern logic.

1974 Matthew Lipman writes *Harry Stottlemeier's Discovery*, a logic textbook for fifth-grade children.

1974 Richard Montague's *Formal Philosophy* explains "Montague grammar" for natural and formal languages.

1974 Alvin Plantinga's *The Nature of Necessity* gives a vigorous defense of modality, modal logic, and Aristotelian essentialism—with applications to philosophy of religion.

1975 The first personal computer, the Altair 8800, sells in kit form for $439 through *Popular Electronics* magazine. It uses Intel's 8080 chip, the first general-purpose microprocessor; the name "Altair" allegedly comes from a Star Trek episode. Soon most of us would own computers that can do useful tasks by executing quickly a huge number of logical functions.

1975–1992   Alan Ross Anderson, Nuel Belnap, and Michael Dunn's *Entailment: The Logic of Relevance and Necessity* (2 vols.) argues for a non-classical approach to the logic of conditionals and entailment.

1977   Michael Dummett's *Elements of Intuitionism* endorses anti-realism and an intuitionist rejection of the law of excluded middle.

1979   David Kaplan's "On the Logic of Demonstratives" investigates terms like "I" and "this."

1979   Douglas Hofstadter's *Gödel, Escher, Bach* becomes a best-seller.

1983   Jon Barwise and John Perry's *Situations and Attitudes* is critical of standard logical approaches and argues for the importance of situations and attitudes in understanding meaning.

1985   Raymond Smullyan's *To Mock a Mockingbird and Other Logic Puzzles* entertains us with logic.

1986   David Lewis's *On the Plurality of Worlds* argues for an intriguing modal realism that says that every possible world is equally real. It is sheer provincialism, he claims, to hold that our world is any more real than any other possible world that we might imagine.

1986   Johan van Benthem's *Essays in Logical Semantics* explores the border between logic and linguistics.

1989   George Boolos and Richard Jeffrey's *Computability and Logic* gives an unusually lucid introduction to technical topics connecting logic, metalogic, and computers.

1989   Solomon Feferman's *The Number Systems* analyzes the construction of numbers from sets.

1993   Graham Priest and Timothy Smiley debate "Can Contradictions Be True?" Priest argues that a statement can be both true and false and that we need a paraconsistent logic that can contain self-contradictions without leading to intellectual chaos. Smiley defends the law of non-contradiction.

1994 The World Wide Web begins, and it becomes increasingly more important year by year. Information (like text, pictures, sound, or video) can now be sent quickly between places—by being translated into 1s and 0s, transferred electrically, and then reassembled by computers that quickly execute a huge number of logical functions.

1994 George Boolos's brief "Gödel's Second Incompleteness Theorem Explained in Words of One Syllable" gives a witty introduction to a difficult subject.

1996 George Edward Hughes and M. J. Cresswell bring out their *New Introduction to Modal Logic*, marking the growth of modal logic since their earlier introduction to the subject.

1996 Susan Haack's *Deviant Logic, Fuzzy Logic* defends classical logic.

2000 Colin McGinn's *Logical Properties: Identity, Existence, Predication, Necessity, Truth* argues, against many contemporary philosophers, for a logical realism that takes logical properties and facts to be part of an independent realm, not reducible to the physical or the mental.

2001 Ian Hacking's *Introduction to Probability and Inductive Logic* shows the importance and usefulness of inductive reasoning for a wide range of human activities.

2001 Graham Priest's *Introduction to Non-Classical Logic* argues that classical logic is inadequate and sketches non-classical alternatives.

Introduction

Logic is about reasoning—about going from premises to a conclusion. Logic was born in ancient Greece and was radically transformed a hundred years ago; it is transforming our lives through the computer revolution. Since medieval times, logic has been important in higher education; people who reason well are apt to do better in many areas of life.

This introduction will give an overall view of logic—the big picture— in order to provide a broader context for the dictionary entries. We will start with the importance of logic, its key concepts, and its earlier history. Then we will sketch the current state of logic, which involves both classical and non-classical deductive logic and also logic in a broader sense.

The Importance of Logic

Logic is important because reasoning is important. Philosophers like to reason about big questions, about things like free will and determinism, the existence of God, and the nature of morality. But we all reason about things, about big questions and little questions, about sports, politics, religion, science, the environment, and every other area of life. People study logic mostly to improve their understanding of reasoning and to become better at it.

Logic is important for philosophy and is generally required of philosophy majors. Philosophy can be defined as *reasoning about the ultimate questions of life*. Philosophers ask questions like "Why accept or reject free will?" "Can one prove or disprove God's existence?" and "How can one justify a moral belief?" If we do not know any logic, we will have only a vague grasp of such issues; and we will lack the tools needed to understand and evaluate philosophical reasoning. Logic can help us express reasoning clearly, determine whether a conclusion follows from the premises, and focus on key premises to defend or criticize. Logic, while not itself resolving the big issues, gives us intellectual tools to understand them better and reason about them in a clearer way.

Logic is often highly recommended for students interested in other disciplines that involve close reasoning and argumentation—like law, mathematics, and computer programming. It can be useful in any area where good reasoning is important, including politics, religion, medicine, science, journalism, business, and education. The ability to reason and argue in a clear way is a big advantage in most areas of life.

Logic is important also because of how it connects with other disciplines. The ability to reason is a central aspect of human life. Aristotle defined humans as "rational animals," animals who can reason. We are animals, yes, but we can reason—and that changes everything. Since reasoning is so important to human existence, every area of study has to come to grips with it, in its own way; and thus every area of study connects, in its own way, with logic.

Let me give examples. To understand humans adequately from a biological perspective, we need to know, in a more detailed way than we do at present, how reasoning connects with brain processes and how we evolved into reasoning beings. To understand humans from an anthropological perspective, we need to grasp the role of reasoning in human culture and the extent to which different cultures agree or disagree on the basics of reasoning. To understand humans from a psychological perspective, we need to appreciate how reasoning develops in the individual and how it can be stimulated or retarded by various social and educational conditions. And to build artificial-intelligence computer programs, we need to spell out the logical steps involved in solving various problems. One could go on and on, relating logic and reasoning to other areas, like religion or ethics or physics or rhetoric or whatever (*see* **logic: and other areas**). But the general point should be clear: part of the importance of logic is how it connects with other areas of study.

So the importance of logic lies in two areas: (1) in helping us to understand reasoning and develop analytical skills and (2) in connecting with other disciplines. In addition, many people find logic intrinsically interesting and stimulating; for many students, logic is a fun subject.

Some Key Logical Concepts

I begin my basic logic course with a multiple-choice pretest (which is on the Web at http://www.jcu.edu/philosophy/gensler/logic.htm). The test has

10 problems, each giving information (premises) and asking what conclusion necessarily follows. The problems are easy, but most students get almost half of them wrong. Here are two problems like the ones on the pretest, with the correct answers boxed:

If your car stalls, then you are late for class. You aren't late for class.	If your car stalls, then you are late for class. Your car didn't stall.
Therefore:	Therefore:
(a) Your car stalled.	(a) You are late for class.
(b) Your car didn't stall.	(b) You aren't late for class.
(c) You are late for class.	(c) Your car stalled.
(d) None of these follows.	(d) None of these follows.

Boxed answers: left problem — (b) Your car didn't stall. Right problem — (d) None of these follows.

While almost everyone gets the first problem right, many wrongly pick (b) for the second. Here "You aren't late for class" does not necessary follow, since you may be late for another reason; maybe you were stopped for speeding or delayed by road construction. Most people, once they grasp this point, will see that (b) for the second problem is wrong.

Untrained logical intuitions are often unreliable. But logical intuitions can be improved. The study of logic develops our logical intuitions and teaches special techniques for testing arguments.

An *argument*, in the sense used in logic, is a set of statements consisting of premises and a conclusion; normally the premises give evidence for the conclusion. Arguments put into words a possible act of *reasoning*. Here is an example of a valid argument ("∴" is for *therefore*):

Valid Argument → If your car stalls, then you are late for class. You aren't late for class. ∴ Your car didn't stall.

An argument is *valid* if it would be contradictory (impossible) to have the premises all true and conclusion false. In calling an argument *valid*, we are not saying whether the premises are true. We are just saying that the conclusion *follows from* the premises—that if the premises were all true, then the conclusion would also have to be true. In saying this, we assume that there is no shift in the meaning or reference of the terms.

Our argument is valid because of its *logical form*—its arrangement of logical notions (like "if-then" and "not") and content phrases (like "your car stalls"). We can display an argument's form by using words or symbols for logical notions and letters for content phrases:

Valid *Argument* ➜	If your car stalls, then you are late for class. You aren't late for class. ∴ Your car didn't stall.	If A then B Not-B ∴ Not-A

Our argument is valid because its form is correct. If we take another argument of the same form, but substituting other ideas for "A" and "B," then this second argument also will be valid. Here is an example:

Valid *Argument* ➜	If you are in France, then you are in Europe. You are not in Europe. ∴ You are not in France.	If A then B Not-B ∴ Not-A

Logic studies forms of reasoning. The content can deal with anything—backpacking, mathematics, cooking, physics, ethics, or whatever. So logic gives us tools of reasoning that can be applied to any subject.

In our invalid example, the second premise denies the *first* part of the IF-THEN instead of the *second*; this change makes all the difference:

Invalid *Argument* ➜	If your car stalls, then you are late for class. You car didn't stall. ∴ Your aren't late for class.	If A then B Not-A ∴ Not-B

You might be late for some other reason—just as, in the following similar argument, you might be in Europe because you are in Italy:

Invalid *Argument* ➜	If you are in France, then you are in Europe. You are not in France. ∴ You are not in Europe.	If A then B Not-A ∴ Not-B

We begin the study of logic with unclear logical intuitions; but we eventually reach clear logical principles on which there is relatively little disagreement. We do this by focusing on the formal structure of arguments and searching for clear formal principles that lead to intuitively correct results in concrete cases without leading to any clear absurdities.

We said a "valid" argument is one in which the conclusion follows from the premises. To prove something, we need more than this; we also need premises that are true. If we have both—true premises from which the conclusion logically follows—then the conclusion has to be true. In this case, the argument is said to be "sound"; so a *sound* argument is a valid one with true premises.

Correspondingly, there are two ways to attack the soundness of an argument: we could try to show that the conclusion does not follow from the premises, or we could try to show that one or more of the premises are false or doubtful. If we succeed in doing either, then the argument fails.

Logic deals with the validity question: does the conclusion follow from the premises? The main branches of logic deal with validity but differ in what logical terms they focus on. Propositional logic focuses on "if-then," "and," "or," and "not." Quantificational logic includes these but adds "all" and "some." Modal logic adds "necessary" and "possible." Other branches of logic focus on other logical terms.

"Logic" can be defined as "the analysis and appraisal of arguments." This emphasizes that logic is an activity, something we do; when we do logic, we try to clarify reasoning and separate good from bad reasoning. Other common definitions include "the study of valid reasoning," "the science of the principles governing the validity of inference," and "the art and science of right reasoning." While these vary in emphasis, they are more similar than different.

Actually, the meaning of the term "logic" is more complicated than this, because the term can be used in a narrow and a broad sense. *Logic in the narrow sense* is the study of deductive reasoning, which is about what logically follows from what. *Logic in the broad sense* includes also various other studies that relate to the analysis and appraisal of arguments; we will talk more about logic in this broad sense later, starting with our discussion of Aristotle in the next section.

The Earlier History of Logic

Back in ancient Greece, a remarkable man by the name of Aristotle invented logic. Of course, there was reasoning before Aristotle. Plato had taught Aristotle how to reason about the big questions of life and how to counter the verbal trickery of the Sophists. And geometry, which involves tight reasoning, was popular in ancient Greece. But no one before Aristotle had spelled out the *principles* involved in reasoning. Aristotle did this, and thus he invented the subject that we have come to call "logic."

Aristotle created syllogistic logic, which studies arguments like these:

Valid *Argument*	➜	All humans are mortal. All Greeks are humans. ∴ All Greeks are mortal.	all H is M all G is H ∴ all G is M

The argument is *valid* because of its formal structure, as given by the formulation on the right. Any argument having this same structure will be valid. If we change the structure a little, we may get an invalid argument, like this one:

Invalid *Argument*	➜	All Romans are mortal. All Greeks are mortal. ∴ All Greeks are Romans.	all R is M all G is M ∴ all G is R

This is invalid because its form is wrong. Aristotle's logic investigated logical forms using "all," "no," and "some." It gave ways to determine whether arguments using these are valid or invalid.

Logic in the narrow sense is about what follows from what; Aristotle's syllogistic logic is an example. Aristotle pursued also other topics relevant to the appraisal of arguments, like how to define terms, how to generalize from experience, practical hints about investigating and debating, and errors of thinking to be avoided. These are about *logic in the broad sense*.

Aristotle proposed two basic principles of thought. The *law of non-contradiction* states that a statement A and its contradictory not-A cannot both be true at the same time, unless we shift what we mean by their terms. And the *law of excluded middle* states that either a statement A is true or its contradictory not-A is true.

After Aristotle, the Stoics developed another form of logic, one that used "if-then," "and," and "or." One form of inference that they used, called *modus tollens*, is crucial in philosophical reasoning, where we often argue against a view by showing that it implies something that is false:

	If your view is correct, then	
Valid ➔	such and such is true.	If C then S
Argument	Such and such is false.	Not-S
	∴ Your view is not correct.	∴ Not-C

For the next two thousand years or so, the logic of Aristotle, with additions from the Stoics, would rule in the Western world.

Roughly at the same time, another tradition of logic rose up, independently, in the Eastern world—in India, China, and Tibet. This is sometimes called *Buddhist logic*, even though it was pursued also by Hindus and others. It studied many of the same topics that were important in the West, including patterns of inference, fallacies, and issues about language.

In the West, Aristotelian logic was continued and developed in the Middle Ages—first by Moslems and later by Christians. Reasoning was important for the medieval philosophers; the writings of St. Thomas Aquinas, for example, are filled with arguments. Medieval logicians like Peter of Spain and William of Ockham wrote logic textbooks, which were used in the new universities. Jean Buridan developed an improved set of rules for testing the validity of syllogisms, a set that is still in common use today; it has the rule, for example, that a valid syllogism cannot have two negative premises. And logicians investigated many other issues about logic, including whether abstract entities (like human nature) exist independently of our minds; this is the problem of universals, and it is still discussed today. One sign of the continuing influence of medieval logic is the persistence of Latin terms (like *modus tollens*) even today.

Aristotelian logic dominated until the end of the 19th century. There were some improvements; for example, John Venn developed a useful graphical method to test syllogisms using "Venn diagrams." But most logicians would have agreed with the philosopher Immanuel Kant, who claimed that Aristotle had invented and perfected logic; nothing else of fundamental importance could be learned or added, although we might improve teaching techniques. Kant would have been shocked to learn about the revolution in logic that came about a hundred years later.

The philosopher Gottfried Leibniz, who was one of the coinventors of calculus, had some insight into future developments. He proposed the idea of a symbolic language that would reduce reasoning to something like arithmetic calculation. If controversies arose, the parties involved could take up their pencils and say, "Let us calculate."

Leibniz's idea was partly realized when the 19th-century mathematician George Boole presented his Boolean algebra. Boole used letters for sets; so

"M" might stand for the set of mortals and "H" for the set of humans. Putting two letters together represents the *intersection* of the sets; so "HM" represents the set of those who are *both human and mortal*. Then we can symbolize "All humans are mortal" as "H = HM," which says that the set of humans = the set of those who are both human and mortal. We can then symbolize a syllogism as a series of equations:

Valid		All humans are mortal.	H = HM
Argument	➜	All Greeks are humans.	G = GH
		∴ All Greeks are mortal.	∴ G = GM

We can derive the conclusion "G = GM" from the premises by substituting equals for equals; see if you can do it without looking at the Boole entry.

Boole, who is considered the father of mathematical logic, thought that logic belonged with mathematicians instead of philosophers. But the effect of his work was to make logic a subject shared by both groups, each getting the slice of the action that fits it better. While Boole's work was important, a far greater revolution in logic was soon to come.

Classical Symbolic Logic

Gottlob Frege's *Begriffsschrift* ("Concept Writing") in 1879 radically transformed logic. This slim book of 88 pages gave the first complete formalization of propositional and quantificational logic, which today are the standard systems of symbolic logic. Frege figured out how to construct a system that lets us combine quantifier words ("all," "some," "no") and propositional connectives ("and," "or," "if-then," "not") in every conceivable way. Thus the gap between the logic of Aristotle and that of the Stoics was finally overcome in a higher synthesis.

Frege saw logic as part of his larger project of showing that arithmetic is reducible to logic; he wanted to show that all the basic concepts of arithmetic (like numbers and addition) are definable in purely logical terms and that all the truths of arithmetic are provable using axioms of logic. Frege's work used set theory. In particular, it used a harmless looking axiom that said that every condition on x (like "x is a cat") picks out a set containing just those elements that satisfy that condition. For example, the condition "x is a cat" picks out the set of cats. But consider that some sets are members of themselves (the set of abstract objects is itself an abstract object) while other sets are not (the set of cats is not itself a cat). By Frege's axiom, "x is not a member of itself" picks out the set containing just those things that are not members of themselves. Call this "set R." So any x is a member of R, if and only if x is not a member of x:

For all x, x ∈ R if and only if x ∉ x.

But, Bertrand Russell asked, what about set R itself? By the principle just given, R is a member of R, if and only if R is not a member of R:

R ∈ R if and only if R ∉ R.

So is R a member of itself? If it is, then it is not—and if it is not, then it is; either way we get a contradiction. Since this contradiction, called "Russell's paradox," was provable in Frege's system, that system was flawed.

Russell, in collaboration with Alfred North Whitehead, worked to develop logic and set theory in a way that avoided the contradiction. He also came up with a more intuitive symbolism, which was based on the work of Giuseppe Peano. The result was the massive *Principia Mathematica*, which was published in 1910–1913. *Principia* had a huge influence and became the standard formulation of the new logic.

Russell, like Frege, sought to reduce arithmetic to logic, or perhaps to logic plus set theory. Kurt Gödel later showed that this project was impossible. Gödel's 1931 paper established "Gödel's theorem," which states that arithmetic cannot be reduced to any formal system; no consistent set of axioms and inference rules would suffice to prove all arithmetic truths. This is perhaps the most striking and surprising result of 20th-century logic.

However, the two central logical systems used by Frege and Russell, namely propositional and quantificational logic, *can* be reduced to tight formal systems. These are the branches of logic that one is apt to study today in an introductory logic course. Propositional logic is the more basic system, with quantificational logic building on it.

Propositional logic studies arguments whose validity depends on "if-then," "and," "or," "not," and similar notions. One way of presenting it uses capital letters for true-or-false statements, parentheses for grouping, and five special symbols for logical notions:

$$\begin{array}{rcl}
{\sim}P & = & \text{Not-P} \\
(P \cdot Q) & = & \text{Both P and Q} \\
(P \vee Q) & = & \text{Either P or Q} \\
(P \supset Q) & = & \text{If P then Q} \\
(P \equiv Q) & = & \text{P if and only if Q}
\end{array}$$

The example we gave to illustrate Stoic logic would translate this way:

Valid
Argument →

If your view is correct, then
such and such is true.
Such and such is false.
∴ Your view is not correct.

$(C \supset S)$
${\sim}S$
∴ ${\sim}C$

The new logic was very powerful, since it allowed one to test arguments using complex combinations, like "If either A or not B, then if C then not both D and E," which translates as "$((A \vee {\sim}B) \supset (C \supset {\sim}(D \cdot E)))$."

Truth tables, which display graphically how the formulas work, were invented by Ludwig Wittgenstein and Emil Post. One of the simplest truth tables, the one for "P and Q," looks like this (where T = true, F = false):

P Q	$(P \cdot Q)$
F F	F
F T	F
T F	F
T T	T

An AND is true if
both parts are true—
and is false if one or
both parts are false.

The left side gives all possible truth combinations for "P" and "Q": maybe both are false, or just the second is true, or just the first is true, or both are true. "$(P \cdot Q)$" is false in the first three cases and true in the third. "$(P \cdot Q)$" claims that both parts are true.

Truth tables can be used to test whether propositional arguments are valid; but they get tedious for complex arguments. Other methods of establishing validity include proofs (where we derive some formulas from others by certain specified inference rules) and truth trees (a more graphical test). People learn such things in introductory logic courses.

Quantificational logic builds on propositional logic but adds ways to symbolize "all" and "some," terms for variables and specific individuals (like "x" the variable and like "a" for "Albert"), and terms for predicates and relations (like "Ix" for "x is Italian" and like "Lxy" for "x loves y").

The example we gave for Aristotle's logic would translate this way:

Valid *Argument* ➔	All humans are mortal.	(x)(Hx ⊃ Mx)
	All Greeks are humans.	(x)(Gx ⊃ Hx)
	∴ All Greeks are mortal.	∴ (x)(Gx ⊃ Mx)

The symbolization of "All humans are mortal" is tricky:

(x)(Hx ⊃ Mx)	=	All humans are mortal.
	=	For all x, *if* x is human *then* x is mortal.

This uses the universal quantifier "(x)," which claims that the formula that comes after it is true in *all* cases. Similarly, the existential quantifier "(∃x)" claims that the formula that comes after it is true in *some* cases. So "All are Italian" is "(x)Ix," and "Some are Italian" is "(∃x)Ix." As before, we can test arguments that use complex combinations. We can also test relational arguments, like this one:

Valid *Argument* ➔	There is someone that everyone loves.	(∃y)(x)Lxy
	∴ Everyone loves someone.	∴ (x)(∃y)Lxy

This becomes invalid if we switch the premise and conclusion, since we might all love different people. The two formulas here differ only in the order of the quantifiers:

	=	There is someone that everyone loves.
(∃y)(x)Lxy	=	There is some one person that everyone loves.
	=	There is some y such that, for all x, x loves y.

	=	Everyone loves someone.
(x)(∃y)Lxy	=	Everyone loves at least one person.
	=	For all x there is some y, such that x loves y.

The inclusion of relational arguments was a major advance.

Logic was important in the development of modern computers. The basic insight behind the computer is that truth functions like "and" and "or" can be simulated electrically by *logic gates*. This idea goes back to the American logician Charles Sanders Peirce in the 1880s and was rediscovered by Claude Shannon in 1938. Logicians like Alan Turing and

John von Neumann helped design the first computers. Since logic is important for computers, it is studied in computer science departments. So now there are three main departments that study logic—philosophy, mathematics, and computer science—each from a different perspective.

Classical symbolic logic includes propositional and quantificational logic. An approach to these is considered "classical" if it accords with the systems of Frege and Russell about which arguments are valid, regardless of differences in symbolization and proof techniques. Classical symbolic logic became the new orthodoxy in the 20th century, replacing the Aristotelian logic that had dominated for over 20 centuries.

Non-Classical Logics

Is classical symbolic logic sufficient? Some think it needs to be supplemented by other logical systems. Others claim that it is wrong on some points and that parts of it need to be changed.

The most important supplement is modal logic, which C. I. Lewis investigated in 1932 (although it goes back to Aristotle and the medievals). Modal logic studies arguments whose validity depends on "possible," "necessary," and similar notions. It adds two new symbols: "◇" and "□" (diamond and box):

◇A = A is possible (true in some possible world).
A = A is true (true in the actual world).
□A = A is necessary (true in all possible worlds).

Calling something *possible* is a weak claim, weaker than calling it *true*. Calling something *necessary* is a strong claim; it says not just that the thing is true, but that it has to be true—it could not be false.

We can rephrase "possible" as *true in some possible world*—and "necessary" as *true in all possible worlds*. A *possible world* is a consistent and complete description of how things might have been or might in fact be. Picture a possible world as a *consistent story*. The story is *consistent*, in that its statements do not entail self-contradictions; it describes a set of possible situations that are all possible together. The story is *complete*, in that it is imagined to include every statement or its negation. The story may or may not be true. The *actual world* is the story that is true—the description of how things in fact are.

At first, modal logic was controversial. Willard Van Orman Quine argued that modal logic was based on a confusion; he thought that logical necessity was unclear and that quantified modal logic led to an objectionable metaphysics of necessary properties. There was lively debate on modal logic for many years. Then in 1959 Saul Kripke presented the possible-worlds way to explain modal logic; this made more sense of it and gave it new respect among logicians. Possible worlds have proved useful in many other areas and are now a common tool in logic; and several logicians have defended a metaphysics of necessary properties. Today, modal logic is a well-established extension of classical logic.

Other extensions have been developed for ethics (about "A ought to be done" or "A is good"), theory of knowledge (about "X believes that A" or "X knows that A"), the part-whole relationship (about "X is a part of Y"), temporal relationships (about "It will be true at some future time that A" and "It was true at some past time that A"), and many other areas. These extensions of classical logic have been an exciting development.

Deviant logics say that classical symbolic logic is wrong on some points and that parts of it need to be changed. Some propose using more than two truth values. Maybe we need a third truth value for "half-true": so "1" = "true," "0" = "false," and "½" = "half-true"; truth tables and alternative propositional logics have been set up on this "multi-valued logic" basis. Or maybe we need a "fuzzy logic" range of truth values, from completely true (1.00) to completely false (0.00); then statements would be more or less true. Or perhaps for some mathematical statements we could have "A" and "not-A" both false (intuitionist logic). Or perhaps we should allow a logic in which contradictions can be true; so then we could have "A" and "not-A" both true (paraconsistent logic). Or perhaps classical logic's treatment of "if-then" is flawed; some suggested relevance-logic alternatives reject the validity of *modus ponens* ("If A then B, A ∴ B") and *modus tollens* ("If A then B, not-B ∴ not-A"). These and other deviant logics have been proposed. Today there is more questioning of basic logical principles than ever before.

The issue of deviant logics is controversial and much discussed. Some see deviant logics as a breath of fresh air; they welcome the fact that logic is becoming, in some circles, as controversial as other areas of philosophy. Others defend orthodox classical logic and see the new trends as dangerous; they wonder what would happen to other areas of philosophy if we could not take for granted that *modus ponens* and *modus tollens* are valid and that contradictions are to be avoided. We can anticipate much further discussion on this issue of deviant logics.

Logic in the Broad Sense

We said earlier that "logic" has a narrow and a broad sense. Logic in the narrow sense is the study of deductive reasoning, which is about what logically follows from what. Logic in the broad sense includes various other studies that relate to the analysis and appraisal of arguments. We will now sketch four major areas of *logic* in this broader sense: informal logic, inductive logic, metalogic, and philosophy of logic.

    1. *Informal logic* covers various non-formal skills that we need to have when we appraise arguments. It includes topics like locating the premises and conclusion in a passage that contains reasoning, supplying implicit premises, symbolizing English arguments, appraising the plausibility of premises, clarifying the meaning of a statement, and recognizing informal fallacies.

    Informal fallacies are common errors of thinking that cannot be expressed neatly in a standard system of logic. Examples include *ad hominem* (where we fallaciously attack the person instead of the issue), ambiguity (where we shift the meaning of a term or phrase within an argument), complex question (where we ask a question that assumes the truth of something false or doubtful: "Are you still beating your wife?"), and straw man (where we misrepresent an opponent's views). Informal fallacies are frequently studied in introductory logic courses.

    2. *Inductive logic* is about forms of reasoning in which we extrapolate from observed patterns to conclude that a given conclusion is *probably* true. Consider these two arguments:

Deductively Valid	Inductively Valid
All who live in Iowa	Most who live in Iowa were born in Iowa.
live in the U.S.	Jones lives in Iowa.
Jones lives in Iowa.	This is all we know about the matter.
∴ Jones lives in the U.S.	∴ Jones was born in Iowa (probably).

The first argument has a tight connection between premises and conclusion; it would be impossible for the premises to be all true but the conclusion false. The second argument has a looser connection between premises and conclusion; here, relative to the premises, the conclusion is only a good guess—it is likely true but could be false. Inductively logic is about arguments like the second one. It deals also with probability (like the probability of drawing two aces randomly from a deck of cards) and scientific reasoning (like how observations logically connect with laws of physics).

3. *Metalogic* is the study of formal systems. Its most important task is to prove that specific systems are consistent (will not prove contradictory formulas), sound (all arguments that are provable are valid), and complete (all valid arguments are provable).

Branches of logic are usually set up as formal systems. A *formal system* is an artificial language in which it is specified, in a notational way (independently of what the formulas may mean), what the well-formed formulas are and what sequences of such formulas are considered proofs. To connect a formal system with claims about reality, there is often added a *semantics*, which is about what the symbols mean; thus we might specify that certain letters represent true-or-false statements, or objects, or properties of objects. In *formal semantics* (a technique started by Alfred Tarski), we give truth conditions for the formulas of the system relative to various interpretations, where an *interpretation* is an abstract model of how language could link to different possible situations. Then an argument of our formal system is *valid*, if and only if every interpretation that makes all the premises true also makes the conclusion true.

The key issue in metalogic is the harmony, in a formal system, between the syntax and the semantics. What arguments are *provable* is specified by syntactic (notational) rules. What arguments are *valid* is specified by the semantic rules about what these formulas mean. The key issue is whether the two fit together: is a given formal system *adequate*, in that all the provable arguments are valid and all the valid arguments are provable? Using metalogical techniques, we can show that standard versions of propositional and quantificational logic are adequate in this sense; their axioms and inference rules match what they should be, given that the logical terms have certain meanings and semantic properties.

Classical metalogic makes classical assumptions about the meanings and semantic properties of the logical terms; for example, it assumes that statements are either true or false, but not both, and that there are no other truth values besides true and false. Non-classical metalogic rejects some of these assumptions; thus the techniques of metalogic can be applied to deviant logics as well.

4. *Philosophy of logic* deals with a whole slew of philosophical issues raised by logic. Many of these issues deal with language and how it relates to the world; here we have topics like abstract entities and ontology, the analytic/synthetic distinction, ambiguity, definite descriptions, definitions, general/singular terms, meaning, truth and paradoxes about truth, and the

sense/reference distinction. Philosophy of logic, in dealing with these issues, overlaps with philosophy of language.

Philosophy of logic deals also more specifically with issues about logic, such as its definition and scope, the analysis of its key concepts, the relationship between logic and ordinary language, the possibility of deviant logics, and the status and justification of basic logical principles.

Summary

Logic is about reasoning, about going from premises to a conclusion. Logic is important because reasoning is important; by studying logic we can come to understand reasoning better and become better at it. Logic was created in ancient Greece by Aristotle, and it was transformed about a hundred years ago by Gottlob Frege and Bertrand Russell. And so now we have the classical symbolic logic that is standard in logic courses. But logic is still growing and developing in the computer revolution, in extensions to classical logic, in controversies about deviant logics, and in a series of related investigations—about informal logic, inductive logic, metalogic, and philosophy of logic.

While logic may seem simple at first, it turns out to be much more complicated when we pursue it further. Although logic has been around for a long time, humanity still has a desperate need for more of it.

The Dictionary

A FORTIORI ARGUMENT (Latin for "even more so"). One that says that since something is true in a first case, it must also be true in a second (since the factors making it true in the first case apply even more in the second). For example, since Jim's little brother can pick up this chair, it must be that Jim can pick it up too—since Jim is stronger; we can construe this reasoning as a *modus ponens* argument with an "even more so" reason, perhaps implicit, for this conditional premise:

> Jim's little brother can pick up this chair.
> If Jim's little brother can pick up this chair, then Jim can pick up this chair. (The ability to pick up the chair depends on strength, and Jim is stronger than his little brother.)
> ∴ Jim can pick up this chair.

A PRIORI/A POSTERIORI (Latin for "from what is earlier/from what is later," equivalent to *rational/empirical*). Knowledge or beliefs are *a priori* if they are based on reason or thinking, while they are *a posteriori* if they are based on sense experience. Here is an example of each kind of knowledge:

> *A Priori:* "All bachelors are unmarried."
> *A Posteriori:* "Some bachelors are happy."

We know the second statement from our experience of bachelors; we have met many and have seen that some of them are happy. In contrast, we know the first statement by grasping what it means and seeing that it must be true; we do not have to investigate bachelors.

A priori knowledge requires some experience. We cannot know that all bachelors are unmarried unless we have learned the concepts involved; this requires some experience of language and of what it is to be married. But it still makes sense to call such knowledge *a priori*. Suppose we gain the concepts using experience. Then, to know that all bachelors are unmarried, we do not have to appeal to any further experience, other than thinking.

Most knowledge is *a posteriori*—based on sense experience. "Sense experience" here covers the five outer senses (sight, hearing, smell, taste, and touch), inner sense (our awareness of our thoughts and feelings), and any other experiential access to the truth that we may have (perhaps mystical experience or extrasensory perception). Logical and mathematical knowledge is usually considered (except by **John Stuart Mill** and **Willard Van Orman Quine**) to be *a priori*. To test an argument's validity, we do not go out and do experiments. Instead, we just think and reason, sometimes writing things out to help our thinking.

Here are further examples of statements known *a priori*:

"2=2"	"If everything is green, this is green."
"1>0"	"If there is rain, there is precipitation."
"All frogs are frogs."	"All red things are colored."

These are also examples of statements that are analytic, or self-contradictory to deny. Some philosophers think that the *a priori/a posteriori* and **analytic/synthetic** distinctions coincide; they think there is only one distinction, although it is drawn in two ways:

$$\begin{array}{rcl} \textit{A Priori} & = & \text{analytic} \\ \textit{A Posteriori} & = & \text{synthetic} \end{array}$$

Is this view true? If it is true at all, it is not true just because of how we defined the terms. By our definitions, the basis for the two distinctions is different. A statement is analytic or synthetic depending on whether its denial is self-contradictory. Knowledge is *a posteriori* or *a priori* depending on whether it rests on sense experience. Our definitions leave it open whether the two distinctions coincide.

These two combinations are common:

analytic *a priori*   synthetic *a posteriori*

Most of our knowledge in math and logic is analytic and *a priori*. Most of our scientific and everyday knowledge about the world is synthetic and *a posteriori*. These next two combinations are more controversial:

analytic *a posteriori*   synthetic *a priori*

"π is a little over 3" is presumably an analytic truth that can be known either by *a priori* calculations (the more precise way to compute π)—or by measuring circles empirically (as the ancient Egyptians did). And "It is raining or not raining" is an analytic truth that can be known either *a priori* (and justified by truth tables) or by deducing it from the empirical statement "It is raining." But perhaps any analytic statement that is known *a posteriori* could also be known *a priori*. While this claim seems plausible, **Saul Kripke** has questioned it (or at least he has questioned the related claim that all logically necessary truths are *a priori*).

A big controversy has raged over **Immanuel Kant**'s question of whether we have any synthetic *a priori* knowledge. This question asks whether there is any statement A such that (1) A is synthetic (neither self-contradictory to affirm nor self-contradictory to deny), (2) we know A to be true, and (3) our knowledge of A is based on reason (not on sense experience). In one sense of the terms, an **empiricist** is one who rejects such knowledge (and thus limits what we can know by reason alone to analytic statements) while a **rationalist** is one who accepts such knowledge (and thus gives a greater scope to reason).

Empiricists deny the possibility of synthetic *a priori* knowledge for two main reasons. First, it is difficult to understand how there could be such knowledge. Analytic *a priori* knowledge is fairly easy to grasp. Suppose a statement is true simply because of the meaning and logical relations of the concepts involved; then we can know it in an *a priori* fashion, by reflecting on these concepts and logical relations. But suppose a statement could without self-contradiction be either true or false. How could we then possibly know by pure thinking which it is?

Second, those who claim to know synthetic *a priori* truths do not agree much on what these truths are. They just seem to follow their prejudices and call them "deliverances of reason."

Rationalists affirm the existence of synthetic *a priori* knowledge for two main reasons. First, the opposite view, at least if it is claimed to be known, is self-refuting. Consider empiricists who claim to know this to be true: "There is no synthetic *a priori* knowledge." This statement itself would have to represent synthetic *a priori* knowledge. For the statement is synthetic (it is not true by virtue of how we defined the terms "synthetic" and "*a priori*"—and it is not self-contradictory to deny). And it would have to be known *a priori* (since surely we cannot know it on the basis of sense experience). So the empiricists' claim would be synthetic *a priori* knowledge—the very thing it rejects.

Second, we seem to have synthetic *a priori* knowledge of various truths—such as the following:

> If you believe that you see an object to be red and you have no special reason to doubt your perception, then it is reasonable for you to believe that you see an actual red object.

This claim is synthetic; it is not true because of how we have defined terms—and skeptics can deny it without self-contradiction. It is known to be true; if we do not know truths like this one, then we cannot justify any empirical beliefs. And it is known *a priori*; it cannot be based on sense experience—instead, knowledge from sense experience is based on truths like this. So our knowledge of this claim is synthetic *a priori*.

The dispute over synthetic *a priori* knowledge concerns the power of pure thinking and whether it can yield anything beyond analytic truths. Our view on this influences how we approach other areas of philosophy, including **ethics**. Are basic ethical principles (which are generally conceded to be synthetic) known *a priori*? Empiricists answer no; so they think knowledge of basic ethical principles is either empirical or non-existent. But rationalists can (and often do) think that we know basic ethical truths *a priori*, from reason alone, either through intuition or through some rational consistency test. *See also* LOGICAL PRINCIPLES, STATUS AND JUSTIFICATION OF.

ABSTRACT ENTITIES. Entities, roughly, that are neither physical (like apples) nor mental (like feelings); alleged examples include numbers, sets, and properties. Abstract entities are controversial.

The study of logic can lead quickly to questions about abstract entities. Consider this valid argument:

> This is green.
> This is an apple.
> ∴ Some apple is green.

Let us focus for the moment on the term "green" in the first premise. First, there is the **set** of green things; this set seems to be not a physical or mental entity, but rather an abstract entity. Second, there is the property of greenness, which can apply either to the color as experienced or to its underlying physical basis; in either case, greenness seems to be not a concrete mental or physical entity, but rather something more abstract that has physical or mental instances. Third, there is the concept of greenness, which is what terms for "green" in various languages mean; this concept could be construed as somehow mental or as something more abstract. Finally, there is the word "green"; this could be seen either as concrete physical marks or as an abstract pattern that is exemplified in written or auditory instances (*see* **type/token**).

Platonists, as logicians use this term, are those who straightforwardly accept the existence of abstract objects, regardless of whether they accept other parts of Plato's philosophy.

Nominalists are unhappy with this proliferation of entities; they typically want to limit what exists to concrete physical or mental entities. **Nominalism**'s challenge is to make sense of logic and mathematics while denying that there are abstract entities.

Various views intermediate between Platonism and nominalism are possible. Maybe we should accept abstract entities, but as creations or fictions of the mind rather than as independently real entities that we can discover. Maybe our talk about abstract entities is just shorthand for talking about concrete entities; maybe abstract entities are **logical constructs** with no independent existence.

Similar issues arise if we consider the whole sentence "This is green." Is this sentence concrete physical marks, or is it a more abstract pattern that has written or auditory instances? Should logic focus instead on the proposition that this is green, which is what is asserted by various ways to say "This is green" in different languages? If so, then is this proposition something mental, or is it something more abstract, like the meaning of "This is green"? How does it relate to the fact that this is green? Is this fact best understood as the this-being-green state of affairs being actual? Are facts and states of affairs to be understood not as concrete physical or mental entities but rather as something more abstract? Platonists are happy with entities like these, while nominalists are unhappy with the more abstract ones.

There are controversies about how abstract entities differ from concrete ones. Some say abstract objects (like numbers) are non-physical, non-mental, non-spatial, non-empirical, non-temporal, do not interact causally with other things, and exist of logical necessity. Others dispute these characterizations. For example, sometimes numbers have been alleged to be mental in nature (perhaps existing as ideas in the mind of God), to have temporal qualities (perhaps negative numbers were invented in a specific time and place), and to enter into causal relationships (perhaps big numbers have caused headaches). Some see the abstract/concrete distinction to be too vague to be useful.

Controversies about abstract entities go back to ancient and **medieval** disputes about forms and universals. They continue to rage today in disputes about **ontology**. *See also* QUINE; STATEMENT; TRUTH.

AD HOC (**Latin for "toward this"**). An implausible view that is accepted only in order to defend another belief. For example, the second-century astronomer Ptolemy believed that the earth was the center of the solar system; to keep this belief from being refuted by observations about planetary motion, he made complex and implausible *ad hoc* proposals about the epicycle paths that planets take around the earth. Later theories, which saw planets as traveling around the sun, explained observations about planetary motion more simply.

AD HOMINEM (**Latin for "toward the person"**). Attacking a person instead of an issue. A personal-attack argument can be either legitimate or fallacious. A legitimate *ad hominem* form goes as follows:

> X holds that A is true.
> In holding this, X violates legitimate rational standards (e.g.,
>   X is inconsistent, biased, or not correctly informed).
> ∴ X is not fully reasonable in holding A.

It is a **fallacy** to appeal to non-rational grounds, as some Nazis did who argued that Albert Einstein's theories must be wrong because he was Jewish. It is also a fallacy to conclude that a belief must be false because the person holding it is irrational; to show a belief to be false, we must argue against the belief, not against the person.

AESTHETICS. Relates to logic and mathematics mainly through the notion of "elegance." Given two definitions, proofs, or systems that give the same results, the more elegant one is the one that is simpler. Elegance in logic and mathematics is a highly prized aesthetic value. *See also* OCKHAM.

AFFIRMING ANTECEDENT/CONSEQUENT. *See* CONDITIONAL; *MODUS PONENS*.

ALGORITHM (decision procedure). Mechanical computational procedure for solving a type of problem that will always, if correctly followed, give the right answer in a finite number of steps. We learned
algorithms in grade school for adding up numbers; and in introductory
logic we learned **truth-table** algorithms for determining whether a
propositional argument is valid. Algorithms can easily be put into
computer programs; they do not require guesswork or ingenuity.

Some problems, even in mathematics and logic, do not admit of algorithms. While there are algorithms for computing the decimal expansion of π (3.14159265358979323846264433832795 . . .) to any desired
number of decimal places, there is no algorithm for deciding whether
an arbitrary string of digits (like "888888") will ever occur in this
decimal expansion. We could build a computer program that keeps
generating further digits in π's expansion and checks if the desired
string of digits occurs. But suppose the program runs for a thousand
years and has not found the sequence; maybe the sequence will show
up later or maybe it will not—we may never know.

A **function** is *effectively computable* just if there is an algorithm to
compute it. The function "the number that is the sum of x and y" is
effectively computable; the function "the decimal place in the expansion of π where the first string of x instances of '8' occurs" is not.

Algorithms contrast with *heuristics*, which are rules of thumb that
give the desired result, at best, only most of the time. While we have no
algorithms for winning at chess, we do have helpful heuristic strategies.
See also CHURCH; QUANTIFICATIONAL LOGIC; TURING.

AMBIGUITY, FALLACY OF. The **fallacy** of shifting the meaning of a
term or phrase within an argument. Here is an example:

All rational animals are men.	all R is M
No woman is a man.	no W is M
∴ No woman is a rational animal.	∴ no W is R

The premises can both be true only if we shift the meaning of "man." If
"man" means "human being," the second premise is false; if it means
"adult male," the first premise is false. *See also* AMPHIBOLY; BOX-
INSIDE/BOX-OUTSIDE AMBIGUITY; SCOPE AMBIGUITY.

AMPHIBOLY. Ambiguity that comes from how words are put together.
Willard Van Orman Quine's example was "pretty little girls camp";

this could be a camp for girls who are little and pretty, or a pretty camp for little girls, or a little and pretty camp for girls. "Pretty little" can also mean "rather little"; but this ambiguity comes from various meanings that "pretty" can have and is not an example of amphiboly.

Another amphiboly is "Not A and B," which in English can mean "Not-both A and B" or "Both not-A and B." Logical systems use devices like parentheses to avoid such **scope ambiguities**.

ANALOGICAL REASONING. Reasoning that two things that are similar in most ways are likely similar in some further way.

Suppose we are exploring our first Las Vegas casino. It is huge and crowded. We see slot machines, tables for blackjack and poker, and a big roulette wheel. There is a bar and an inexpensive all-you-can-eat buffet. We then go into our second Las Vegas casino and notice most of the same things: the size of the casino, the crowd, the slot machines, the blackjack and poker tables, the roulette wheel, and the bar. We are hungry. Recalling what we saw in our first casino, we conclude, "This place likely has an inexpensive all-you-can-eat buffet, just like the first casino." This is an argument by analogy:

Analogy Syllogism

Most things true of X are also true of Y.
X is A.
This is all we know about the matter.
∴ Probably Y is A.

The first premise here is rough, since many similarities are irrelevant. It would not weaken our reasoning if the two casinos had a different number of letters in their name; but it would weaken it if the places differed greatly in size. So we do not just *count* similarities—since many of these are trivial and unimportant. Rather, we look to *relevant* similarities. But how do we decide which similarities are relevant? We somehow appeal to our background information about what features are likely to go together. It is difficult to give rules here—even vague ones. So our analogy-syllogism formulation is a rough sketch of a subtle form of reasoning. Analogical reasoning, like other forms of **inductive** reasoning, is elusive and difficult to put into strict rules.

The *analogy argument for other minds* tries to justify a belief that we each have, that we are not the only conscious being in the world:

> Most things true of me are also true of Jones. (We are both alike in general behavior, nervous system, and so on.)
> I generally feel pain when showing outward pain behavior.
> This is all I know about the matter.
> ∴ Probably Jones, too, generally feels pain when showing outward pain behavior.

Critics raise several objections to the argument. First, Jones and I differ in many ways; such differences, if relevant, weaken the argument. Second, since I cannot directly feel Jones's pain, I cannot have direct access to the truth of the conclusion; this makes the argument peculiar and may weaken it. Third, we all have a sample-projection argument (*see* **inductive logic**) against the belief in other conscious beings:

> All the conscious events that I have experienced are mine.
> I have examined a large and varied group of conscious events.
> ∴ Probably all conscious events are mine.

The conclusion is another way to say that I am the only conscious being. Finally, the analogical argument, being weakened by such objections, at most makes it only somewhat probable that there are other conscious beings. But normally we take this belief to be solidly based.

The analogical argument for other minds highlights a problem with inductive arguments: it is often controversial whether and to what extent the premises, if true, give good reason to accept the conclusion.

ANALYTIC/SYNTHETIC. An *analytic statement* is one that is self-contradictory to deny—for example, "All bachelors are unmarried." Analytic statements are sometimes called *necessary truths* or *logical truths*. Such truths are based on logic, the meaning of concepts, or necessary connections between properties; they are sometimes said to be "true by definition" or "true by virtue of the meaning of the words." Here are further examples of analytic statements:

"2=2"	"If everything is green, this is green."
"1>0"	"If there is rain, there is precipitation."
"All frogs are frogs."	"All red things are colored."

In contrast, a *synthetic statement* is one that is neither analytic nor self-contradictory; *contingent* is another term for the same idea. Statements divide into analytic, synthetic, and self-contradictory:

Analytic:	"All bachelors are unmarried."
Synthetic:	"Daniel is a bachelor."
Self-contradictory:	"Daniel is a married bachelor."

There are only two kinds of truth: analytic and synthetic; self-contradictory statements are necessarily false. **Modal logic** symbolizes "A is necessary (analytic)" as "□A," "A is possible" as "◇A," "A is contingent" as "(◇A · ◇~A)," and "A is self-contradictory" as "~◇A."

The explanation just given takes "analytic" in a broad sense, to include truths of the following categories:

1. "All A is B" statements in which the predicate is contained in the **definition** of the subject, like "All bachelors [unmarried men] are unmarried."
2. Truths of **classical symbolic logic**, like "If it is raining, then it is raining."
3. Statements that become truths of classical symbolic logic when we substitute definitions for terms, like "If you do not have a brother, then either you have a sister or you are an *only child*" (which becomes a truth of classical logic if for "only child" we substitute "someone with no brother and no sister").
4. Statements that are true because of necessary connections between properties, like "All *red* things are *colored*."

Terminology varies. **Immanuel Kant**'s definition of "analytic statement" seemed to include just category 1. Some other philosophers use "analytic statement" to include just the first three categories, or just categories 1 and 3; they may then use "necessary truth" to cover all four categories. And some thinkers use "logical truth" to cover just category 2. *See also A PRIORI/A POSTERIORI*; PLANTINGA; QUINE; TAUTOLOGY.

ANCIENT LOGIC BEFORE ARISTOTLE. While logic as a systematic study started with **Aristotle** (384–322 BC), a growing interest in reasoning in ancient Greece prepared the way for Aristotle's work.

Pythagoras (c. 580–500 BC) and his followers applied close reasoning to **geometry**; we still speak of their "Pythagorean theorem." Geometry reached its definitive ancient formulation only after Aristotle, when Euclid wrote his *Elements* about 300 BC.

Pre-Socratic philosophers (sixth and fifth century BC), in the enigmatic fragments that they left, showed interest in issues that relate to logic. Heraclites insisted that we both do and do not step into the same river and that we both are and are not; Aristotle later took him to deny the **law of non-contradiction**. Parmenides criticized Heraclites and others who say being and non-being are the same and not the same; instead, he argued that being must be one and unchanging. His disciple Zeno of Elea argued that nothing moves: if something moved then it would have to move where it is or where it is not—and both are impossible. He similarly argued that Achilles cannot pass the tortoise, since this would require traversing an infinite number of points. Finally, the Sophist Gorgias constructed elaborate reasoning to show that nothing exists; if something existed, it would have to be being, non-being, or both—and for various reasons these are excluded as impossible.

Sophists emerged in the fifth century BC as paid teachers of wealthy young men. The Sophists taught **rhetoric**—how to speak and argue effectively; this was an important path to power in ancient Greece.

Socrates (c. 470–399 BC) and Plato (c. 428–347 BC) tried to purify reasoned inquiry from sophistical elements. Plato saw geometry, which deduced conclusions from *a priori* principles, as the model of rational thinking. His dialogues were unprecedented models of careful and plausible philosophical reasoning; these dialogues tested ideas by trying to derive absurdities from them and sought out beliefs that could be held consistently after a careful examination.

ANCIENT LOGIC SINCE ARISTOTLE. Aristotle's disciple Theophrastus (c. 371–286 BC) carried on logic in his master's tradition. He added a fourth figure of syllogism (*see* **Barbara, Celarent**) but, perhaps out of respect for his teacher, regarded it as a variation on the first figure. He cleaned up Aristotle's **modal logic** by proposing that a valid conclusion prefixed by "it is necessary that" normally requires that both premises be so prefixed; he said that the conclusion tends to follow the weaker premise—where "necessary" is the strongest, then "true," then "possible." He developed two-premise arguments using only **conditionals**, patterning these after his teacher's syllogisms:

Aristotle	Theophrastus
All A is B	If A then B
All C is A	If C then A
∴ All C is B	∴ If C then B

Later ancient logicians in the Aristotelian tradition include three from the second century AD: Lucius Apuleius; Alexander of Aphrodisias; and Galen, a medical doctor whose *Institutio Logica* is the oldest surviving logic textbook.

The Megarian school of philosophers (fourth century BC) followed Socrates and were often critical of Aristotle. They studied modal logic, conditionals, and the **liar paradox**. Most of the Megarians interpreted "possible" as "true at some time," and "necessary" as "true at all times." However, Diodorus Chronos interpreted "possible" as "true now or at some future time," and "necessary" as "true now and at all future times." Chronos proposed a *master argument* for fatalism (*see* **temporal logic**) and saw "If A then B" as true, if and only if it is *never* at any time the case that A is true while B is false. In contrast, Philo of Megara saw "If A then B" as true, if and only if it is not *now* the case that A is true and B is false; this, which fits the modern **truth table** for "if-then," led to much controversy (*see* **relevance logic**).

The Stoic Chrysippus (c. 279–206 BC) developed a form of **propositional logic**. Using numerals for statements, he accepted five basic inference rules:

(a)   If 1 then 2; but 1; therefore 2.
(b)   If 1 then 2; but not-2; therefore not-1.
(c)   Not both 1 and 2; but 1; therefore not-2.
(d)   Either 1 or 2 (but not both); but 1; therefore not-2.
(e)   Either 1 or 2 (but not both); but not 2; therefore 1.

We could use these rules to derive further results—for example, to derive "Not-B" from the first three lines below:

1   If both A and B, then C
2   Not-C
3   A
4   ∴ Not both A and B   {from 1 and 2 by rule (b)}
5   ∴ Not-B   {from 3 and 4 by rule (c)}

Unlike Aristotle, Chrysippus treated each premise and conclusion as a separate statement; he did not combine them into one big conditional. It is truly unfortunate that we have only small fragments of the writings of Chrysippus and the other logicians of this period.

Zeno of Citium, the founder of Stoicism, divided philosophy into three parts: **physics, ethics**, and **logic**. The Stoic Chrysippus believed that students should start with logic, and then study ethics, and lastly study physics; Sextus Empiricus agreed, adding that we need to know some logic to do ethics properly. Other Stoic logicians include Diogenes Laertius and Cicero.

At first, the logics of Aristotle and of the Stoic Chrysippus were looked upon as incompatible rivals. Later commentators and textbook writers combined the two into the **traditional logic** approach that dominated in higher education until the 20th century.

AND. *See* CONJUNCTION.

ANECDOTAL EVIDENCE. Evidence based on a few personal examples instead of a large and random sample. *See also* INDUCTIVE LOGIC.

ANTECEDENT. *See* CONDITIONAL.

ANTHROPOLOGY. Raises questions like these about logic and reasoning: Is reasoning based more on culture or on biology? Is logical thinking universal in all the cultures of the world, or is it peculiar to Western cultures? Do most or all languages have equivalents for the key logical notions of **classical symbolic logic**—notions like "and," "or," "if-then," "not," "all," and "some"—and assume similar principles for their use? Do different cultures have conflicting views about central principles of logic, like *modus ponens*, **Barbara**, the **law of non-contradiction**, and the **law of excluded middle**? What sort of cultural conditions would favor or retard logical thinking or the study of logic? *See also* BIOLOGY; BUDDHIST LOGIC.

ANTILOGISM. *See* LADD-FRANKLIN.

ANY/ALL. While "Any man is mortal" means "All men are mortal," "any" and "all" are often not interchangeable. Here are two rules for translating "any" into **quantificational logic**:

1. Paraphrase a sentence with "any" into an equivalent sentence without "any"; then translate this second sentence.
2. Translate "any" by a "(x)" that begins the formula (going outside all parentheses), regardless of the English word order.

These two methods often give different, but equivalent, translations. Here are examples of translations using both rules:

　Not anyone is rich
　　$=$　No one is rich　$=$　$\sim(\exists x)Rx$
　　$=$　For all x, x is not rich　$=$　$(x)\sim Rx$

If anyone is just, there will be peace
　　$=$　If someone is just, there will be peace　$=$　$((\exists x)Jx \supset P)$
　　$=$　For all x, if x is just there will be peace　$=$　$(x)(Jx \supset P)$

Not any Italians are lovers
　　$=$　No Italians are lovers　$=$　$\sim(\exists x)(Ix \cdot Lx)$
　　$=$　For all x, x is not both Italian and a lover　$=$　$(x)\sim(Ix \cdot Lx)$

In this example from **quantified modal logic**, only rule 2 works:

Anyone could be above average
　　$=$　For all x, it could be that x is above average　$=$　$(x)\lozenge Ax$

This differs from "It could be that everyone is above average," "$\lozenge(x)Ax$," which is false in non-empty universes.

APPEAL TO AUTHORITY. Basing a belief on what some authority says. A legitimate form of appeal to authority goes as follows:

> X holds that A is true.
> X is an authority on the subject.
> The consensus of authorities agrees with X.
> $\therefore$ There is a presumption that A is true.

It is a **fallacy** if we appeal to someone who is not an authority on the subject, if the authorities widely disagree, or if we say something *must* be true (and is not just *probably* true) because authorities support it.

APPEAL TO EMOTION. The **fallacy** of stirring up feelings instead of arguing in a logical manner. There is an important difference between using slanted (derogatory or laudatory) language about an issue and presenting an argument (premises and a conclusion) about it.

APPEAL TO FORCE. The **fallacy** of using threats or intimidation to get a conclusion accepted.

APPEAL TO IGNORANCE. The **fallacy** of arguing that what has not been proved must thus be false, or that what has not been disproved must thus be true.

APPEAL TO THE CROWD. The **fallacy** of arguing that something that most people believe must thus be true.

ARAB LOGIC. *See* MEDIEVAL LOGIC.

ARGUMENT. Set of statements consisting of *premises* and a *conclusion*. Normally we use the premises to give evidence for the conclusion; but sometimes we are just exploring what the premises lead to. Arguments put into words a possible act of reasoning. An argument is **valid** if it would be impossible for the premises to all be true while the conclusion was false; it is **sound** if it is valid and has only true premises.

Logicians like to express arguments clearly, with each premise beginning a new line and the conclusion prefixed by "∴" or "therefore." Arguments in real life are seldom so neat and clean. Instead we often find convoluted wording or extraneous material; and important parts of the argument may be omitted or only hinted at. It can take much work to reconstruct a clearly stated argument from a passage. Here is a helpful procedure for analyzing arguments (for example, arguments that one might find in a work of philosophy):

1. *Formulate the argument clearly in English.* Identify and write out the premises and conclusion. Try to arrive at a valid argument expressed as clearly and directly as possible. Using the principle of **charity**, interpret unclear reasoning in the way that gives the best argument. Supply implicit premises where needed, avoid emotional terms, and phrase similar ideas in similar words. This step can be difficult if the passage is unclear.

2. *Translate into some logical system and test for validity.* If the

argument is invalid, we might return to the previous step and try a different formulation. If we cannot get a valid argument, we can skip the next two steps.

3. *Identify difficulties.* Look for premises that are controversial and terms that are obscure or ambiguous.

4. *Appraise the premises.* Try to decide if the premises are true. Look for informal **fallacies**, especially **circularity** and **ambiguity**. Look for further arguments for or against the premises.

The first step in untangling obscure reasoning is to identify the premises and conclusion. While logicians like to put the conclusion last, ordinary people sometimes put it first or in the middle of the argument. These terms often indicate premises or conclusions:

These Often Indicate Premises:	Because, for, since, after all . . . I assume that, as we know . . . For these reasons . . .
These Often Indicate Conclusions:	Hence, thus, so, therefore . . . It must be, it cannot be . . . This proves (or shows) that . . .

When we do not have this help, we need to ask what is argued *from* (these are the premises) and what is argued *to* (this is the conclusion).

Sometimes people use **idiomatic** ways to express central logical concepts like "all" and "if-then." When we symbolize statements, we need to know how to translate from such idioms (like "Only As are Bs" and "A unless B") into the more central logical concepts.

The third step is to identify difficulties. This is especially important in presenting our own arguments. A good technique is to imagine ourselves in the place of our opponent and to keep asking, "How would an opponent object to this?"

A *good argument* in a broad sense is one that is logically correct and fulfills the purposes for which we use arguments. A good argument should be **deductively** valid (or **inductively** strong) and have only true premises; have this validity and truth be as evident as possible to the parties involved; be clearly stated; avoid circularity, ambiguity, and emotional language; and be relevant to the issue at hand.

Arguments can be useful even if they fall short of these ideals. We would like to use premises that are so obvious that everyone will immediately accept them; but in practice this is too high a standard. We sometimes appeal to premises that only some will accept—perhaps those of similar philosophical, religious, or political views. And we sometimes appeal to personal hunches, like "I can get to the gun before the thief does"; this may be the best we can do at a given moment.

A good argument need not spell everything out; it is often fine to omit premises that are obvious to the parties involved. If we are hiking the Appalachian Trail, we might say to our hiking partner, "We cannot still be on the right trail, since we do not see white blazes on the trees." This is fine if our friend knows the truth of the italicized premise below—since then the full argument would be pedantic:

> We do not see white blazes on the trees.
> *If we were still on the right trail, then we*
> *would see white blazes on the trees.*
> ∴ We are not still on the right trail.

In philosophy, it is wise to spell out *all* our premises, since implicit ideas are often crucial but unexamined. Suppose someone argues, "We cannot be free, since all human actions are determined." This assumes the italicized premise:

> All human actions are determined.
> *No determined action is free.*
> ∴ No human actions are free.

We should be aware that we are assuming this controversial premise.

A good argument normally convinces others, but it need not. Some people are not open to rational argument on certain issues. Some believe that the earth is flat—despite good arguments to the contrary. On the other hand, bad arguments, including fallacies, sometimes convince people. Studying logic can help to protect us from bad reasoning.

Logicians normally allow arguments with no premises; a premiseless argument is *valid* if and only if the conclusion is a logical truth. While logicians normally allow only premises that are true or false, defenders of **imperative logic** want to allow imperative premises, which tell what to do instead of making true or false assertions. And while logicians normally allow only premises and conclusions that are of finite length, **infinitary logic** allows ones that are infinitely long.

There is a second sense of "argument" that means "input to a **function**"; in the function "x + y," for example, the variables "x" and "y" are said to be *arguments*. *See also* INDUCTIVE/DEDUCTIVE; LANGUAGE, USES OF; PROOF.

ARISTOTELIAN ESSENTIALISM. The view (which **Aristotle** may or may not have held) that there are properties that some beings have of necessity that some other beings totally lack. *See also* KRIPKE; PLANTINGA; QUANTIFIED MODAL LOGIC; QUINE.

ARISTOTLE (384–322 BC). Philosopher and scientist of ancient Greece. With Socrates and Plato, he was one of the three great philosophers of the ancient world. He did influential work in many areas, including astronomy, biology, ethics, history, literary theory, logic, metaphysics, political theory, psychology, and rhetoric. Aristotle's ideas largely shaped Western thought for the next two thousand years; he was arguably the most influential secular thinker ever.

Aristotle invented logic as a systematic study. While there was much previous interest in reasoning (*see* **ancient logic before Aristotle**), he was the first, as far as we know, to formulate a correct principle of logic, to use letters for terms, and to construct an axiomatic system (*see* **Barbara, Celarent**). His main work was in **syllogistic logic**; here is an example of a **syllogism**:

All humans are mortal.	all H is M
All Greeks are humans.	all G is H
∴ All Greeks are mortal.	∴ all G is M

Aristotle's logic was about *forms* of reasoning that apply to any subject matter. While the above example is about humans, mortals, and Greeks, the form of reasoning is **valid** universally and remains valid if we substitute any other terms for "H," "M," and "G." This idea of logical **form** and **formal logic** started with Aristotle.

Aristotle developed syllogistic logic in his *Prior Analytics*. He took some syllogism forms to be self-evidently valid and derived others from these. He refuted invalid forms by giving cases where they had all true premises and a false conclusion. In evaluating syllogisms, he assumed they do not use empty terms (*see* **square of opposition**).

Several factors make his presentation difficult for modern readers.

First, he put syllogisms inside paragraphs instead of putting premises and conclusions on separate lines. Second, he liked to say "B is predicated of all A" instead of "all A is B." Third, he formulated a syllogism not as three statements but as a long conditional, as in "If all humans are mortal and all Greeks are humans, then all Greeks are mortal."

Aristotle worked out a **modal** version of syllogisms, in which premises or conclusions are prefaced by "it is necessary that" or "it is possible that" ("□" and "◇" in modern symbolism). His discussion is often confusing. He often switches between two senses of "possible": "not necessarily false" ("not impossible") and "neither necessarily false nor necessarily true" ("contingent"). And he seems to say that to get a conclusion prefixed with "it is necessary that" we need only have a similar prefix on one of the two premises, but we need it on the right one; logicians today think you normally need it on both premises. Aristotle's follower Theophrastus (c. 371–286 BC) later resolved some of these problems (*see* **ancient logic since Aristotle**).

Aristotle wrote six works about logic in the broad sense:

- *Categories:* about substance, quantity, quality, relation, place, time, position, state, action, and passion
- *On Interpretation:* about language and **semantics**, including the square of opposition
- *Prior Analytics:* about syllogistic logic
- *Posterior Analytics:* about science and theory of knowledge
- *Topics:* about **inductive reasoning**, informal hints about debating, and essential properties (*see* **Aristotelian essentialism**)
- *Sophistical Refutations:* about informal **fallacies**

Aristotle saw logic (then called "analytics" or "dialectic") more as a thinking tool than as a separate branch of knowledge; so his logical works later came to be called the "Organon" ("tool").

On Interpretation discusses future contingents. Suppose that there may be a sea battle tomorrow. If "There will be a sea battle tomorrow" (abbreviated as "S" below) is *now* either true or false, this seems to imply that whether the battle occurs is a matter of necessity:

> Either it is true that S or it is false that S.
> If it is true that S, then it is necessary that S.
> If it is false that S, then it is necessary that not-S.
> ∴ Either it is necessary that S or it is necessary that not-S.

Aristotle rejected the conclusion, since he thought there was no necessity either way. He seemed to deny the first premise and thus the universal validity of the **law of excluded middle** (which he defends in book 4 of his *Metaphysics*); if we interpret him this way (which is controversial) then he anticipated **many-valued logic** in positing a third **truth value** besides **true** and **false** (*see* **bivalence**). Many today think the second and third premises have the **box-inside/box-outside ambiguity**: taking them as "(A ⊃ □B)" makes them doubtful while taking them as "□(A ⊃ B)" makes the argument invalid.

Book 4 of Aristotle's *Metaphysics* defends the **law of non-contradiction**, that the same property cannot at the same time both belong and not belong to the same object in the same respect. So "S is P" and "S is not P" cannot both be true at the same time—unless we take "S" or "P" differently in the two statements. Aristotle saw the law as so certain that it could not be proved by anything more certain; he thought not all knowledge can be proved, since otherwise we would need an infinity of arguments, whereby every premise of every argument is proved by a further argument. Those who deny the law (like Heraclitus and modern **dialethists**) assume it in their deliberations; to drive the point home, we might pretend to agree with them that contradictions are fine and then bombard them with contradictions until they plead for us to stop.

Aristotle did not investigate inferences using "If A then B" and "Either A or B." Theophrastus and the Stoics dealt with these later; *see* **ancient logic since Aristotle**.

Aristotle's logic, with additions and refinements, was dominant for the next two thousand years. **Immanuel Kant** (1724–1804) claimed that Aristotle had invented and perfected logic; nothing else of fundamental importance could be learned or added. But logic started changing radically in the 19th century (*see* **logic: history of**) and now has gone much beyond Aristotle.

ARITHMETIC. The branch of **mathematics** that deals with numbers. Numbers are of various sorts, including fractions (like ¼), negatives (like -4), and real numbers (like π); "2" is a *cardinal*, while "2nd" is an *ordinal*. Here we will focus on the *natural numbers* (0, 1, 2, . . .) and *three numerical operations* (addition, multiplication, and exponentiation). **Gottlob Frege** and **Bertrand Russell**, the founders of modern logic, thought arithmetic could be reduced to logic; this view, called *logicism*, made two claims:

- *Analysis thesis:* The natural numbers and three numerical operations can be defined using **classical symbolic logic** (perhaps adding **identity logic, set theory**, or **second-order logic**).
- *Provability thesis:* The truths of arithmetic can be captured by a **formal system**, so that all the truths of arithmetic but none of its falsehoods would be provable in the system.

Formalists like David Hilbert (1862–1943) rejected the analysis thesis (since they saw arithmetic as a game about symbols and were not concerned with what numbers were) but accepted the provability thesis (since they saw "truth" in arithmetic as equivalent to "provability").

Regarding the *analysis thesis*, logicists tended to see a number as a second-order set. The number 2, for example, was identified with the set containing just those sets that contain exactly two members; so 2 = {{Tom, Suzy}, {Canada, Mexico}, {this dog, this cat}, . . .}. While this seems circular, we can paraphrase out the "two" in "contain exactly *two* members" using identity logic (read "$\in$" as "is a member of"):

$$2 = \text{the set containing just those sets A that satisfy this condition: for some x and y, } x \neq y, x \in A, y \in A, \text{ and there is no z such that } z \neq x, z \neq y, \text{ and } z \in A.$$

Then "I have two hands" would be analyzed as "the set of my hands $\in$ 2." Having thus analyzed the natural numbers in terms of sets, we could then analyze "plus," "times," and "to the power of" in similar terms. And then we could, presumably, give an axiomatization of logic and set theory that would suffice to prove all the truths of arithmetic; this would establish the provability thesis. Such was the hope of logicism, and it spurred much of the early development of logic.

However, there are other definitions of numbers in terms of sets. Let "Λ" be the null set, the set that is empty and thus has no members. Then "$\{\Lambda\}$" is the set that contains only the null set, "$\{\{\Lambda\}, \Lambda\}$" is the set that contains only these two sets, "$\{\{\{\Lambda\}, \Lambda\}, \{\Lambda\}, \Lambda\}$" is the set containing only these three sets, and so forth. Then the numbers 0, 1, 2, 3, and so forth can be identified with these sets:

0	=	Λ	(which has 0 members)
1	=	$\{\Lambda\}$	(which has 1 member)
2	=	$\{\{\Lambda\}, \Lambda\}$	(which has 2 members)
3	=	$\{\{\{\Lambda\}, \Lambda\}, \{\Lambda\}, \Lambda\}$	(which has 3 members)

Then "I have two hands" can be analyzed as "the members of the set of my hands can be matched one to one with the members of the set 2." Everything else would work out much like before. So there are alternative definitions of numbers in terms of sets that seem to work; this raises doubts about whether any of these definitions gives a uniquely correct analysis of numbers.

Regarding the *provability thesis*, logicists claimed to give consistent axiomatic systems that could prove any truth of arithmetic, including those that quantify over numbers (*see* **quantificational logic**). Kurt Gödel in 1931, in an historic result known as *Gödel's theorem*, showed that this idea will not work. *See also* CANTOR; GÖDEL; PEANO.

ARROW, KENNETH JOSEPH (1921–). An American social scientist most famous for *Arrow's theorem*, which says roughly that every voting system is flawed. Suppose your department is deciding between candidates A, B, and C. Three people show up to vote. Your ranking is ABC (A first, then B, then C). One of your colleagues ranks the candidates BCA, and the other ranks them CAB. So most prefer A to B, most prefer B to C, and most prefer C to A. Who wins is then determined by how the voting procedure is set up. For example, suppose you first take two candidates and have everyone vote, and then have the winner go against the remaining candidate. Then the candidate not in the first vote will win. For example, if you first decide between A and B, then A will win; but then A will go against C, and C will win. Note that the argument here is logical, not empirical.

ASSUMPTION. Depending on context, "assumption" can be mean "premise," "uncertain belief," or "line in a proof that does not follow from previous lines but that is used in some other way" (for example, in a formal **proof** you might assume something in order to derive a contradiction from it and thus prove the opposite, using **RAA**).

AXIOLOGICAL LOGIC. A branch of logic that studies arguments whose validity depends on "good," "bad," "better," and similar notions. One approach takes the **relation** "a is better than (preferable to) b" ("aPb") as basic, where a and b are states of affairs; an example is "That I experience pleasure is better than that I experience pain." The "better than" relation is assumed to be asymmetrical (if a is better than b, then b is not better than a) and transitive (if a is better than b, and b is better than c, then a is better than c). Then "a is good" is defined as "a is better than not-a" and "a is bad" as "not-a is better than a." *See also* DEONTIC LOGIC; ETHICS.

AXIOM. Basic principle that is not proved but can be used to prove other things; these "other things" proved from axioms are called **theorems**. In **formal systems**, an axiom is a formula that can be put on any line of a formal **proof**, regardless of earlier lines. An axiom may be a specific formula (like "0=0") or a *schema* that specifies that any formula of a given form is an axiom (like any formula that consists of a specific numeral, then "=," and then that same numeral again). Axioms are distinguished from **inference rules**, which say that, given certain previous lines in a proof, we can derive a given further line.

AXIOM OF CHOICE. A controversial principle of **set theory** that says that, given an infinite collection of non-empty sets with no common members, there is a further set with exactly one member from each set.

– B –

BARBARA, CELARENT. This is a **medieval** mnemonic verse, going back to William of Sherwood and Peter of Spain of the 13th century, that showed how **Aristotle** axiomatized **syllogistic logic**:

> Barbara, Celarent, Darii, Ferioque, prioris;
> Cesare, Camestres, Festino, Baroco, secundae;
> tertia, Darapti, Disamis, Datisi, Felapton,
> Bocardo, Ferison, habet; quarta insuper addit
> Bramantip, Camenes, Dimaris, Fesapo, Fresison.

Each capitalized name represents a valid syllogism. The vowels "A," "I," "E," and "O" in the names signify specific sentence forms ("negative" is misspelled below to make the mnemonic work in English):

A	all – is –	A ffirmative universal
I	some – is –	aff I rmative particular
E	no – is –	n E gotive universal
O	some – is not –	neg O tive particular

So "Barbara," with AAA vowels, has three "all" statements:

$$\begin{array}{c} \underline{\text{Bar}\underline{\text{ba}}\text{ra}} \\ \text{all all all} \end{array} \quad \begin{array}{c} \text{all M is P} \\ \text{all S is M} \\ \therefore \text{ all S is P} \end{array} \quad \begin{array}{c} \text{MP} \\ \hline \text{SM} \end{array} = \text{figure 1}$$

Aristotelian syllogisms have two premises. It is traditional to use "M" (*middle* term) for the term common to both premises, "P" (conclusion *predicate*) as the other term in the first (major) premise, and "S" (conclusion *subject*) for the other term in the second (minor) premise. There are four possible figures, or arrangements of the premise letters:

1 (prioris)	2 (secundae)	3 (tertia)	4 (quarta)
MP	PM	MP	PM
SM	SM	MS	MS

To remember these, think of the "M" letters as tracing a big "W." Aristotle recognized only the first three figures; but his follower Theophrastus (c. 371–286 BC) added the fourth (which he considered a variation on the first).

The four axioms of Aristotle's system are the valid forms of the first figure—Barbara, Celarent, Darii, and Ferio:

Barbara	Celarent	Darii	Ferio
all M is P	no M is P	all M is P	no M is P
all S is M	all S is M	some S is M	some S is M
∴ all S is P	∴ no S is P	∴ some S is P	∴ some S is not P

The other 15 forms (four in figure 2, six in figure 3, and five in figure 4) can be derived as theorems. Derivations use four rules of inference, here labeled with the letters "s," "m," "p," and "c":

s   Switch the order of letters in a "some A is B" or "no A is B" statement. This is called "simple conversion."

m   Switch the order of premises.

p   Strengthen a premise or weaken the conclusion—where "all A is B" is stronger than "some B is A," and "no A is B" is stronger than "some B is not A." This is called "accidental conversion."

c Switch the contradictory of a premise with the contradictory of the conclusion (where "all A is B" and "some A is not B" are contradictories, as are "no A is B" and "some A is B").

Use the axiom that starts with the same letter as the form you want to derive. The letters "s," "m," "p," and "c" after vowels tell which rules to use on the axiom. Here is the first of three sample derivations:

(1)   Ferio   Festino

        no M is P   → by s →   no P is M
        some S is M   some S is M
        ∴ some S is not P   ∴ some S is not P

We start with Ferio (axiom), switch the order of letters in "no M is P" (by rule s), and thus get Festino of figure 2. Our second derivation is more difficult:

(2)   Barbara   by m   by p   Bramantip

       all M is P   all S is M   all S is M   all P is M
       all S is M   all M is P   all M is P   all M is S
       ∴ all S is P   ∴ all S is P ∴ some P is S ∴ some S is P

We start with Barbara (axiom), switch the premises (by rule m), weaken the conclusion (by rule p), switch "P" and "S" (by an assumed further rule) to keep "S" as subject and "P" as predicate of the conclusion, and thus get Bramantip of figure 4. Here is our third derivation:

(3)   Barbara   by c   Baroco

       all M is P   all M is P   all P is M
       all S is M   some S is not P   some S is not M
       ∴ all S is P   ∴ some S is not M   ∴ some S is not P

We again start with Barbara (axiom), switch the contradictory of the second premise with the contradictory of the conclusion (by rule c), switch "P" and "M" to keep "M" as the term common to both premises and "P" as the conclusion predicate, and thus get Baroco of figure 2.

We need only Barbara and Celarent as axioms. To get Ferio, take Celarent, reverse the letters in the first premise, switch the contradictory of the second premise with the contradictory of the conclusion, and switch "M" and "P." To get Darii, take Celarent, switch the premises, switch the contradictory of the second premise with the contradictory of the conclusion, reverse the letters in the second premise and conclusion, and write "M" for "S," "S" for "P," and "P" for "M."

Given rule p (which assumes that we use no empty terms—*see* **square of opposition**), we can derive five further valid syllogisms with weakened conclusions: Barbari and Celaront of figure 1, Cesaro and Camestrop of figure 2, and Camenop of figure 4. These forms were largely ignored because they violate the idea that we should derive the strongest conclusion that follows from our premises.

The classic syllogism is "All men are mortal; Socrates is a man; therefore Socrates is mortal." Many medievals took "Socrates is a man" to be an A-statement; then our example would have the Barbara form: "All men are mortal; all Socrates is a man; therefore all Socrates is mortal." **Quantificational logic** takes "Socrates is a man" to ascribe a property to an entity ("Fa")—while it takes "all F is G" to claim that, for every entity x, if x is F then x is G ("(x)(Fx ⊃ Gx)").

BARCAN, RUTH. *See* MARCUS, RUTH BARCAN.

BAYES, THOMAS (1702–1761). English theologian and mathematician who is best known today for his ideas about **probability**.
Bayes's theorem, or at least a simple form of it, says,

$$\text{The probability of A, given B} = \frac{\text{Prior probability of A}}{\text{Prior probability of B}} \times \text{The probability of B, given A}$$

This can be derived from the definition of "The probability of A, given B" as "The probability of (A and B) divided by the probability of B."

Suppose Mary's father has a disease, and it is 50 percent likely that Mary carries the disease. If Mary is a carrier, it is 50 percent likely that her child will get the disease; if she is not a carrier, there is no chance that her child will get it. Based on these facts, it is 75 percent likely that a child of Mary will be normal (in the sense of not getting the disease).

Now suppose Mary has a normal child; that makes it less likely that she is a carrier. But how much less likely does it make it? According to Bayes's theorem, it is now only 1/3 likely. Let C = Mary is a carrier, and N = her child is normal. Then Bayes's theorem says,

$$\begin{array}{l} \text{The probability} \\ \text{of C, given N} \end{array} = \frac{\text{Prior probability of C}}{\text{Prior probability of N}} \times \begin{array}{l} \text{The probability} \\ \text{of N, given C} \end{array}$$

The prior probability of her being a carrier is .5; the prior probability of her having a normal child is .75; and the probability of her having a normal child, given that she is a carrier, is .5. So the probability of her being a carrier, given that she has a normal child = $(.5 \div .75) \times .5 = .33$.

Let us now instead suppose she has a diseased child. Then it is 100 percent probable that she is a carrier. Bayes's theorem here says,

$$\begin{array}{l} \text{The probability} \\ \text{of C, given D} \end{array} = \frac{\text{Prior probability of C}}{\text{Prior probability of D}} \times \begin{array}{l} \text{The probability} \\ \text{of D, given C} \end{array}$$

The prior probability of her being a carrier is .5; the prior probability of her having a diseased child is .25; and the probability of her having a diseased child, given that she is a carrier, is .5. So the probability of her being a carrier, given that she has a diseased child = $(.5 \div .25) \times .5 = 1$.

Often *Bayesians* are contrasted with *frequentists*; the former see probability as a measure of one's confidence in a statement, while the latter see it as a ratio of observed frequencies.

BEGGING THE QUESTION. *See* CIRCULAR REASONING.

BEING. *See* IS.

BELIEF LOGIC (doxastic logic). The study of patterns of consistent believing; belief logic is "logic" in an extended sense, since it does not study what follows from what. While some belief logics *describe* how consistent individuals would believe, the version sketched here *prescribes* that we be consistent. This approach builds on **imperative, propositional, quantificational, modal,** and **deontic logic.**

Our belief logic adds the symbol ":" and two ways to form **wffs:**

1. The result of writing a small letter and then ":" and then a wff is a descriptive wff.
2. The result of writing an underlined small letter and then ":" and then a wff is an imperative wff.

Statements about beliefs translate into *descriptive* belief formulas:

$$u:A \quad = \quad \text{You believe that A is true.}$$
$$u:\sim A \quad = \quad \text{You believe that A is false.}$$
$$\sim u:A \quad = \quad \text{You do not believe that A is true.}$$

If you do not believe that A is true (you refrain from believing A), then you need not believe that A is false; maybe you take no position on A, neither believing A nor believing not-A: "($\sim u:A \cdot \sim u:\sim A$)." *Imperative* belief formulas have the small letter before ":" underlined:

$$\underline{u}:A \quad = \quad \text{Believe that A is true.}$$
$$\underline{u}:\sim A \quad = \quad \text{Believe that A is false.}$$
$$\sim \underline{u}:A \quad = \quad \text{Do not believe that A is true.}$$

There are three approaches that we might take to belief logic. The first would study what belief formulas validly follow from what other belief formulas. We might try to prove arguments such as this one:

You believe A.   u:A
∴ You do not believe not-A.   ∴ ~u:~A

But this is invalid, since people can be confused and illogical. Students and politicians can assert A and assert not-A almost in the same breath; given that someone believes A, we can deduce little or nothing about what else the person believes. So this first approach will not work.

A second approach would study how people would believe if they were completely consistent believers (a highly idealized notion). A person X is a *completely consistent believer* if and only if:

- X believes some things,
- the set S of things that X believes is logically consistent, and
- X believes anything that follows logically from set S.

Our previous argument would be valid if we added, as an additional premise, that you are a completely consistent believer:

You are a completely consistent believer.
You believe A.
∴ You do not believe not-A.

A belief logic of this sort would take "You are a completely consistent believer" as an implicit premise of its arguments. This premise would be assumed, even though it is false, to help us explore what belief patterns a consistent believer would follow. **Jaakko Hintikka** used roughly this approach, but symbolizing "s believes that P" as "B_sP."

The third approach, that of Mark Fisher and Harry Gensler, is to construct a belief logic that generates consistency imperatives like this:

Do not combine believing A with believing not-A.
= ~(u̲:A · u̲:~A)

This approach tells us to avoid inconsistent combinations; it assumes an implicit "One ought to be consistent" principle. Here are two further consistency norms that this approach can generate:

Do not combine inconsistent beliefs.
= If A is inconsistent with B, then do not combine *believing A* with *believing B*.
= (~◇(A · B) ⊃ ~(u̲:A · u̲:B))

Do not believe something without believing what follows from it.
= If A logically entails B, then do not combine *believing A* with *not believing B*.
= (□(A ⊃ B) ⊃ ~(u̲:A · ~u̲:B))

We can expand belief logic to cover willing as well as believing. We can do this by treating "willing" as *accepting an imperative*—just as we previously treated "believing" as *accepting an indicative*:

u:A $\quad$ = You accept (say in your heart) "A is true."
$\quad$ = You believe that A.

u:A̲ $\quad$ = You accept (say in your heart) "Let act A be done."
$\quad$ = You will that act A be done.

In translating "u:A̲," we can use terms more specific than "will"—terms like "act," "resolve to act," or "desire." Which of these fits depends on whether the imperative is present or future, and whether it applies to oneself or to another.

We would lose important distinctions if we symbolized *desiring* as an operator on indicatives. Consider these three wffs:

u:(∃x)(Kx · R<u>x</u>)	=	You desire that some who kill *repent*.
	=	You say in your heart "Would that some who kill *repent*."
u:(∃x)(K<u>x</u> · Rx)	=	You desire that some *kill* who repent.
	=	You say in your heart "Would that some *kill* who repent."
u:(∃x)(K<u>x</u> · R<u>x</u>)	=	You desire that some both *kill* and *repent*.
	=	You say in your heart "Would that some both *kill* and *repent*."

The three are different. The underlining shows which parts are desired: repenting, or killing, or killing-and-repenting. If we attached "desire" to indicative formulas, all three would translate the same way, as "You desire that (∃x)(Kx · Rx)" ("You desire that there is someone who both kills and repents"). So "desire" is better symbolized in terms of accepting an imperative.

This expanded version of belief logic could include principles about consistent willing like these:

> Do not combine *believing* that you ought to do A with *not acting* to do A.
> = ~(<u>u</u>:OA<u>u</u> · ~<u>u</u>:A<u>u</u>)

> Do not accept "It is wrong for anyone to kill" without being resolved that if killing were needed to save your family, then you would not kill.
> = ~(<u>u</u>:(x)O~K<u>x</u> · ~<u>u</u>:(N ⊃ ~K<u>u</u>))

> You ought not to combine *believing* that it is wrong for you to do A with *acting* to do A.
> = O~(<u>u</u>:O~A<u>u</u> · <u>u</u>:A<u>u</u>)

These ethical principles are *formal* in that they can be formulated using the abstract notions of belief logic plus variables (like "u" and "A") that stand for any person and action (*see* **formal ethics**). Belief logic can be expanded to include other such principles, including the golden rule.

The "One ought to be consistent" principle assumed here is subject to some qualifications. For the most part, we do have a duty to be consistent. But, assuming that "ought" implies "can," this duty can be nullified when we are unable to be consistent; such inability can come from emotional turmoil or our incapacity to grasp complex inferences. And the obligation to be consistent can be overridden by other factors; if Dr. Evil would destroy the world unless we were inconsistent in some way, then our duty to be consistent would be overridden.

Another problem is whether to accept this *conjunctivity principle*:

> You ought not to combine believing A and
> believing B and not believing (A and B).
> =   O~((u:A · u:B) · ~u:(A · B))

Consider the "lottery paradox." Suppose eight people have an equal chance to win a lottery. Perhaps you could reasonably combine believing that person 1 will lose, that person 2 will lose, . . . , and that person 8 will lose—without believing the conjunction of these (equivalent to the claim that everyone will lose). If this is reasonable, then we would have to reject or qualify the conjunctivity principle.

BESIDE THE POINT. The **fallacy** of arguing for a conclusion irrelevant to the issue at hand. For example, Hitler, when facing a group opposed to the forceful imposition of dictatorships, sidetracked their attention by attacking pacifism; his arguments, even if sound, were beside the point. Such arguments are sometimes called "red herrings," after a practice used to train hunting dogs: a red herring fish would be dragged across the trail to distract the dog from tracking an animal. In arguing, we must keep the issue clearly in mind and not be misled by smelly fish.

BICONDITIONAL. Statement of the form "P if and only if Q." **Propositional logic** symbolizes this as "(P ≡ Q)." An English example is "I went to Paris if and only if I went to Quebec." Logicians sometimes use "iff" for "if and only if."

"(P ≡ Q)" has this **truth table** (where T = true, F = false):

P Q	(P ≡ Q)	
F F	T	An IF-AND-ONLY-IF is
F T	F	true if both parts have the same
T F	F	truth value—and is false if both
T T	T	parts differ in truth value.

"(P ≡ Q)" claims that both parts have the *same* truth value: both are true or both are false. So "I went to Paris if and only if I went to Quebec" is true just in case I went to *both* places, or I went to *neither* place. So "≡" is much like "equals."

Our truth table can produce unintuitive results. Take this example:

$$\text{I had eggs for breakfast if and} \atop \text{only if the world will end at noon.} \qquad (E \equiv W)$$

Suppose both parts are false: I did not have eggs for the breakfast and the world will not end at noon. By our table, the biconditional is then true—since both parts have the same truth value. This is strange. We would normally take the conditional to be false, since the two parts do not have much relationship to each other. So translating "if and only if" as "≡" does not seem totally right.

Our "≡" is a simplified "if and only if" that ignores how the parts relate to each other. "(P ≡ Q)" has a simple meaning; it just claims that both parts have the same truth value—both are true or both are false. An IF-AND-ONLY-IF understood this way is called a *material equivalence*. Most logicians contend that translating "if and only of" this way is an odd but useful simplification, since it captures the part of "if and only if" that normally determines validity. But some logicians disagree and try to develop alternative models for **conditionals** and biconditionals (*see* **conditional, kinds of** and **relevance logic**).

Various inference rules govern biconditionals. Given an IF-AND-ONLY-IF, we can infer the truth of one part from the truth of the other—and the falsity of one part from the falsity of the other:

(P ≡ Q)	(P ≡ Q)	(P ≡ Q)	(P ≡ Q)
P	Q	~P	~Q
Q	P	~Q	~P

Given a false IF-AND-ONLY-IF, we can infer the falsity of one part from the truth of the other—and the truth of one part from the falsity of the other:

~(P ≡ Q)	~(P ≡ Q)	~(P ≡ Q)	~(P ≡ Q)
P	Q	~P	~Q
~Q	~P	Q	P

An IF-AND-ONLY-IF is false if the parts have opposite truth value—and is true if both parts have the same truth value (defenders of relevance logic reject these last two as *paradoxes of material equivalence*):

$$\frac{P, \sim Q}{\sim(P \equiv Q)} \qquad \frac{\sim P, Q}{\sim(P \equiv Q)} \qquad \frac{P, Q}{(P \equiv Q)} \qquad \frac{\sim P, \sim Q}{(P \equiv Q)}$$

And we can reverse the parts of an IF-AND-ONLY-IF:

$$\frac{(P \equiv Q)}{(Q \equiv P)}$$

This last one can fail in ordinary English, where the connection between the parts can be important: "I get sick if and only if I eat broccoli" does not seem equivalent to "I eat broccoli if and only if I get sick." Again, "≡" is a simplified "if and only if" that ignores how the parts relate to each other. *See also* IDIOMS; TRUTH FUNCTIONS.

BIOLOGY. Raises questions like these about logic and reasoning: How does reasoning connect with brain processes? How did we evolve into reasoning beings? Is reasoning based more on biology or on culture? To what extent do brains resemble computers? What is the logical structure of DNA? To what extent are other animals able to reason? What kind of reasoning can be used to defend theories like evolution?

BITING THE BULLET. Accepting an implausible consequence of a view in order to avoid admitting that you have been refuted. Suppose you say that "good" means "socially approved," and someone objects that this entails the absurdity that racism would become good if it were socially approved. Saying "Yes, then it would be good," even though this seems implausible, would be an example of biting the bullet.

BIVALENCE. The principle that there are only two **truth values**, namely, **true** and **false**. This is consistent with there being truth-value gaps for sentences that are meaningless (like "Glurklies glurkle") or vague ("Her shirt is white," when it is between white and gray). *See also* ARISTOTLE; INTUITONIST LOGIC; LAW OF EXCLUDED MIDDLE; MANY-VALUED LOGIC.

BLACK-AND-WHITE THINKING. The **fallacy** of oversimplifying a complex situation by assuming that one or another of two extreme cases must be true. One commits this fallacy in thinking that people must be logical or emotional but cannot be both. People who think in a black-and-white manner like simple dichotomies, like logical-emotional, capitalist-communist, or intellectual-jock. Such people have a hard time seeing that the world is more complicated than that.

BOOLE, GEORGE (1815–1864). An English mathematician and logician. He invented *Boolean algebra*, which symbolizes logical notions using an algebraic notation and uses something akin to mathematical calculation to check the correctness of inferences.

On Boole's approach, letters can represent **sets**; so "A" might stand for the set of animals and "C" for the set of cats. Putting two letters together represents the *intersection* of the sets; so "CA" represents the set of things that are *both cats and animals.* We can use these ideas to symbolize "All cats are animals" as "C = CA," which says that the set of things that are cats = the set of things that are both cats and animals. Then we can express a **syllogism** as a series of equations:

all M is P	M = MP
all S is M	S = SM
∴ all S is P	∴ S = SP

In the algebraic version, we can derive the conclusion from the premises by substituting equals for equals. Start with the second premise: S = SM. Cross out "M" and write "MP" (the first premise says M = MP); this give us S = SMP. Then cross out "SM" and write "S" (the second premise says S = SM); this gives us S = SP.

If we add "0" to represent the *null set* (the set that contains nothing), then we can symbolize other statements as follows:

- "No dogs are cats" is "DC = 0": the set of things that are both dogs and cats = the null set.
- "Some animals are dogs" is "AD ≠ 0": the set of things that are both animals and dogs is not the null set.
- "Some animals are not dogs" is "A ≠ AD": the set of things that are animals is not the same as the set of things that are both animals and dogs.

Boole used "1" for the *universal set* (the set that contains everything). He represented the *complement* of a set "A" by "(1 − A)"; so if A is the set of animals, then (1 − A) is the set of non-animals. So then we can symbolize these facts about sets:

$$A(1 - A) = 0 \qquad\qquad AB = BA$$

The first formula says that the intersection of the set of animals and the set of non-animals = the null set; the second says that the intersection of the set of animals and the set of birds = the intersection of the set of birds and the set of animals. Boole's formulas, except for a few like AA = A, follow standard algebraic laws.

We can take Boolean formulas to be about sets or about statements. On the second interpretation, letters represent statements; "AB" means "A and B," "(1 − A)" represents "not-A," "1" represents "true," and "0" represents "false." On this interpretation, the first boxed formula above means "(A and not-A) is false": the **law of non-contradiction**; and the second boxed formula means "(A and B) is equivalent to (B and A)": the law of commutation. When we speak of *Boolean operators* today, we often have in mind the second interpretation; such Boolean operators include "and," "or," "not," and other **propositional** connectives.

Boole, who is considered the father of **mathematical logic**, thought that logic belonged with mathematicians instead of philosophers. But the effect of his work was to make logic a subject shared by both groups, each getting the slice of the action that fits it better.

BOX-INSIDE/BOX-OUTSIDE AMBIGUITY. In English and many other languages, sentences like "If A, then it is necessary (must be) that B" are ambiguous between two **modal logic** formulas:

$(A \supset \Box B)$ = If A, then B (by itself) is necessary.

$\Box(A \supset B)$ = It is necessary that if A then B.

So "If you are a bachelor, then you must be unmarried" could have either of these two meanings:

	=	"If you are a bachelor, then you are
$(B \supset \Box U)$		*inherently unmarriable* (in no possible world would anyone ever marry you)."
	=	If B, then U (by itself) is necessary.
	=	"It is necessary that *if* you are a bachelor
$\Box(B \supset U)$		*then* you are unmarried."
	=	It is necessary that if B then U.

The box-inside "$(B \supset \Box U)$" posits an *inherent necessity*: given that the antecedent is true, then "You are unmarried" is inherently necessary. This is insulting and false. The box-outside "$\Box(B \supset U)$" posits a *relative necessity*: what is necessary is not "You are a bachelor" or "You are unmarried" by itself, but only the connection between the two. This version is true because "bachelor" means *unmarried man*.

Likewise, "If A, then it is impossible (could not be) that B" is ambiguous between "$(A \supset \Box \sim B)$" and "$\Box(A \supset \sim B)$."

The medievals called the box-inside form the "necessity of the *consequent*" (the second part being necessary); they called the box-outside form the "necessity of the *consequence*" (the IF-THEN being necessary). The ambiguity is important; some intriguing but fallacious philosophical arguments depend on the ambiguity for their plausibility.

Here is an example of an argument involving the ambiguity:

> If you are a bachelor, then you must be unmarried.
> You are a bachelor.
> ∴ It is logically necessary that you are unmarried.

Here the first premise could have either of these two meanings:

| $(B \supset \Box U)$ | "If you are a bachelor, then you are *inherently unmarriable*—in no possible world would anyone ever marry you." (This is false.) |

| $\Box(B \supset U)$ | "It is necessary that *if* you are a bachelor *then* you are unmarried." (This is true because "bachelor" means *unmarried man*.) |

So the argument is ambiguous between these two symbolizations:

Box Inside	(B ⊃ □U) B ∴ □U	□(B ⊃ U) B ∴ □U	*Box Outside*

The box-inside interpretation, as we saw, has a false premise and thus is unsound. The box-outside interpretation is invalid; to get the "□U" conclusion requires a "□B" premise—the "B" premise gets us only a "U" conclusion. To refute the box-outside argument, we can use this galaxy of possible worlds—which has "B" and "U" both true in the actual world but both false in some alternative possible world W:

	B, U
W	~B, ~U

In this galaxy, the box-outside argument has true premises and a false conclusion, making it invalid. So the argument that you are inherently unmarriable ("□U") fails, since the box-inside form has a false premise while the box-outside form is invalid.

This example of the fallacy is more philosophically interesting:

> If you know, then you could not be mistaken.
> You could be mistaken.
> ∴ You do not know.

This, when generalized to deny all cases, is an argument for skepticism, which is the view that we never know anything. Since we cannot have knowledge if we are mistaken, and we always could be mistaken, it seems to follow that we cannot ever have knowledge. But again, the first premise is ambiguous; it could mean either of these:

(K ⊃ □~M)	"If you know, then you are *inherently incapable of being mistaken*—in no possible world are you mistaken." (This is false, since it reserves knowledge for infallible beings.)
□(K ⊃ ~M)	"It is necessary that *if* you know *then* you are not mistaken." (This is true because "S *knows* that P" entails "P is true.")

So the argument is ambiguous between these two symbolizations:

$$\textit{Box Inside} \quad \begin{array}{c} (K \supset \Box \sim M) \\ \Diamond M \\ \hline \therefore \sim K \end{array} \qquad \begin{array}{c} \Box (K \supset \sim M) \\ \Diamond M \\ \hline \therefore \sim K \end{array} \quad \textit{Box Outside}$$

The box-inside interpretation, as we saw, has a false premise and thus is unsound. The box-outside interpretation is invalid; to get the "~K" conclusion requires a "M" premise (that you *are* mistaken)—the premise "◇M" premise gets us only a "◇~K" conclusion (that it is logically possible that you do not know). To refute the box-outside argument, we can use this galaxy of possible worlds—which has "You know and are not mistaken" true in the actual world but "You are mistaken and do not know" true in some alternative possible world W:

$$W \quad \boxed{\begin{array}{c} K, \sim M \\ \hline M, \sim K \end{array}}$$

In this galaxy, the box-outside argument has true premises and a false conclusion, making it invalid. So the argument fails, since the box-inside form has a false premise while the box-outside form is invalid.

Other arguments in this dictionary that involve the box-inside/box-outside ambiguity include an argument from Boethius about whether divine foreknowledge precludes human freedom (*see* **ancient logic since Aristotle**), an argument from **Aristotle** about whether future contingent statements are true or false, and a **quantified modal** argument about **Aristotelian essentialism**.

BUDDHIST LOGIC. From about the third century BC, there has existed in India and neighboring areas (including China and Tibet) an Eastern logic tradition parallel to and independent of the Western tradition that started with **Aristotle** (384–322 BC). This Eastern tradition includes mostly Buddhists but also some Hindus and others. It studies topics like patterns of inference, grammar, debate rules, fallacies, reference, contradictions, and universals (with realists debating nominalists). A characteristic inference pattern is the five-step syllogism:

1   Here there is fire,   {conclusion}
2   because there is smoke.   {premise}
3   Wherever there is smoke there is fire, as in
     a kitchen.   {rule and example}
4   Here there is smoke.   {premise}
5   ∴ Here there is fire.   {conclusion}

The first two steps here repeat the last two; the last three steps mirror a **quantificational logic** inference:

3   All cases of smoke are cases of fire. $(x)(Sx \supset Fx)$
4   This is a case of smoke. St
5   ∴ This is a case of fire.   ∴ Ft

This reconstruction omits "as in a kitchen," which points to an **inductive** justification of the rule; in our experience of smoke and fire (as in the kitchen), smoke always seems to involve fire.

In about the second century BC, the Nyāya (logical) school of Hindu philosophy began in India. It taught four ways to reach knowledge: direct experience, inference, analogy, and testimony.

The Eastern logic tradition is as yet poorly understood in the West; it includes many thinkers over many centuries, with few texts translated into Western languages, and with many texts being difficult to interpret. Some Western commentators emphasize similarities between East and West; they see human thinking as essentially the same everywhere. Others emphasize differences and caution against imposing a Western framework on Eastern thought. And some **dialethic** logicians see the Eastern tradition as more congenial to their view than is the **Aristotelian** approach dominant in the West.

In the East, as in the West, there are differences of opinion; we should be critical of oversimplified stereotypes of Eastern and Western thought. So Buddhism, besides having this logic tradition, in its Zen form has what seems to be an anti-logic tradition. The Zen masters delighted in using *koans*, or paradoxical riddles, to move students beyond logical thinking toward a mystical enlightenment; we are to meditate, for example, on the sound of one hand clapping. Western logicians, too, delight in **paradoxes** but typically regard them as problems to be solved rather than as mysteries to be encountered.

BURDEN OF PROOF. Some beliefs (like those of common sense or those about a person's legal innocence) are generally presumed to be true unless there is strong contrary evidence. Whoever denies such beliefs has the responsibility (or "burden of proof") to provide this contrary evidence.

– C –

CALCULUS. *See* FORMAL SYSTEM.

CANTOR, GEORG (1845–1918). German mathematician who created modern **set theory** and the theory of *transfinite numbers* (which are infinite numbers that may differ in size).

Cantor proposed that two sets *have the same number of members* if the members of one can be matched up one-to-one with the members of the other. For example, the set of Gospels and the set of Beatles have the same number of members:

$$\begin{array}{lll} \text{Set of Gospels:} & \{\text{Matthew, Mark, Luke, John}\} \\ \text{Set of Beatles:} & \{\text{John, Paul, George, Ringo}\} \end{array}$$

More surprisingly, there are the same number of positive integers as there are *even* positive integers:

$$\begin{array}{ll} \text{Set of Positive Integers:} & \{1, 2, 3, \ldots\} \\ \text{Set of Even Positive Integers:} & \{2, 4, 6, \ldots\} \end{array}$$

A set with the same number of items as the set of positive integers is said to be *denumerable*, or *countably infinite*. Using a Hebrew letter, the set of positive integers is said to have $\aleph_0$ (aleph-null) members.

Rational numbers are numbers that can be expressed as fractions made out of positive integers (like 2/7 or 459/3). It can be shown that the number of rational numbers is denumerable; so there are the same number of rational numbers as there are positive integers.

Real numbers are numbers that can be expressed as infinite decimal expansions. An example of a real number that is not a rational number is π (the ratio between the circumference and diameter of a circle), which is roughly 3.1415926535 and is precisely the sum of the infinite series 4/1 − 4/3 + 4/5 − 4/7 + 4/9 − 4/11 + 4/13 − . . . . Another real number that is not a rational is √2 (the square root of two), which is about 1.4142135623. Cantor used a *diagonalization argument* to prove that there are more real numbers than there are positive integers; so the set of real numbers is said to be *undenumerable*, or *uncountably infinite*. Intuitively, there are more points in a line (where a point corresponds to a real number) than there are positive integers. Cantor came up with infinite sets even larger than the set of real numbers; in fact, he came up with an ascending series of ever larger infinite sets.

Cantor's diagonalization argument to prove that there are more real numbers than positive integers is clever. Suppose the real numbers between 1.0 and 0.0, each written as infinite series of digits, can be put into a list—so we have a first such number, a second, and so forth—thus matching up each such real number one-to-one with a positive integer, as below:

.03456792345 . . .
.54396857124 . . .   ←   We create a new
.76792039485 . . .   number that is not
          . . .   on the list.

Then we can define a real number in this range that is *not* on the list. Let the nth digit of this unlisted real number be based on, but different from, the nth digit in the nth formula; let us say, to be definite, that if the nth digit in the nth formula is "4" then we will use "1" for the nth digit of our unlisted number, and otherwise we will use "4." In the case above, the unlisted number will start ".414 . . ." (based on the digits "0," "4," and "7"). This unlisted real number between 1.0 and 0.0 will differ in at least one digit from the each member of the list, which by hypothesis contains all the real numbers between 1.0 and 0.0. Thus the hypothesis that these real numbers can be put into a list, so that each real number is matched up one-to-one with a positive integer, is false. So there are more real numbers than positive integers.

So we have the number of positive integers and the number of real numbers. Both numbers are infinite, but the second is larger than the first. The *continuum hypothesis* says that there are no infinite numbers intermediate between the two; no one has yet been able to prove or disprove the continuum hypothesis.

Cantor's approach to transfinite numbers strikes many people as bizarre and unintuitive; and it can easily lead to contradictions if we are not careful. While many mathematicians delight in Cantor's results, **intuitionist** mathematicians reject the whole thing.

CARROLL, LEWIS (1832–1898). Pseudonym for Charles L. Dodgson, an eccentric English logician. Carroll delighted in silly **syllogisms**, like "No misers are unselfish; none but misers save egg-shells; so no unselfish people save egg shells." He wrote *Alice in Wonderland* and other children's books that entertain and drive home points about logic. He loved logic puzzles, of the kind that torment students taking the LSAT and begin like "All whom I have taught, except my own sons, know something." He introduced a box-alternative to **Venn** diagrams. And he developed ways to symbolize logical relationships, including these:

$$
\begin{aligned}
a' &= \text{Not-a} \\
(a \dagger b) &= \text{a and b} \\
(a \P b) &= \text{If a then b} \\
a_1 &= \text{Some a exists} \\
a_0 &= \text{No a exists} \\
ab_1 &= \text{Some a is b} \\
ab'_0 &= \text{No a is not-b}
\end{aligned}
$$

He saw "All a is b" as the conjunction of the last two, thus assuming that each general term refers (*see* **square of opposition**).

CATEGORICAL STATEMENT. Statement of one of these four forms: "all A is B," "no A is B," "some A is B," and "some A is not B." *See also* SYLLOGISTIC LOGIC.

CAUSAL LOGIC. A form of **modal logic** developed by Arthur Burks, in which "☐" is used for "it is causally necessary that." It can perhaps be used to analyze subjunctive conditionals (*see* **conditional, kinds of**).

CHARITY, PRINCIPLE OF. The principle that we ought to interpret another's unclear ideas in such a way that they become as sensible and plausible as possible. This may involve understanding ambiguous terms one way rather than another, supplying implicit premises where needed, or adding needed qualifications to premises.

CHILDREN. Fifth-graders have been introduced to logic, seemingly with great success, as part of Matthew Lipman's philosophy for children program. *Harry Stottlemeier's Discovery* is the classic logic text for fifth-grade students.

CHURCH, ALONZO (1903–1995). American logician who helped found the Association for Symbolic Logic in 1935 and who edited its *Journal of Symbolic Logic* from 1936 to 1979. He showed (*Church's theorem*) that the problem of determining validity in **quantificational logic** cannot be reduced to an **algorithm** (a finite mechanical procedure). His related *Church's thesis* proposes that the intuitive idea of an algorithm matches a more precise mathematical definition in terms of "recursive functions"; his student **Alan Turing** later showed that the latter is equivalent to what could be done by a simple kind of computer called a *Turing machine*. Church also developed the *lambda calculus*, which symbolizes **functions** using the Greek letter λ.

CIRCULAR REASONING (begging the question). The **fallacy** of presuming the truth of what is to be proved. A simple example is "The soul is immortal, because it cannot die"; the premise here just repeats the conclusion in different words. A series of arguments is circular if it uses a given premise to prove a conclusion, and then uses that conclusion to prove the premise; for example, one might say "We know that the Bible tells the truth, because it is God's word; we know the Bible is God's word, because the Bible says so and it tells the truth."

CLASS. Some use "class" as synonymous with "**set**." For others, a *class* is a collection of objects none of which are themselves classes; then sets can contain other sets, but classes cannot contain other classes.

CLASSICAL SYMBOLIC LOGIC (standard logic). Includes **propositional logic**, first-order **quantificational logic**, and often **identity logic**. An approach to these systems is considered "classical" if it accords with the systems of **Gottlob Frege** and **Bertrand Russell** about which arguments are valid, regardless of differences in symbolization and proof techniques. *Classical symbolic logic* is opposed to **traditional logic**; to **deviant logics**, which disagree on which arguments are valid; and to *extensions of classical logic*, which extend but do not conflict with classical logic. *See also* LOGIC: DEDUCTIVE SYSTEMS.

COGNITIVE SCIENCE. An interdisciplinary approach to thought that includes areas like linguistics, **psychology**, **biology** (especially areas dealing with brain and sensory systems), **computers** (especially artificial intelligence), and philosophy (especially **logic, epistemology**, and philosophy of **mind**).

COHERENCE CRITERION. Other things being equal, we ought to prefer a theory that harmonizes with existing well-established beliefs. *See also* SCIENTIFIC REASONING.

COMPACTNESS THEOREM. The theorem, which can be proved for **propositional** or **quantificational logic**, that if an infinite set of sentences is inconsistent, then some finite subset of this set is inconsistent.

COMPLETE SYSTEM. Formal system in which every valid argument expressible in the system is provable in the system. *See also* META-LOGIC; SEMANTICS/SYNTAX.

COMPLEX QUESTION. The **fallacy** of asking a question that assumes the truth of something false or doubtful. The classic example is, "Are you still beating your wife?" A "yes" implies that you still beat your wife, while a "no" implies that you used to beat her. The question combines a statement with a question: "You have a wife and used to beat her; do you still beat her?" The proper response may be: "Your question presumes something false—namely that I have a wife and used to beat her." Sometimes it is misleading to give a "yes" or "no" answer.

COMPUTERS. Relate to logic in various ways. **Logic gates** are electrical or electronic devices whose input-output functions mirror **truth tables**; modern computers use a huge number of such logic gates, supplemented by memory and input-output devices. Logicians **John von Neumann** and Arthur Burks were important members of the team that designed the first large-scale electronic computer, the ENIAC, which was completed in 1946. The abilities and limitations of computers are often studied using logical models, such as **Turing** machines.

Programming languages somewhat resemble **formal systems** and generally incorporate basic notions of **propositional logic** like "and," "or," and "not"; books on programming languages often use truth tables to explain how these notions work. *Logic programming* languages, like Prolog, mirror **quantificational logic** and include *facts* (like "Socrates is a man") and *rules* (like "All men are mortal"); such languages are widely used in *artificial intelligence*, which tries to simulate the intelligent activities of human beings, like playing chess, writing poetry, or finding websites on a given topic.

Computerized **software for learning logic** can help students learn areas like proofs, translations, truth tables, and informal fallacies. *See also* MACHINES, LOGIC.

CONCLUSION. Statement that is supposed to be supported by other statements. *See also* ARGUMENT; INDUCTIVE/DEDUCTIVE.

CONDITIONAL (hypothetical). Statement of the form "If P then Q." **Propositional logic** symbolizes this as "(P ⊃ Q)." An English example is "If I went to Paris, then I went to Quebec." The if-part is called the *antecedent*, the then-part the *consequent*.

The word "if" functions in part as a left-hand parenthesis. So "It is false that if P then Q" translates as "~(P ⊃ Q)" and denies the whole conditional—while "If it is false that P, then Q" is "(~P ⊃ Q)" and says that if P is false then Q is true.

"(P ⊃ Q)" has this **truth table** (where T = true, F = false):

P Q	(P ⊃ Q)
F F	T
F T	T
T F	F
T T	T

An IF-THEN is true in every case except where we have the first part true and the second false.

"(P ⊃ Q)" claims that what we do not have is the first part true and the second false. Suppose you say "If I went to Paris, then I went to Quebec." By our table, you speak truly if you went to neither place, or to both, or to Quebec but not Paris. You speak falsely if you went to Paris but not Quebec.

Our truth table can produce unintuitive results. Take this example:

If I had eggs for breakfast, then   (E ⊃ W)
       the world will end at noon.

Suppose I did not have eggs, so E is false. By our table, the conditional is then true—since if E is false then "(E ⊃ W)" is true. This is strange. We would normally take the conditional to be false, since we would take it to claim that my having eggs would *cause* the world to end. So translating "if-then" as "⊃" does not seem totally right.

Our "⊃" is a simplified "if-then" that ignores elements like causal connections and temporal sequence. "(P ⊃ Q)" has a simple meaning; it just denies that we have P-true-and-Q-false:

(P ⊃ Q) If P is true, then Q is true.	=	~(P · ~Q) We do not have P true and Q false.

An IF-THEN understood this way is called a *material implication*. Most logicians contend that translating "if-then" this way is an odd but useful simplification, since it captures the part of "if-then" that normally determines validity. This understanding of "if-then" goes back to Philo of Megara (*see* **ancient logic since Aristotle**) and thus did not begin with modern logic. Some logicians dislike material implication and use alternative models for conditionals (*see* **relevance logic**).

Various inference rules govern conditionals. Most important are *modus ponens* (Latin for "affirming mode") and *modus tollens* (Latin for "denying mode"). Given an IF-THEN, if we have the antecedent true we can conclude that the consequent is true—and if we have the consequent false we can conclude that the antecedent is false:

$$\begin{array}{cc} \textit{Modus} & \dfrac{\begin{array}{c}(P \supset Q)\\ P\end{array}}{Q} \\ \textit{Ponens} & \end{array} \qquad \begin{array}{cc} \dfrac{\begin{array}{c}(P \supset Q)\\ \sim Q\end{array}}{\sim P} & \textit{Modus}\\ & \textit{Tollens} \end{array}$$

For example, assume that if you are a dog then you are an animal:

- If we add that you are a dog, we can conclude that you are an animal. This is *modus ponens* (also called "affirming the antecedent"): (D ⊃ A), D ∴ A.
- If we add that you are not an animal, we can conclude that you are not a dog. This is *modus tollens* (also called "denying the consequent"): (D ⊃ A), ~A ∴ ~D.
- If we add that you are not a dog, we cannot conclude that you are not an animal (since you might be a cat-animal). This is the *fallacy of denying the antecedent*: (D ⊃ A), ~D ∴ ~A.
- If we add that you are an animal, we cannot conclude that you are

a dog (since again you might be a cat-animal). This is the *fallacy of affirming the consequent*: (D ⊃ A), A ∴ D.

Here are two other common inference rules:

$$\text{Contra-} \quad \frac{(P \supset Q)}{(\sim Q \supset \sim P)} \qquad \frac{\begin{array}{c}(P \supset Q)\\(Q \supset R)\end{array}}{(P \supset R)} \quad \text{Hypothetical}$$
position   *Syllogism*

"If P than Q" and "If not-Q then not-P" are equivalent and are called **contrapositives** of each other.

Given that "(P ⊃ Q)" is equivalent to "∼(P · ∼Q)," then "∼(P ⊃ Q)" is equivalent to "(P · ∼Q)." So these simplification rules hold:

$$\frac{P, \sim Q}{\sim(P \supset Q)} \qquad \text{Conditional} \qquad \frac{\sim(P \supset Q)}{P, \sim Q}$$
                            Simplification

Defenders of relevance logic reject the second form. They also reject the *paradoxes of material implication* that are part of classical logic: that from "Not-P" we can infer "If P then Q"—and that from "Q" we can infer "If P then Q." *See also* BICONDITIONAL; CONDITIONAL, KINDS OF; ENTAILMENT; IDIOMS; TRUTH FUNCTIONS.

CONDITIONAL PROOF. A form of reasoning that derives an IF-THEN conclusion by assuming the antecedent and then deriving the consequent. *See also* DEDUCTION THEOREM; PROOF.

CONDITIONAL, KINDS OF. Conditionals and their relatives form a diverse family—going from logical entailment, through various weaker flavors of "if-then," and down to mere suggestion or insinuation. There is much controversy about how to analyze conditionals.

Classical symbolic logic analyzes the "If P then Q" conditional in a simple way, as just denying that we have P-true-and-Q-false:

(P ⊃ Q) If P is true, then Q is true.	=	∼(P · ∼Q) We do not have P true and Q false.

An IF-THEN understood this way is called a *material implication*. While most logicians contend that translating "if-then" this way is an odd but useful simplification, some logicians dislike this and propose alternative models for conditionals (*see* **relevance logic**).

Indicative conditionals differ from *counterfactual* ones—since the second example below is much less plausible than the first:

- *Indicative:* If Oswald did not kill Kennedy, someone else did.
- *Counterfactual:* If Oswald had not killed Kennedy, someone else would have.

A material-implication analysis clearly does not work for counterfactuals. A counterfactual entails that its antecedent is false; but this does not automatically make the whole IF-THEN true, as with a material implication. Some propose analyzing "If P had been true then Q would have been true" as something like "P is false, but in the **possible world** most similar to the present one in which P is true it also is the case that Q is true"; the last part is often symbolized as "(P □→ Q)."

There are also *subjunctive conditionals*, like "If this glass were to fall to the concrete floor, then it would break." Some suggest analyzing these using "It is *causally necessary* that if P then Q" ("ⒸP ⊃ Q)" in **causal logic**), where this means something like "The conjunction of P with the laws of nature (plus perhaps certain facts about circumstances) would logically entail Q."

The strongest member of the conditionals family is *logical entailment* (sometimes called *strict implication*). Standard **modal logic** analyzes "A logically entails B" as "It is logically necessary that if A is true then B is true" and symbolizes this as "□(A ⊃ B)"; this claims that in no possible situation is A true but B false. Defenders of **relevance logic** object to this analysis.

Much weaker than these is a statement of conditional **probability**, such as "It is likely that you were born in Iowa, given that you live on a farm there." This expresses a probabilistic connection between statements. It does not satisfy *modus ponens*, since adding "You live on a farm in Iowa" does not let us validly deduce "You were born there."

The weakest implication is *conversational implication*, where your speech merely suggests or insinuates something. Suppose you say "*Some* of my students work hard." This insinuates that *not all* of your students work hard, or at least that you do not think that they all work hard; if you had thought the latter, you would have said so, at least in normal circumstances. So when you make a weak claim instead of a stronger one, your speech often suggests that you take the stronger claim to be false (*see* **Paul Grice**).

CONJUNCTION. Statement of the form "P and Q." **Propositional logic** symbolizes this as "(P · Q)." An English example is "I have been to Paris and I have been to Quebec." The parts of a conjunction are called *conjuncts.*

The word "both" functions in part as a left-hand parenthesis and helps to avoid ambiguity. For example, "Not P and Q" has **scope ambiguity**; it could mean either of these:

$$(\sim P \cdot Q) \quad = \quad \text{Both not-P and Q}$$
$$\sim(P \cdot Q) \quad = \quad \text{Not-both P and Q}$$

The first is definite and says that P is false and Q is true. The second just says that not both are true (at least one is false).

"(P · Q)" has this **truth table** (where T = true, F = false):

P Q	(P · Q)	
F F	F	An AND is true if
F T	F	both parts are true—
T F	F	and is false if one or
T T	T	both parts are false.

The left side gives all possible truth combinations for "P" and "Q": maybe both are false, or just the second is true, or just the first is true, or both are true. "(P · Q)" is false in the first three cases and true in the third. "(P · Q)" claims that both parts are true.

Various inference rules govern conjunctions. From a conjunction, we can infer both conjuncts—and from two statements we can infer their conjunction:

$$\begin{array}{cc} \textit{Simplifi-} & \dfrac{(P \cdot Q)}{P, Q} \qquad\qquad \dfrac{P, Q}{(P \cdot Q)} & \textit{Conjunc-} \\ \textit{cation} & & \textit{tion} \end{array}$$

If not both are true, but this one is true, then the other must be false:

$$\frac{\begin{array}{c}\sim(P\cdot Q)\\ P\end{array}}{\sim Q} \qquad \begin{array}{c}\textit{Conjunctive}\\ \textit{Syllogism}\end{array} \qquad \frac{\begin{array}{c}\sim(P\cdot Q)\\ Q\end{array}}{\sim P}$$

The falsity of a conjunction follows from the falsity of a conjunct:

$$\frac{\sim P}{\sim(P\cdot Q)} \qquad\qquad \frac{\sim Q}{\sim(P\cdot Q)}$$

Order and grouping do not matter with conjunctions:

$$\textit{Commu-}\quad \frac{(P\cdot Q)}{(Q\cdot P)} \qquad \frac{((P\cdot Q)\cdot R)}{(P\cdot(Q\cdot R))} \qquad \frac{(P\cdot(Q\cdot R))}{((P\cdot Q)\cdot R)} \quad \textit{Associ-}$$
$$\textit{tation}\qquad\qquad\qquad\qquad\qquad\qquad\qquad\qquad\qquad \textit{ation}$$

People sometimes omit inner parentheses in a string of conjunctions.

Our symbolic "$\cdot$" is simpler than the English "and," which some-times can mean "and then" (as it does in "Suzy got married and had a baby"—which differs from "Suzy had a baby and got married"). Our "$\cdot$" just claims that both parts are true.

The "$\cdot$" must link whole statements. So it would be wrong to trans-late "Bob and Lauren got married to each other" as "(B $\cdot$ L)." This is wrong because the English sentence does not mean "Bob got married and Lauren got married" (which omits "to each other"). While **quanti-ficational logic** can translate the English sentence in as "Mbl" ("Bob married Lauren"), the best that propositional logic can do is translate it as a single letter, like "M." *See also* IDIOMS; TRUTH FUNCTIONS.

CONJUNCTIVITY PRINCIPLE. *See* BELIEF LOGIC.

CONNOTATION. *See* SENSE/REFERENCE.

CONSEQUENT. *See* CONDITIONAL.

CONSISTENT. Can have various meanings, depending on context, and can be applied to various sorts of things.

A *statement* is *logically consistent* if there is no self-contradiction in its being true; "I ran a mile in thirty seconds this morning," while physically impossible, is logically consistent (*see* **analytic/synthetic**). By extension, a *set of statements* is *logically consistent* if there is no self-contradiction in all its members being true at once. These definitions must be expanded if we want to apply "logically consistent" to imperatives, which are not true or false (*see* **imperative logic**).

A *statement* is *pragmatically consistent* if there is no logical absurdity in its being asserted; "I do not exist," while logically consistent (since it could have been true), is pragmatically inconsistent (since it cannot be asserted truly).

A **formal system** is *consistent* if there cannot be derived from it both a statement and also its contradictory. Proving the consistency of systems is an important part of **metalogic**.

A formal system containing **arithmetic** is *omega-consistent*, provided that, whenever every numerical instance of a formula is provable (e.g., "0=0," "1=1," "2=2," and so on), it is not provable that there is some number for which the formula is false (e.g., "(∃x)~x=x").

A *person* is *a logically consistent believer* if the person believes some things, believes a total set S of things that is logically consistent, and believes anything that follows logically from set S. Since this last phrase is highly idealized, it could be weakened to something like "and believes anything that can reasonably be seen as logically following from set S." *See also* BELIEF LOGIC; FORMAL ETHICS; SELF-REFUTING STATEMENT.

CONSTANT. A term with a fixed value—as opposed to a variable, which can assume different values. In the formula "x = 2," "x" is a variable while "2" is a constant. *See also* VARIABLE/CONSTANT.

CONTINGENT STATEMENT. One that without self-contradiction could have been either true or false; in **modal logic**, the claim that a statement A is *contingent* is symbolized as "($\Diamond$A · $\Diamond$~A)." In **propositional logic**, a formula is contingent if its **truth table** has some cases true and some false. *See also* ANALYTIC/SYNTHETIC.

CONTINUUM HYPOTHESIS. *See* CANTOR.

CONTRADICTION (logical falsehood). Any statement of the form "A and not-A" or, more generally, any statement that is not logically **consistent**. *See also* ANALYTIC/SYNTHETIC.

CONTRADICTORIES. A pair of statements so related that one must be true if and only if the other is false. In **propositional logic**, two formu-

las are *explicit contradictories* if they are exactly alike except that one
starts with an additional "~." *See also* SQUARE OF OPPOSITION.

CONTRAPOSITIVE. The contrapositive of a **conditional** is the result of
switching and negating both parts. So the contrapositive of "If you are a
dog, then you are an animal" is "If you are not an animal, then you are
not a dog"; both mean the same thing. Similarly the contrapositive of
"All A is B" is "All non-B is non-A," and both mean the same thing.

CONTRARIES. A pair of statements that cannot both be true, although it
could be that both are false; "This is white" and "This is black" are
contraries. *See also* SQUARE OF OPPOSITION.

CONVERSATIONAL IMPLICATION. *See* GRICE.

COROLLARY. A claim that follows easily from something just proved.

COUNTEREXAMPLE. An example that shows the falsity of a universal
claim. We can refute "All A is B" by finding something that is A but
not B; and we can refute "No A is B" by finding something that is A
and also B. *See also* EXCEPTIONS PROVING RULES.

CRAIG, WILLIAM. A logician best known for *Craig's theorem* (1953),
which tries to show the dispensability of theoretical terms in science.
Suppose we have a **formal system** that expresses a given scientific
theory, uses certain theoretical terms (e.g., "electron"), and has certain
observational implications (about what will be observed under certain
test conditions—where these are specified without using the theoretical
terms). Craig's theorem shows that there must then be a second formal
system that has all the same observational implications but contains
none of the theoretical terms.

– D –

DE MORGAN, AUGUSTUS (1806–1871). An English logician who is
best known for the *De Morgan laws* for **propositional logic:**

$\sim(A \cdot B)$	=	$(\sim A \lor \sim B)$
Not both A and B	=	Either not-A or not-B
$\sim(A \lor B)$	=	$(\sim A \cdot \sim B)$
Not either A or B	=	Both not-A and not-B

Set theory has similar laws; for example, "$-(A \cap B) = (-A \cup -B)$."

De Morgan complained that the logic of his day could not handle relational arguments like "All dogs are animals; therefore all heads of dogs are heads of animals." He introduced the idea of a limited **universe of discourse**. And, like his contemporary **George Boole**, he introduced new ways to symbolize logical relationships. He used capital letters for sets and corresponding small letters for their complements; so if "A" stands for the set of animals, then "a" stands for the set of non-animals. A period is used for "not." Parentheses are used to show *distribution* (*see* **syllogistic logic**); a letter is distributed if the parenthesis next to it has its open end facing the letter. So "all A is B," in which A is distributed but not B, is symbolized by "A))B"; likewise "some A is B," in which neither letter is distributed, is "A()B." This symbolism, fortunately, did not become popular.

DE RE/DE DICTO (**Latin for "about the thing/about the saying").** *De dicto* predication asserts something about a statement, as in "It is necessary that Socrates is mortal" or "I believe that the last Olympic gold medalist swimmer is a fine swimmer"; *de re* predication asserts something about a person or thing, as in "Socrates has the necessary property of being mortal" or "The person who is the last Olympic gold medalist swimmer is believed by me to be a fine swimmer." Using symbols, we can show the difference by where the modal or belief operator is placed; in *de dicto* predication the operator governs the whole statement, as in "□Ms" or "B$_i$Sg," while in *de re* predication the operator governs a expression with a variable that is "quantified into" from outside the operator, as in "$(\exists x)(x=s \cdot \Box Mx)$" or "$(\exists x)(x=g \cdot B_iSx)$." There is controversy about how to understand *de re* predication; **Willard Van Orman Quine** rejected it as unclear, while **Jaakko Hintikka** and **Alvin Plantinga** defended it. *See also* ARISTOTELIAN ESSENTIALISM; FREE LOGIC; QUANTIFIED MODAL LOGIC.

DECISION PROCEDURE. *See* ALGORITHM.

DEDUCTION THEOREM. The claim that if a conclusion C is derivable in a given system from a series of premises $P_1, P_2, \ldots, P_n$, then likewise the conditional $(P_n \supset C)$ is derivable from premises $P_1, P_2, \ldots, P_{n-1}$. The deduction theorem can be proved for most formal systems and gives a justification for **conditional proof**.

DEDUCTIVE. *See* INDUCTIVE/DEDUCTIVE.

DEFINITE DESCRIPTION. Phrase of the form "the so and so." Such phrases are meant to pick out a definite single person or thing. The classic analysis of definite descriptions is from **Bertrand Russell**.

Consider these two sentences and how they are commonly symbolized in **quantificational logic**:

Socrates is bald.	The king of France is bald.
= Bs	= Bk

The first sentence has a proper name ("Socrates") while the second has a definite description ("the king of France"); both seem to ascribe a property (baldness) to a particular object or entity. Russell thought this object-property analysis was misleading in both cases, but especially in the second; sentences with definite descriptions (like "the king of France") were in reality more complicated and should be analyzed in terms of a complex of predicates and quantifiers:

> The king of France is bald.
> = There is exactly one king of France, and he is bald.
> = For some x, x is king of France, there is no y such that y≠x and y is king of France, and x is bald.
> = $(\exists x)((Kx \cdot \sim(\exists y)(\sim y=x \cdot Ky)) \cdot Bx)$

Russell sometimes used "$(\iota x)Kx$" for "the entity that is K"; then "The King of France is bald" would be "$B(\iota x)Kx$." But here we will keep to the longer analysis, since it is more revealing.

Russell saw his analysis as having several advantages. First, "The king of France is bald" might be false for three reasons:

1.   There is no king of France;
2.   there is more than one king of France; or
3.   there is exactly one king of France, and he has hair.

In fact, "The king of France is bald" is false for the first reason: France is a republic and has no king. By contrast, the object-property analysis suggests that if "The king of France is bald" is false, then "The king of France is not bald" would have to be true—and so the king of France would have to have hair! So Russell's analysis seems to express better the logical complexity of definite descriptions.

Second, the object-property analysis of definite descriptions can lead us into metaphysical errors, like positing existing things that are not real. The philosopher Alexius Meinong allegedly argued as follows:

> "The round square does not exist" is a true
> statement about the round square.
> If there is a true statement about something,
> then that something has to exist.
> ∴ The round square exists.
> But the round square is not a real thing.
> ∴ Some things that exist are not real things.

For a time, Russell accepted this argument. Later he saw the belief in non-real existing things as foolish; he rejected Meinong's first premise and appealed to his theory of descriptions to clear up the confusion.

According to Russell, Meinong's error comes from his naïve object-property understanding of statements like "The round square does not exist." This, Russell contended, is not a true statement ascribing non-existence to some object called "the round square." If it were a true statement about the round square, then the round square would have to exist—which the statement denies. Instead, the statement just denies that there is exactly one round square. So Russell's analysis keeps us from having to accept that there are existing things that are not real.

Russell's approach has had its critics. Peter Strawson thought that it was truer to ordinary language to say that, in case there is not exactly one king of France, the sentence "The king of France is bald" is neither true nor false; instead, the question of whether the king of France is bald then "does not arise" or make sense. Russell responded that he was trying to clean up ordinary language and that it was an advantage to make the statement be true or false regardless of whether there is exactly one king of France. In this dispute, Strawson trusts ordinary language more than logical analysis—and Russell trusts logical analysis more than ordinary language. Russell saw ordinary language as often vague, ambiguous, and prone to lead us into philosophical errors; he searched for a logically ideal language to overcome these defects. *See also* QUANTIFIED MODAL LOGIC.

DEFINITION. Rule of paraphrase intended to explain **meaning**. More precisely, a definition of a word or phrase is a rule saying how to eliminate this word or phrase in any sentence using it and produce a second sentence that means the same thing—the purpose of this being to explain or clarify the meaning of the word or phrase.

Definitions may be *lexical* (explaining current usage) or *stipulative* (specifying our own usage). Here is a correct lexical definition: "bachelor" means "unmarried man." This says we can interchange "bachelor" and "unmarried man" in any sentence; the resulting sentence will mean the same as the original, according to current usage. This leads to the interchange test for lexical definitions:

> *Interchange Test:* To test a lexical definition claiming that A means B, try switching A and B in a variety of sentences. If some resulting pair of sentences do not mean the same thing, then the definition is incorrect.

According to our definition of "bachelor" as "unmarried man," for example, these two sentences would mean the same thing: "Albert is a *bachelor*" and "Albert is an *unmarried man*." These seem to mean the same thing. To refute the definition, we would have to find a pair of sentences that are alike, except for interchanging "bachelor" and "unmarried man," but do not mean the same thing.

Here is an incorrect lexical definition: "Bachelor" means "happy man." If the definition were correct, then these sentences would mean the same thing: "Albert is a *bachelor*" and "Albert is a *happy man*." But they do not mean the same thing, since we could have one true but not the other. So the definition is wrong.

The interchange test is subject to at least two restrictions. First, definitions are often intended to cover just one sense of a word that has various meanings; we should then use the interchange test only on sentences using the intended sense. Second, we should not use the test on sentences where the word defined appears in quotes.

To see further how the interchange test works, consider cultural relativism's definition of "good": "X is good" means "X is approved by my society." To evaluate this, we would try switching "good" and "approved by my society" in various sentences. For example, "Slavery is *good"* becomes "Slavery is *approved by my society.*" These two clearly do not mean the same thing, since it is consistent to affirm one but deny the other. Those who disagree with social norms often say things like "X is approved by my society, but it is not good"—and they are not contradicting themselves.

Here are five traditional rules for good lexical definitions:

1.   A good lexical definition is not too broad or too narrow.

Defining "bachelor" as "man" is too broad, since some men are not bachelors. And defining "bachelor" as "unmarried male astronaut" is too narrow, since some bachelors are not astronauts.

2.   A good lexical definition avoids circularity and poorly understood terms.

Defining "true" as "known to be true" is circular, since it defines "true" using "true." And defining "good" as "having positive aretaic value" uses poorly understood terms, since "aretaic" is less clear than "good."

3.   A good lexical definition matches in vagueness the term defined.

Defining "bachelor" as "unmarried male over 18 years old" is overly precise. The ordinary sense of "bachelor" is vague, since it is unclear on semantic grounds at what age the term begins to apply. "Man" or "adult" would be better, since these match "bachelor" in vagueness.

4.   A good lexical definition matches, as far as possible, the emotional tone (positive, negative, or neutral) of the term defined.

It will not do to define "bachelor" as "*fortunate* man who never married" or "*unfortunate* man who never married." These have positive and negative overtones; the original term "bachelor" is fairly neutral.

5.   A good lexical definition includes only properties essential to the term.

Suppose all bachelors live on planet earth. Even so, living on planet earth is not a property *essential* to the term "bachelor"—since we could imagine a bachelor who lives on the moon. So it is wrong to include "living on planet earth" in the definition of "bachelor."

Lexical definitions are important in philosophy. Many philosophers, from Socrates to the present, have sought lexical definitions for central concepts of human existence—like *knowledge, truth, virtue, goodness,* and *justice.* Such definitions are important in understanding and applying the concepts. Defining "good" as "what society approves of" would lead us to base our ethical beliefs on what is socially approved; we would reject this method if we defined "good" as "what I like" or "what God desires" or if we regarded "good" as indefinable.

Some *reductionist* philosophers analyze one sphere of reality in terms of another; for example, they analyze material objects into sensations, minds into behavior, or moral claims into what is socially approved. Other philosophers, like **Ludwig Wittgenstein** in his later years, try to show that many or most terms do not admit of strict definitions; such terms have to be elucidated in some other way, perhaps by describing how they fit into a form of life.

Now we will consider *stipulative definitions*; these specify how we are going to use a term. Since our use may differ from conventional usage, it would be unfair to criticize a stipulative definition for clashing with the latter. Stipulative definitions should be judged not as correct or incorrect but rather as useful or useless.

Logic uses stipulative definitions for terms like "logic," "argument," "valid," "wff," and so forth. These definitions specify the meaning we are going to use for the terms (which sometimes is close to their standard meaning) and create a technical vocabulary.

A *clarifying definition* is one that stipulates a clearer meaning for a vague term. For example, a scientist might stipulate a technical sense of "pure water" in terms of bacteria; this sense, while related to the normal one, is more precise. Likewise, courts might stipulate a more precise definition of "death" to resolve certain legal disputes; the definition might be chosen on moral and legal grounds to clarify the law.

Philosophers often use stipulative definitions. Here is an example: "In this discussion, I use 'rational' to mean 'always adopting the means believed necessary to achieve one's goals.'" This definition signals that the author will use "rational" to abbreviate a longer phrase; there is no claim that this exactly reflects the ordinary meaning of the term. Other philosophers may use "rational" in quite different senses, such as "logically consistent," "emotionless," or "always forming beliefs solely by the methods of science." These philosophers need not be disagreeing; they may just be specifying their technical vocabulary differently.

We could use subscripts for different senses; "rational$_1$" might mean "logically consistent," and "rational$_2$" might mean "emotionless." We should not think that because being rational in one sense is desirable, therefore being rational in some other sense also must be desirable.

While stipulative definitions need not accord with current usage, they should

- use clear terms that the parties involved will understand,
- avoid circularity,
- allow us to paraphrase out the defined term,
- accord with how the person giving it will use the term, and
- help our understanding and discussion of the subject matter.

Regarding the last norm, a stipulative definition is a device to abbreviate language. The introduction to this dictionary uses a stipulative definition: "*Logic* is the analysis and appraisal of arguments"; this lets us use one word in place of six. Without the definition, our explanations would be wordier and harder to grasp; so the definition is useful. Stipulative definitions should promote understanding. It is seldom useful to stipulate that a well-established term will be used in a radical new sense (for example, that "biology" will be used to mean "the study of earthquakes"); this would create confusion. And it is seldom useful to multiply stipulative definitions for terms that we seldom use. But at times we find ourselves repeating a cumbersome phrase over and over; then a stipulative definition can be helpful.

Some of our definitions seem to violate the "avoid circularity" norm. For example, the **propositional logic** entry defines "**wffs**" as sequences that we can construct using these rules:

1.   Any capital letter is a wff.
2.   The result of prefixing any wff with "~" is a wff.
3.   The result of joining any two wffs by "·" or "∨" or "⊃" or "≡" and enclosing the result in parentheses is a wff.

This definition is *recursive*: it first specifies some things that "wff" applies to and then specifies that if "wff" applies to certain things then it also applies to certain other things. So it defines "wff" in terms of "wff," which seems circular. And the definition does not seem to let us paraphrase out the term "wff"; we do not seem able to take a sentence using "wff" and say the same thing without "wff."

Actually, our definition is perfectly fine. We can rephrase it in the following way to avoid the apparent circularity and show how to paraphrase out the term "wff":

> "Wff" means "member of every set S of strings that satisfies these conditions: (1) Every capital letter is a member of set S; (2) the result of prefixing any member of set S with '~' is a member of set S; and (3) the result of joining any two members of set S by '·' or '∨' or '⊃' or '≡' and enclosing the result in parentheses is a member of set S."

See also ANALYTIC/SYNTHETIC; SENSE/REFERENCE.

DENOTATION. See SENSE/REFERENCE.

DENYING ANTECEDENT/CONSEQUENT. See CONDITIONAL.

DEONTIC LOGIC. A branch of logic that studies arguments whose validity depends on "ought," "permissible," and similar notions. This branch is less established, with much diversity. Here we will sketch the approach of Hector-Neri Castañeda, which builds on **imperative, propositional, quantificational,** and **modal logic.**

In Castañeda's approach, underlining turns indicatives into imperatives. So "A̲" and "Au̲" might mean "Do A," and "Ax̲y" might mean "X, do A to Y." Deontic logic adds two operators: "O" (for "ought") and "R" (for "all right" or "permissible"—many use "P" for this); these attach to imperatives to form deontic **wffs:**

> The result of writing "O" or "R," and then
> an imperative wff, is a deontic wff.

This rule lets us construct wffs like these:

OA̲	=	Act A is obligatory (required, a duty).
OAx̲	=	X ought to do A.
OAx̲y	=	X ought to do A to Y.
RA̲	=	Act A is all right (permissible).
RAx̲	=	It is all right for X to do A.
RAx̲y	=	It is all right for X to do A to Y.

"Ought" here is intended in the all-things-considered, normative sense that we often use in discussing moral issues. The deontic operators "O" and "R" are somewhat like the modal operators "□" and "◇."

Here are some further translations:

	=		Act A is not all right.
~R$\underline{A}$	=	O~$\underline{A}$	=	Act A ought not to be done.
	=		Act A is wrong.
O($\underline{A}$ · $\underline{B}$)	=	It ought to be that A and B.
(A ⊃ O~$\underline{B}$)	=	If you do A, then you ought not to do B.
O~($\underline{A}$ · $\underline{B}$)	=	You ought not to combine doing A with doing B.

Here are examples using quantifiers:

O(x)A$\underline{x}$	=	It is obligatory that everyone do A.
~O(x)A$\underline{x}$	=	It is not obligatory that everyone do A.
O~(x)A$\underline{x}$	=	It is obligatory that not everyone do A.
O(x)~A$\underline{x}$	=	It is obligatory that everyone refrain from doing A.

These two are importantly different:

O(∃x)A$\underline{x}$	=	It is obligatory that someone answer the phone.
(∃x)OA$\underline{x}$	=	There is someone who has an obligation to answer the phone.

The first might be true while the second is false; it may be obligatory (on the group) that someone or other answer the phone—while yet no specific person has an obligation to answer it. To prevent the "Let the other person do it" problem, we sometimes need to assign duties.

Compare these three:

O(∃x)(Kx · R$\underline{x}$)	=	It is obligatory that some who kill repent.
O(∃x)(K$\underline{x}$ · Rx)	=	It is obligatory that some kill who repent.
O(∃x)(K$\underline{x}$ · R$\underline{x}$)	=	It is obligatory that some both kill and repent.

The three wffs are different; the underlining shows which parts are obligatory: repenting, killing, or killing-and-repenting. If we attached "O" to indicatives, our formulas could not distinguish the forms; all three would translate as "O(∃x)(Kx · Rx)." Because of such examples, we need to attach "O" to imperative wffs, not to indicative ones. We cannot distinguish the three as "(∃x)(Kx · ORx)," "(∃x)(OKx · Rx)," and "(∃x)O(Kx · Rx)"—since putting "(∃x)" outside the "O" changes the meaning, as we saw in the previous paragraph.

We now add four inference rules for deontic **proofs**. The first two, following the modal and quantificational pattern, are for reversing squiggles. These hold regardless of what pair of contradictory imperative wffs replaces "A̲"/"~A̲" (here "→" means we can infer whole lines from left to right):

Reverse
Squiggle

~OA̲   →   R~A̲
~RA̲   →   O~A̲

These let us go from "not obligatory to do" to "permissible not to do"—and from "not permissible to do" to "obligatory not to do." These rules can be based on the interdefinability of the two deontic operators:

$$OA̲ \quad = \quad ~R~A̲$$
Act A is obligatory.   =   It is not all right to omit doing A.

$$RA̲ \quad = \quad ~O~A̲$$
Act A is all right.   =   It is not obligatory to omit doing A.

To present the next two rules, we need to expand the approach used in modal logic for **possible worlds** and world prefixes. For deontic logic, a *possible world* is a consistent and complete set of indicatives and imperatives. And a *deontic world* is a possible world (in this expanded sense) in which (a) the indicative statements are all true and (b) the imperatives prescribe some jointly permissible combination of actions. So then these equivalences hold:

OA̲   =   Act A is obligatory.
    =   "Do A" is in *all* deontic worlds.

RA̲   =   Act A is permissible.
    =   "Do A" is in *some* deontic worlds.

A *world prefix* is now a string of zero or more instances of the letters "W" or "D." As before, world prefixes represent possible worlds. "D,"

"DD," and so on represent deontic worlds; we can use these in derived steps and assumptions, such as:

D ∴ A (So deontic world D has A.)

DD asm: A (Assume deontic world D has A.)

We drop deontic operators using the next two rules (which hold regardless of what imperative wff replaces "$\underline{A}$"). This is the drop-"R" rule:

Drop "R" | R$\underline{A}$ → D ∴ $\underline{A}$, use a *new* string of Ds

Here the line with "R$\underline{A}$" can use any world prefix—and the line with "∴ $\underline{A}$" must use a world prefix that is the same except that it adds a *new* string (a string not occurring in earlier lines) of one or more Ds at the end. If act A is permissible, then "Do A" is in some deontic world; we may give this world an arbitrary and hence *new* name—corresponding to a new string of Ds. And this is the drop-"O" rule:

Drop "O" | O$\underline{A}$ → D ∴ $\underline{A}$, use a blank or any string of Ds

Here the line with "O$\underline{A}$" can use any world prefix—and the line with "∴ $\underline{A}$" must use a world prefix that is either the same or else the same except that it adds one or more Ds at the end. If act A is obligatory, then "Do A" is in all deontic worlds. So if we have "O$\underline{A}$" in the actual world, then we can derive "∴ $\underline{A}$," "D ∴ $\underline{A}$," "DD ∴ $\underline{A}$," and so on. This rule can be used to derive "Hare's law" (named after R. M. Hare):

Hare's Law

□(O$\underline{A}$ ⊃ $\underline{A}$)

An ought judgment entails the corresponding imperative: "You ought to do A" entails "Do A."

Hare's law (also called "prescriptivity") equivalently claims that "You ought to do it; but do not do it" is inconsistent. This law fails for some weaker *prima facie* or descriptive senses of "ought"; there is no inconsistency in this: "You ought (according to company policy) to do it; but do not do it." But the law seems to hold for the all-things-considered, normative sense of "ought"; this seems inconsistent: "All things considered, you ought to do it; but do not do it." However, some philosophers reject Hare's law; those who reject it would want to specify that in applying the drop-"O" rule the world prefix of the derived step cannot be the same as that of the earlier step.

Here is a deontic proof using these rules:

$$
\begin{array}{lll}
1 & \text{O}\underline{A} \\
2 & \text{O}\underline{B} \\
& [\therefore \text{O}(\underline{A} \cdot \underline{B}) \\
3 & \text{asm: } \sim\text{O}(\underline{A} \cdot \underline{B}) \\
4 & \therefore \text{R}\sim(\underline{A} \cdot \underline{B}) \quad \{\text{from 3}\} \\
5 & \text{D} \therefore \sim(\underline{A} \cdot \underline{B}) \quad \{\text{from 4}\} \\
6 & \text{D} \therefore \underline{A} \quad \{\text{from 1}\} \\
7 & \text{D} \therefore \underline{B} \quad \{\text{from 2}\} \\
8 & \text{D} \therefore \sim\underline{B} \quad \{\text{from 5 and 6}\} \\
9 & \therefore \text{O}\sim\underline{B} \quad \{\text{from 3; 7 contradicts 8}\}
\end{array}
$$

This is like a modal proof, except for underlining and having "O," "R," and "D" in place of "□," "◇," and "W."

This principle "O$\underline{A}$, O$\underline{B}$ ∴ O($\underline{A} \cdot \underline{B}$)" has been disputed using the premise that we have individual all-things-considered obligations, to do A and to do B, that cannot be combined. If we were convinced of this premise (as most are not), then we could modify the drop-"O" rule so that it could be used only once to put things in a given world; this would outlaw step 7 in the above proof.

These further quasi-logical "laws" are all somewhat controversial:

- *Kant's Law:* "Ought" implies "can."
- *Hume's Law:* We cannot deduce an "ought" from an "is."
- *Poincaré's Law:* We cannot deduce an imperative from an "is."

Here is a careful formulation of the first, named for **Immanuel Kant**:

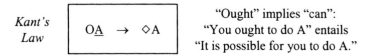

| *Kant's Law* | O$\underline{A}$ → ◇A | "Ought" implies "can": "You ought to do A" entails "It is possible for you to do A." |

This would hold regardless of what imperative wff replaces "<u>A</u>" and what indicative wff replaces "A"—provided that the former is like the latter except for underlining and every wff out of which the former is constructed is an imperative. This proviso outlaws "O(∃x)(Lx · ~L<u>x</u>) ∴ ◇(∃x)(Lx · ~Lx)" ("It is obligatory that someone who is lying not lie ∴ It is possible that someone both lie and not lie"), since "Lx" in the premise is not an imperative wff.

Kant's law equivalently claims that "You ought to do it, but it is impossible" is inconsistent. This law fails for some weaker *prima facie* or descriptive senses of "ought"; if company policy requires impossible things, then it could happen that we "ought" (according to company policy) to do these impossible things. But it is difficult to believe that we could have an all-things-considered moral obligation to do the impossible; this seems inconsistent: "All things considered, you ought to do it; but it is impossible to do it."

Our rule is a weak form of Kant's law. Kant thought that what we ought to do is not just *logically possible* but is also what we are *capable of doing* (physically and psychologically). Our rule expresses only the "logically possible" part; but we could for some arguments interpret "◇" in a stronger way, in terms of what we are capable of doing.

Hume's law (named for **David Hume**) claims that we cannot validly deduce what we *ought* to do from premises that do not contain "ought" or similar notions. Hume's law fails for some weak senses of "ought"; given descriptions of company policy and the situation, we can sometimes validly deduce what ought (according to company policy) to be done. Hume's law seems to hold for the all-things-considered, normative sense of "ought"; but a few thinkers disagree and say we can deduce moral conclusions using only premises about social conventions, personal feelings, God's will, or something similar. Here is a more careful wording of Hume's law:

Hume's Law ~□(B ⊃ O<u>A</u>)	We cannot deduce an "ought" from an "is": If B is a consistent non-evaluative statement and A is a simple contingent action, then B does not entail "Act A ought to be done."

This wording sidesteps trivial cases (like "b=c ∴ (OF<u>ab</u> ⊃ OF<u>ac</u>)")

where we clearly *can* deduce an "ought" from an "is."

Poincaré's law (named for Jules Henri Poincaré) claims that we cannot validly deduce an imperative from indicative premises that do not contain "ought" or similar notions. Here is a more careful wording:

Poincaré's Law $\sim\Box(B \supset \underline{A})$	We cannot deduce an imperative from an "is": If B is a consistent non-evaluative statement and A is a simple contingent action, then B does not entail the imperative "Do act A."

Again, the qualifications block trivial objections. *See also* AXIOLOGICAL LOGIC; BELIEF LOGIC; ETHICS; FORMAL ETHICS.

DEVIANT LOGIC. Any approach to logic that conflicts with **classical symbolic logic** on which arguments are valid. Deviant systems include **free, fuzzy, intuitionist, many-valued, paraconsistent, relative identity,** and **relevance** logics. There are three perspectives on such approaches: (1) classical logic is right and so these deviant logics are wrong; (2) at least some of these deviant logics are right and so classical logic is partly wrong; or (3) classical logic is like the theory of frictionless bodies in physics in that it is a useful simplification—so deviant logics serve not to replace classical logic but to show us its limitations and how it needs to be used with care.

If we take seriously the idea that **formal systems** are artificial languages, and thus things that we create, we might playfully consider the wide range of alternative systems. With **propositional logic**, for example, we could allow connectives (perhaps for "if-then") that are only partly **truth functional**, or formulas or proofs that are infinitely long, or formulas that are ambiguous because they do not have enough parentheses. We could have more than two **truth values** (maybe three or six or an infinite number), or just one (cynics might want it to be falsity), and change **truth tables** accordingly. We could allow a statement to have more than one truth value (maybe being both true and false), or a range of truth values, or no truth value; and we could define "valid" so that what is preserved in a valid inference is a range of truth values, or even falsity. We could follow English slang in taking a double negation (like "I don't say nothing") to be equivalent to a single negation ("I say nothing"). We could make theoremhood a matter of degree, or introduce chance factors into proofs ("flip a coin and if . . ."). While most of these ideas are just fun things to think about on a rainy day, a few of them might have some application.

There is much controversy about deviant logics. Some logicians are very radical and question such things as **modus ponens**, **modus tollens**, and the **law of non-contradiction**. Today there is more questioning of basic logical principles than ever before. Some see this as a breath of fresh air; they welcome the fact that logic is becoming, in some circles, as controversial as other areas of philosophy. Others defend orthodox classical logic and see the new trends as dangerous; they wonder what would happen to other areas of philosophy if we could not take for granted that *modus ponens* and *modus tollens* are valid and that contradictions are to be avoided.

DIAGONALIZATION ARGUMENT. *See* CANTOR.

DIALECTICAL LOGIC. *See* RENAISSANCE-TO-NINETEENTH-CENTURY LOGIC.

DIALETHISM. The view that a statement and its contradictory are sometimes both true. **Graham Priest** is a contemporary defender of this view; he rejects the **law of non-contradiction**, which has been part of orthodox logic since **Aristotle**. Priest does not say that *all* statements and their denials are true—but just that *some* are.

Here are examples where a statement and its negation might both be claimed to be true:

- "We do and do not step into the same river." (Heraclitus)
- "God is spirit and is not spirit." (the mystic Pseudo-Dionysius)
- "The moving ball is here and not here." (Hegel and Marx)
- "The round square is both round and not-round." (Meinong)
- "The one hand claps and does not clap." (Eastern paradox)
- "Sara is a child and not a child." (paradoxical speech)
- "What I am telling you now is false." (**liar paradox**)
- "The electron did and did not go in the hole." (quantum physics)

Most logicians would say that these are not genuine cases of a statement and its negation both being true. They contend that sensible cases of "A and not-A" must take "not" in a deviant sense or take the second

"A" to mean something different from the first. For example, "Sara is a child and not a child" can be sensible (and consistent) only if it really means something like "Sara is a *child-in-age* but not a *child-in-sophistication*." Paradoxical ways of speaking, although sometimes useful in provoking our thinking, seem not to make sense if taken literally. Dialethists need to show that at least some of their allegedly true self-contradictions resist such analyses.

If a single self-contradiction were true, then classical **propositional logic** could deduce from this that *every* statement and its opposite is true—which would bring chaos to human speech and thought. Dialethists respond by rejecting classical logic; instead, they defend a **paraconsistent logic** that lets us contain an occasional self-contradiction without leading to an "anything goes" logical nihilism.

DISCOVERY/JUSTIFICATION. We can ask how scientific laws are *discovered*, or we can ask how they are *verified*. History can tell us how Georg Simon Ohm discovered Ohm's law in 1827; philosophy is more concerned with how such laws are verified, or shown to be true, regardless of their origins. Basically, scientific laws are verified by a combination of observation and argument; but the details about **scientific reasoning** get complicated.

DISJUNCTION. Statement of the form "P or Q." **Propositional logic** symbolizes this as "(P ∨ Q)." An English example is "I went to Paris or I went to Quebec." The parts here are called *disjuncts*.

The word "either" functions in part as a left-hand parenthesis and helps to avoid ambiguity. For example, "Not P or Q" is ambiguous; it could mean either of these:

$$\sim(P \vee Q) \quad = \quad \text{Not-either P or Q}$$
$$(\sim P \vee Q) \quad = \quad \text{Either not-P or Q}$$

The first is definite and says that not either is true (both are false). The second just says that either P is false or Q is true.

"(P ∨ Q)" has this **truth table** (where T = true, F = false):

P Q	(P ∨ Q)
F F	F
F T	T
T F	T
T T	T

An OR is true if at
least one part is true
—and is false if both
parts are false.

The left side gives all possible truth combinations for "P" and "Q": maybe both are false, or just the second is true, or just the first is true, or both are true. "(P ∨ Q)" is false in the first case but true in the other three cases. "(P ∨ Q)" claims that *at least one* part is true.

The English "or" can have two senses. Our "∨" symbolizes the *inclusive* sense, which claims that one or the other *or both* are true. By contrast, the *exclusive* sense claims that at least one is true *but not both*; our symbolism can express this as "((P ∨ Q) • ~(P • Q))."

Various inference rules govern disjunctions. If not either is true, then both are false—and if both are false, then not either is true:

$$\frac{\sim(P \lor Q)}{\sim P, \sim Q} \qquad \textit{Disjunctive} \qquad \frac{\sim P, \sim Q}{\sim(P \lor Q)}$$
$$\textit{Simplification}$$

If at least one is true, but this one is false, then the other must be true:

$$\frac{(P \lor Q)}{Q}\,\frac{}{\sim P} \qquad \textit{Disjunctive} \qquad \frac{(P \lor Q)}{P}\,\frac{}{\sim Q}$$
$$\textit{Syllogism}$$

(Some **paraconsistent logics** reject disjunctive syllogism.) The truth of a disjunction follows from the truth of a disjunct:

$$\frac{P}{(P \lor Q)} \qquad \textit{Addition} \qquad \frac{Q}{(P \lor Q)}$$

Order and grouping do not matter with disjunctions:

$$\textit{Commu-} \quad \frac{(P \lor Q)}{(Q \lor P)} \qquad \frac{((P \lor Q) \lor R)}{(P \lor (Q \lor R))} \quad \frac{(P \lor (Q \lor R))}{((P \lor Q) \lor R)} \quad \textit{Associ-}$$
$$\textit{tation} \qquad\qquad\qquad\qquad\qquad\qquad\qquad\qquad\qquad \textit{ation}$$

People sometimes omit inner parentheses in a string of disjunctions.

See also IDIOMS; TRUTH FUNCTIONS.

DISTRIBUTED. *See* SYLLOGISTIC LOGIC.

DIVISION-COMPOSITION. The **fallacy** of arguing that something true of the whole must be true of all the parts or that something true of all the parts must be true of the whole. A great team need not have great players (since it could have average players who play well together)— and having great players does not necessarily produce a great team.

DOMAIN. *See* UNIVERSE OF DISCOURSE.

DOXASTIC LOGIC. *See* BELIEF LOGIC.

– **E** –

ELEMENTARY LOGIC. *See* QUANTIFICATIONAL LOGIC.

EMPIRICAL. *See* A PRIORI/A POSTERIORI.

EMPIRICIST. One who believes that we have no synthetic *a priori* knowledge—or, more broadly, one who emphasizes *a posteriori* knowledge and denies or minimizes *a priori* knowledge.

ENDLESS LOOP. Doing the same sequence of actions over and over, endlessly. *See* **loop, endless**.

You get into an endless loop if you follow the previous paragraph literally, since then you would go back and forth endlessly between the entries on "endless loop" and "loop, endless."

Endless loops are important in **computer** programming, since including one by mistake may cause a program to stall. Endless loops are also important in proof strategies. A common strategy for **quantificational** proofs involves (1) introducing a new letter for each existentially quantified formula and (2) dropping universal quantifiers for each letter. But this leads to an endless loop in relational arguments using "(x)(∃y)Rxy"; here for each new letter we must apply (2) and then (1), which gives us another new letter, and so we must again apply (2) and then (1), which gives us a new letter, and so on endlessly. The same problem can occur with any strictly mechanical or **algorithmic** way to construct quantificational proofs. *See also* CHURCH.

ENTAILMENT (strict implication). "A logically entails (or strictly implies) B" means "It is logically necessary that if A then B." **Modal logic** symbolizes this as "□(A ⊃ B)" or "(A ⊰ B)."

ENTHYMEME. Argument with an implicit but unstated premise.

ENUMERATIVE INDUCTION. A *sample-projection syllogism* of **inductive logic**.

EPISTEMIC LOGIC. The logic of knowledge. *See also* HINTIKKA.

EPISTEMOLOGY (theory of knowledge). Connects with logic in many ways. *Foundationalists* claim that some truths are given to us by experience and reason and that other truths must be "based on" these; this raises the question of what it is for A to be "based on" B—whether this is a deductive or inductive relationship, or perhaps something more complex with deductive or inductive components. In contrast, *coherentists* claim that the rationality of a set of beliefs depends not on how it relates to the given but on internal "coherence"; this raises the question of what it is for a set of beliefs to be "coherent"—whether this is a matter of logical consistency or some inductive relationship, or something more complex with deductive or inductive components.

Some philosophers think that logic imposes a structure on what can be known. With **Immanuel Kant**, this structure comes from **traditional logic** and leads to specific thought categories (like substance/accident and cause/effect) that we impose on experience. With **Bertrand Russell**, what is given is experiential terms and atomic facts, and what is believed may be compounded from these using ideas from **classical symbolic logic** (like "and," "or," "if-then," "not," "some," and "all").

There are questions about how truths of logic are known—whether, for example, logical truths are *a priori* or empirical, and whether they are based on language conventions or on the nature of things. *See also* LOGICAL PRINCIPLES, STATUS AND JUSTIFICATION OF; PHILOSOPHY OF LOGIC.

ESSENCE/EXISTENCE. *See* IS.

ESSENTIALISM. *See* ARISTOTELIAN ESSENTIALISM.

ETHICS. Connects with logic in many ways. First, there is the question of what logical laws cover terms like "ought" and "good"; **deontic** and **axiological logic** focus on this.

Second, there is the question of what consistency principles apply to ethical beliefs; **formal ethics** focuses on this.

Third, there is the question of how we can establish ethical conclusions. Can we deduce an "ought" from an "is"? Can we take strictly factual premises (perhaps about what produces pleasure or what is socially accepted or what the creator commands) and, without assuming any further premises about what is "good" or "ought to be done," validly deduce conclusions about what is "good" or "ought to be done"? **Hume**'s law says we cannot. A related question is how moral principles connect with concrete moral conclusions. Can we reason like this: "All actions that are F are wrong; this action is F; therefore this action is wrong"? (Here F might stand for a type of action, like intentionally killing an innocent human being, or a property about consequences, like not maximizing the balance of pleasure over pain for sentient beings.) Or must we reason like this: "Any action that is F is wrong unless there is strong reason for doing F; this action is F; there is no strong reason for doing F; therefore this action is wrong"?

EULER, LEONHARD (1707–1783). A Swiss mathematician who made many important contributions. In logic, he is best known for using *Euler circles* to diagram the categorical statements of **syllogistic logic.** "All A is B" is a smaller A-circle inside a larger B-circle. "No A is B" is an A-circle and a B-circle that are entirely apart. "Some A is B" is an A-circle and a B-circle that overlap; but the problem here is that this also seems to diagram "Some A is not B" and "Some B is not A":

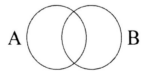

Venn diagrams are clearer on this point. They shade an area to show that it is empty and "×" an area to show that it has something in it; so "Some A is B" and "Some A is not B" are distinguished as follows:

Some A is B   Some A is not B

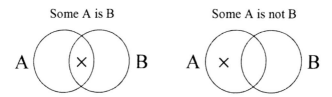

EXCEPTIONS PROVING RULES. What is the saying "The exception proves the rule" supposed to mean? There are three views on this.

1. It is just an inane thing you can say to perplex your opponent and avoid admitting that you have been refuted. Suppose you say "All As are Bs." Your opponent says, "This is an A but it is not a B—so you are wrong." Instead of saying "I guess I am wrong," and perhaps adding, "I should have said '*Most* As are Bs,'" you say, "The exception proves the rule." This seems to mean that you can prove "All As are Bs" from "This is an A but it is not a B"; but think of all the silly things that you could "prove" using that rule.

2. "Prove" in the saying really means "test." So the saying means that you can *test* a rule by looking for exceptions. If you find an exception (a **counterexample**), then the rule does not hold universally.

3. The rule in question is a legal one, like "Do not park on the lawn," which forbids something unless you get an exception. Asking for an exception presupposes that there is a general (but not exceptionless) rule to the contrary.

EXISTENTIAL QUANTIFIER. *See* QUANTIFICATIONAL LOGIC.

EXPECTED GAIN OF AN ALTERNATIVE. Sum of **probability** times gain of the various possible outcomes.

EXTENSION. *See* SENSE/REFERENCE.

EXTENSIONAL/INTENSIONAL. A sentence is *extensional* if its truth depends not on the meaning of the parts but only on their reference or **truth value.** "A and B" is extensional, since its truth depends solely on whether the parts are true, regardless of what they mean. By contrast, "It is necessary that A" is *intensional*, since its truth depends on what A means and not just on the truth of A. **Classical symbolic logic** is extensional, while **modal logic** is intensional.

– F –

FALLACY. A common error of thinking, especially one apt to mislead.

Formal fallacies are ones that can be expressed neatly in a standard system of logic, like **propositional logic**. Examples of formal fallacies include affirming the consequent (*see* **conditional**), the **box-inside/box-outside ambiguity**, the **quantifier-shift fallacy**, and inconsistency (*see* **belief logic, formal ethics**, and **self-refuting statement**).

Informal fallacies are ones that cannot be expressed neatly in a standard system of logic. Examples include **ad hominem, ambiguity, appeal to authority, appeal to emotion, appeal to force, appeal to ignorance, appeal to the crowd, beside the point, black-and-white thinking, circular reasoning, complex question, division-composition, false stereotype, gambler fallacy, genetic fallacy, hasty generalization, logic chopping, opposition,** *post hoc ergo propter hoc,* **pro-con argument,** and **straw man.** While this book tries to cover the more common informal fallacies, there are many more that are not included.

What is normally a fallacy may not be so in specific cases. Consider the "opposition fallacy," which argues that what your opponent says must thus be false. While this is normally a fallacy, it could happen that everything a specific opponent says is false; and we might have good evidence for thinking that the person is in fact always wrong. If this unlikely situation were to occur, then this would be sound reasoning: "Everything that my opponent says is false; my opponent says A; therefore A is false." So sometimes an accusation of fallacy can be rebutted (*see also* **form**).

FALSE STEREOTYPE. The **fallacy** of assuming that members of a certain group are more alike than they actually are. People commit this fallacy in thinking that all Italians exist only on spaghetti, that all New Yorkers are uncaring, or that all who read Karl Marx want to overthrow the government.

FALSITY. "A is *false*" means "Not-A is **true**" (where "Not-A" is the **negation** or **contradictory** of "A"). This is different from saying that "false" means "not true"—since imperatives and meaningless or vague claims can be neither true nor false. *See also* BIVALENCE; MANY-VALUED LOGIC; PARACONSISTENT LOGIC; TRUTH TABLES.

FEELINGS. While feelings are often contrasted with *logical thinking* (*see* **appeal to emotion** and **language, uses of**), people in fact usually appraise the validity of an **argument** by whether it "feels right." While untrained feelings about validity are often unreliable, we can train our feelings about validity to be highly reliable.

FEMINISM. *See* GENDER.

FIRST-ORDER LOGIC. *See* QUANTIFICATIONAL LOGIC; SECOND-ORDER LOGIC.

FORM. The *form* of an argument is how it is structured in terms of logical notions (like "all" and "not") and content phrases (like "Greek," "mortal," and "Socrates"). Consider the argument on the left:

> All Greeks are mortal.   all G is M
> Socrates is Greek.   s is G
> ∴ Socrates is mortal.   ∴ s is M

This argument is valid because of its form. We display its form on the right, by using letters for the content phrases and words for the logical notions; any argument with this same form will be valid. Thus "logical form" is a fill-in-the-blanks notion.

Here is an example of an invalid argument and the corresponding invalid logical form:

> All Greeks are mortal.   all G is M
> Socrates is mortal.   s is M
> ∴ Socrates is Greek.   ∴ s is G

This is invalid because the form is defective. In most cases, an argument's logical form determines whether it is valid.

In some cases, an argument might have a defective form and still be valid—as in these two examples:

> If John can do it, every-   If 2+2=4, then   If A then B
>   one can do it.   snow is white.   B
> Everyone can do it.   Snow is white.   ∴ A
> ∴ John can do it.   ∴ 2+2=4.

Both have the invalid **conditional** form of affirming the consequent. But the first argument is valid because its second premise by itself entails the conclusion by a **quantificational** form: "(x)Dx ∴ Dj." The second argument is valid because, since it is logically impossible to have the conclusion false, it is logically possible to have the premises true and conclusion false; the second argument might be said to be *materially valid*—valid because of its content instead of its form. Another example of a materially valid argument is "This is completely red. ∴ This is not completely green."

There are controversies about logical form. We spoke of "logical notions" and gave "all" and "not" as examples. But there are disputes about which notions are "logical"; for example, do the logical notions include "is a member of," "necessarily," or "it ought to be that"? Perhaps logical form is relative to what system of logic we are using.

It is sometimes unclear how fine-grained an analysis of form to give. Consider "If all Greek philosophers are mortal, then Socrates is mortal." Is the form here "If A then B" or "If all A is B then s is B" or "If all A that is B is C, then s is C"? In practice, we break a specific argument down enough to give an adequate analysis of its validity.

FORMAL ETHICS. The study of *formal ethical principles*. These are ethical principles that can be formulated using abstract logical or quasi-logical notions (including terms like "believing" and "desiring") and variables (for things like persons and actions). Formal ethical principles often are consistency requirements and presuppose the norm that *one ought to be consistent*, which is subject to some implicit qualifications (*see* **belief logic**). Formal ethics is a bridge between logic and **ethics**; some principles of formal ethics are part of belief logic and others can be derived by adding further principles to belief logic.

These formal ethical principles prescribe consistency among beliefs:

> Do not combine inconsistent beliefs.
> = If A is inconsistent with B, then do not combine *believing A* with *believing B*.

> Do not believe something without believing what follows from it.
> = If A logically entails B, then do not combine *believing A* with *not believing B*.

The appeal to consistency in beliefs is important in ethical discussions. When we appeal to a moral principle to defend our beliefs, we can be criticized if we do not apply it consistently. So a white racist who claims that people of lower intelligence ought to be treated poorly can be criticized if he does not also believe that *white* people of lower intelligence ought to be treated poorly.

While the consistency principles above are formulated as "Do not" imperatives, they could equally well be formulated using "One ought not." While the former is simpler, the latter more clearly brings out the ethical nature of the principles.

Here are some further formal ethical principles—all of which forbid combinations that are in some sense inconsistent:

> *Ends–Means Consistency:* Keep your means in harmony with your ends. Do not combine *wanting* to attain end E, *believing* that taking means M is needed to attain this end, and *not acting* to take means M.

> *Conscientiousness:* Keep your actions, resolutions, and desires in harmony with your moral beliefs. For example, do not combine *believing* that you ought to do A with *not acting* to do A.

> *Impartiality:* Make similar evaluations about similar actions, regardless of the individuals involved. Do not believe that you ought to do A without believing that any such act ought to be done in similar circumstances, regardless of where you or others are in the situation.

> *Golden Rule:* Treat others only as you consent to being treated in the same situation. Do not act to do A to X without being willing that if the situation were reversed then A would be done to you.

> *Formula of Universal Law:* Act only as you are willing for anyone to act in the same situation—regardless of imagined variations of time or person.

While these can be symbolized and proved rigorously in an expanded version of belief logic, the details are too complex to go into here. An important part of the expansion is to add logical machinery to cover the **universalizability principle**:

Universalizability: If it is all right for X to do A, then it would be all right for anyone else to do A in the same situation.

=   If act A is permissible, then there is some universal property (or conjunction of such properties) F, such that: (1) act A is F, and (2) in any actual or hypothetical case every act that is F is permissible.

= $(R\underline{A} \supset (\exists F)(F\underline{A} \cdot \blacksquare(\underline{X})(F\underline{X} \supset R\underline{X})))$

While the "all right" form is given here, universalizability also holds with other ethical terms, like "ought" or "wrong."

FORMAL LOGIC. *See* INFORMAL/FORMAL LOGIC.

FORMAL PROOF. *See* PROOF.

FORMAL SYSTEM (calculus). Artificial language with notational grammar rules and notational rules for determining validity or theoremhood. Branches of logic and math are often presented as formal systems; *see* **propositional logic** for an example.

In presenting a formal system, one must specify which symbols will be used, which sequences of symbols will be considered **wffs** (well-formed formulas), and which sequences of wffs will be considered **proofs**. These must be specified in a purely **syntactic** or notational way, in terms of rules (**algorithms**) about the manipulation of symbols, in abstraction from what the symbols might be taken to mean.

To connect a formal system with claims about reality, there is often added a **semantics**, which is about what the symbols means. Thus we might specify that certain letters represent true-or-false statements, or objects, or properties of objects; and a system of arithmetic might specify that certain symbols refer to natural numbers 0, 1, 2, 3, and so on.

FREE LOGIC. Version of **quantificational logic** that is free of the assumption that individual constants must refer to existing beings. Free logic allows empty singular terms, like "s" for "Santa Claus," and so would reject this inference as invalid:

> (x)Fx   Every existing being is F.
> ∴ Fs   ∴ Santa Claus is F.

For this to be valid, free logic requires a further premise that Santa is an

existing being: "(∃x)x=s" ("There exists an x such that x is Santa").
Free logic weakens the drop-universal rule to this:

(x)Fx, (∃x)x=a → Fa, where "a" is *any* constant	Every existing being is F. a is an existing being. ∴ a is F.

And it strengthens the drop-existential rule to this:

(∃x)Fx → Fa, (∃x)x=a, where "a" is a *new* constant	Some existing being is F. ∴ a is F (for some new "a"). ∴ a is an existing being.

When we drop an existential, we get an existence claim, like "(∃x)x=a," to use in dropping universals. The resulting system can prove almost everything that could be proved before, but with longer proofs. The main effect is to block a few proofs, like the one above about Santa Claus; "(x)Fx ∴ Fs" becomes invalid.

This version of free logic would allow empty **possible worlds** with no entities. In such worlds, all **wffs** starting with existential quantifiers are false and all wffs starting with universal quantifiers are true. So then existence claims like "(∃x)(Fx ∨ ~Fx)" would not be provable.

We might also add an inference rule to the effect that whatever has positive properties must exist; in a consistent story where Harry Gensler does not exist, Gensler could not be a logician or a backpacker. So from "a has property F" we could infer "a exists":

Fa → (∃x)x=a	a has property F. ∴ a is an existing being.

This rule holds regardless of what constant replaces "a," what variable replaces "x," and what wff containing only a capital letter and "a" and perhaps other small letters (but nothing else) replaces "Fa." By this rule, "Descartes thinks" entails "Descartes exists." The example on the left below is not an instance of this rule (since the wff substituted for "Fa" cannot contain "~"):

	Wrong:		*Right:*
~Fa	a is not F	Fa	a is F
∴ (∃x)x=a	∴ a exists	∴ (∃x)x=a	∴ a exists

This point is confusing because "a is not F" in English can have two different senses. "Descartes does not think" could mean either of these:

(∃x)(x=d · ~Td)	=	Descartes is an existing being who does not think.
~(∃x)(x=d · Td)	=	It is false that Descartes is an existing being who thinks.

The first form is *de re* (about the thing); it affirms the property of being a non-thinker of the entity Descartes. Taken this first way, "Descartes does not think" entails "Descartes exists." The second form is *de dicto* (about the saying); it denies the statement "Descartes thinks," which may be false either because Descartes is a non-thinking entity or because Descartes does not exist. Taken this second way, "Descartes does not think" does not entail "Descartes exists."

One might object to this rule on the grounds that Santa Claus has properties (such as being fat) but does not exist. But various stories predicate conflicting properties to Santa; they differ, for example, on what day he delivers presents. Does Santa have contradictory properties? Or is one Santa story uniquely "true"? What would that mean? When we say "Santa is fat," we mean that in such and such a story (or possible world) there is a being called Santa who is fat. We should not think of Santa as a non-existing being in our actual world who has properties such as being fat. Rather, what exists in our actual world is stories about there being someone with certain properties—and children who believe these stories. So Santa need not make us give up the rule that "a has property F" entails "a exists." *See also* GENERAL/ SINGULAR TERM; QUANTIFIED MODAL LOGIC.

FREGE, GOTTLOB (1848–1925). German philosopher and mathematician who created modern logic.

Frege's 1879 *Begriffsschrift* ("Concept Writing") marked the beginning of **classical symbolic logic**. This slim book of 88 pages introduced a strange symbolism that, for the first time, allowed one to combine

quantifier words ("all," "some," "no") and **truth-functional** connectives ("and," "or," "if-then," "not") in every conceivable way. So we could now symbolize arguments using not just simple forms like "All A is B" and "If A then B" but also complex ones like "If everything that is A or B is then C if and only if it is D, then either something is not-A or everything that is B is A." Thus the gap between the logic of **Aristotle** and the logic of the Stoics (*see* **ancient logic since Aristotle**) was finally overcome in a higher synthesis. In addition, Frege's logic allowed one to analyze arguments involving **relations** (like "x loves y") and multiple-quantifiers. So we could now show that "There is someone that everyone loves" entails "Everyone loves someone"—but not conversely (*see* **quantificational logic**). Frege presented his logic as a genuine **formal system**, with purely notational rules for determining the grammaticality of formulas and the correctness of proofs.

Frege's work, despite its importance, was largely ignored until **Bertrand Russell** came to praise it in the early years of the 20th century. Part of the problem was that Frege used a symbolism that seemed strange and unintuitive; few people took the time to master his many pages of complex, tree-like diagrams. Frege used lines for "not," "if-then," and "all":

Not-A	If A then B	For all x
$\dashv\!\!-A$	$\displaystyle \overline{} \begin{array}{l} B \\ A \end{array}$	$\neg\!\!-x\!\!-$

These can be combined to symbolize forms like "Not all A is non-B," which is "$\sim(x)(Ax \supset \sim Bx)$" in our symbolism:

Not all A is non-B $\quad = \quad \neg\!\!-x\!\!- \begin{array}{l} Bx \\ Ax \end{array}$

This was also his way to write "Some A is B," or "$(\exists x)(Ax \cdot Bx)$" in our symbolism; he had no simpler notation for "some" or "and."

In 1884 Frege published his *Die Grundlagen der Arithmetik* ("The Foundations of Arithmetic"), which sketched his project of trying to analyze numbers in terms of sets. His *logistic thesis* proposed that **arithmetic** can be reduced to logic: every truth of arithmetic can be formulated using just notions of logic and proved using just axioms and inference rules of logic. His two-volume *Grundgesetze der Arithmetik*

("Basic Laws of Arithmetic"), which came out in 1893 and 1903, tried to fill in the details. But in 1902, between the publishing of the two volumes, Bertrand Russell sent him a letter showing a contradiction in Frege's system; this contradiction came to be called "Russell's paradox" (*see* **set theory** for more details). Frege was crushed, since he saw his life work as collapsing; he tried to deal with the problem in *Grundgesetze* volume 2, but did not meet with complete success. He never fully recovered intellectually. (How might his life have been different if he had used a **paraconsistent logic**?)

Frege was also important for his work on **philosophy of logic**, particularly for his extension of the traditional **sense/reference** distinction. He saw the sense of "human" as the properties it would ascribe to an object and the reference of "human" as the set of objects that the term can be truly predicated of. More controversially, he saw the sense of "Socrates is human" as a proposition and its reference as the true— both being considered **abstract entities**. *See also* GÖDEL; HUSSERL; MATHEMATICS; PEIRCE; RENAISSANCE-TO-NINETEENTH-CENTURY LOGIC.

FUNCTION. A term using variables that refers to exactly one entity for each value of the variables. "The father of x" is a function if each of us has exactly one father; here the **universe of discourse** is implicitly restricted to people. "The parent of x" is not a function, since people can have more than one parent. *See also* ALGORITHM.

FUZZY LOGIC. Can refer either to **many-valued logic**, in which truth comes in degrees, or to a kind of **set theory**, in which set membership comes in degrees. On the first approach, "This shirt is dry" is true to degree n; on the second, this shirt is to degree n a member of the set of dry shirts. In both cases, n can typically be any real number between 1 and 0; but on some approaches n has a fuzzier value, like "very" or "middlish."

Fuzzy logic is often used in devices like clothes dryers to permit more precise control. A crisp-logic dryer might have a rule that if the shirts are dry then the heat is turned off; a fuzzy-logic dryer might have a rule that if the shirts are dry to degree n then the heat is turned down to degree n. We could get the same results using **classical symbolic logic** and a relation "Dxn" that means "shirt x is dry to degree n."

– G –

GAMBLER FALLACY. The **fallacy** of arguing, for example, that, since the first five coin tosses have been heads, thus the next one must be tails. But the coin has no memory. *See also* PROBABILITY.

GENDER. Raises questions like these about logic and reasoning: Do women reason in a different way from men? If they do, then how should we characterize the difference? Is the difference genetic or cultural—and is one way better than another, or are both complementary or incommensurable (and thus cannot be compared in worth in any objective way)? Are women less logical than men?

The common stereotype of women until fairly recently was that they were intellectually inferior, since they were less logical and appealed more to emotion and intuition than to rationality. Such ideas were important in keeping women out of higher education and leadership roles in government and business—and in keeping them from voting. We can put these ideas into an argument:

>Women are less logical than men.
>Those who are less logical are intellectually inferior.
>∴ Women are intellectually inferior to men.

There are three ways to attack this argument, the first two of which reflect opposite poles of thought within the feminist movement.

1. We could dispute the first premise. Empirically, woman are as logical as men with the same educational background; experiments with college students show no significant difference in logical skills between men and women. (In the author's logic courses, women generally do a little *better* than men—taking students from 2003–2004, the 53 men averaged 91.0 on all exams while the 63 women averaged 93.8; perhaps this is because college-age women tend to be more mature for their age and thus approach study with more seriousness.)

2. We could dispute the second premise. Logical skills are only one part of intellectual competence; woman may have other qualities, such as emotional sensitivity and a more holistic intuition, that are just as important in the intellectual realm, if not more so.

3. We could point out that the argument is obscure and would need to be recast to become formally valid. For starters, is the conclusion a universal claim: "*Every* women is intellectually inferior to any man"—

"(x)(y)((Wx · My) ⊃ Ixy)"? If so, the first premise would have to be an implausible universal claim: "*Every* woman is less logical than any man"—"(x)(y)((Wx · My) ⊃ Lxy)." Or is the conclusion "The average intelligence of women is less than the average intelligence of men"? This, even if it were true, would not justify banning *every* woman from higher education, leadership roles in government and business, or voting; for it still could be that *many* women were of greater intelligence than most men who participate in these activities.

In logic, as in many areas, women have for many years been underrepresented; but the gap is narrowing. Many women are mentioned in the bibliography at the end of this dictionary; prominent women logicians include, among others, Alice Ambrose, Susan Haack, Virginia Klenk, Martha Kneale, **Christine Ladd-Franklin**, Susanne Langer, **Ruth Barcan Marcus**, and Susan Stebbing. If we take these women logicians as a group, it is difficult to find clear differences between their work and that of their male colleagues.

GENERAL/SINGULAR TERM. A *general term* is one that describes or puts in a category (like "brave," "drives a car," or "a Canadian"). A *singular term* is one that picks out a specific individual (like "Albert Einstein," "the present prime minister of Canada," or "this child").

Should logic include as general and singular terms those that are *empty*, in that they refer to no existing objects? Is "unicorn" a legitimate general term—and is "Santa Claus" a legitimate singular term? Traditional **syllogistic logic** assumes that general and singular terms are not empty, **quantificational logic** accepts empty general terms but not empty singular terms, and **free logic** accepts empty terms of both sorts.

GENETIC FALLACY. The **fallacy** of arguing that a belief must be false if we can explain its origin. For example, one might dismiss another's religious or political views on the basis of some explanation of how the person came to accept those views in childhood. But to show a belief to be false, we must argue against the *content* of the belief; it is not enough to explain how the belief came to be.

GEOMETRY. From the beginning, geometry was closely related to logic and other areas of philosophy. Geometry was one of the most advanced areas of study in ancient Greece; it was developed by the Pythagoreans and put into a deductive system in Euclid's *Elements* about 300 BC. Geometry was Plato's model for strict rational knowledge; reflection on things like perfect circles, which we can know even though they are not given in sense experience, led him to his views about forms and *a priori* knowledge. Geometry's stress on deductive reasoning was likely one of the factors that encouraged **Aristotle** (384–322 BC) to invent logic as a systematic study. In the Middle Ages, geometry and logic were both included in the seven liberal arts that were part of the university core curriculum. René Descartes (1596–1650), who created analytic geometry, saw the strict proofs of geometry as the model for philosophical thinking. **Immanuel Kant** (1724–1804) later puzzled about how geometry can be *a priori* and also give us information about the physical world; he concluded that geometric principles regulate the *a priori* spatial intuition that we use to construct the world out of sense experiences. Nikolay Lobachevsky (1792–1856) developed non-Euclidean geometries; after that, geometry was increasingly looked upon as a logical derivation from definitions and axioms (with alternative axioms being possible) and not as a description of the properties of physical space (which must be investigated empirically). Logicians like **Gottlob Frege** (1848–1925) and David Hilbert (1862–1943) were interested in seeing what logical principles are presupposed in geometric proofs, in axiomatizing geometry using tools of modern logic, and in proving the consistency of various Euclidean and non-Euclidean systems.

High school geometry gives many students their first exposure to rigorous deductive reasoning. How one does in geometry is often seen as a predictor for future interest and success in philosophy and logic.

GOD. Raises questions like these about logic: Did God create the rules of logic? Is God able to do everything that is not self-contradictory? Can God bring about contradictions? Can we prove (or at least give a strong argument for) the existence of God—or the non-existence of God? Does the reasonableness of belief in God require that we have a proof (or strong argument) for his existence? Is the concept of God logically consistent? Are specific beliefs about God (especially about the Incarnation and the Trinity) logically consistent?

Many famous arguments deal with God. Here are examples:

If God exists in the understanding and not in reality, then there can be conceived a being greater than God (namely, a similar being that also exists in reality).

"There can be conceived a being greater than God" is false (since "God" is defined as "a being than which no greater can be conceived").

God exists in the understanding.

∴ God exists in reality.

Ontological Argument

$((U \cdot \sim R) \supset G)$
$\sim G$
U
∴ R

Some things are caused (brought into existence).

Anything caused is caused by another.

If some things are caused and anything caused is caused by another, then either there is a first cause or there is an infinite series of past causes.

There is no infinite series of past causes.

∴ There is a first cause.

First-Cause Argument

S
A
$((S \cdot A) \supset (F \lor I))$
$\sim I$
∴ F

If God does not want to prevent evil, then he is not all good.

If God is not able to prevent evil, then he is not all powerful.

Either God does not want to prevent evil, or he is not able.

∴ Either God is not all powerful, or he is not all good.

Problem of Evil

$(\sim W \supset \sim G)$
$(\sim A \supset \sim P)$
$(\sim W \lor \sim A)$
∴ $(\sim P \lor \sim G)$

While these three are clearly valid, there is much dispute about the premises. *See also* MODAL SYSTEMS.

GÖDEL, KURT (1906–1978). An Austrian logician and mathematician who later moved to the United States (mostly because of Nazi oppression) and taught at Princeton. While he was part of the Vienna Circle for a time, he deviated from their **logical positivist** doctrines; for example, he had a strong Platonist belief in the **abstract entities** of logic and mathematics, and he developed an ontological argument for the existence of **God**. While Gödel did important work in many areas of logic, he is best known for his 1931 paper on **metalogic** that showed that **arithmetic** is not reducible to a **formal system**. This and related results are called *Gödel's theorem*.

Gödel showed that, for any attempted axiomatization of arithmetic, one of two bad things will happen: either some true statements of arithmetic will not be provable (making the system incomplete), or some false statements of arithmetic will be provable (making the system unsound). So any formal system for arithmetic will be incomplete or unsound. This result is surprising. Arithmetic seems to be an area where everything can be proved one way or the other. But Gödel showed that this was wrong. While his ingenious reasoning about this is difficult, this entry will try to give a glimpse of how it works.

What exactly is this "arithmetic" that we cannot systematize? *Arithmetic* here is roughly like high school algebra, but limited to positive whole numbers. It includes truths like these three:

$$2+2=4$$
$$\text{If } x+y=z, \text{ then } y+x=z.$$
$$\text{If } xy=8 \text{ and } x=2y, \text{ then } x=4 \text{ and } y=2.$$

More precisely, *arithmetic* is the set of truths and falsehoods that can be expressed using symbols for the vocabulary items in these boxes:

Mathematical Vocabulary	Logical Vocabulary
positive numbers: 1, 2, 3, . . .	not, and, or, if-then
plus, times	variables: x, y, z, . . .
to the power of	every, some
parentheses, equals	parentheses, equals

Gödel's theorem claims that no formal system with symbols for all the items in these two boxes can be both sound and complete.

The notions in our mathematical box can be reduced to a sound and complete formal system, one that we will call the "number calculus." And the notions in our logical box can be reduced to a sound and complete formal system, once that includes **propositional**, **quantificational**, and **identity logic**. But combining these two systems produces a monster that cannot be put into a sound and complete formal system.

We will now construct a number calculus (NC) with seven symbols:

$$/ \quad + \quad \cdot \quad \wedge \quad (\quad) \quad =$$

"/" means "one" ("1"). We will write 2 as "//" ("one one"), 3 as "///" ("one one one"), and so on. "+" is for "plus," "·" for "times," and "^" for "to the power of." Our seven symbols cover all the notions in our mathematical box.

Meaningful sequences of NC symbols are *numerals*, *terms*, and **wffs**:

1.   Any string of one or more instances of "/" is a *numeral*.
2.   Every numeral is a *term*.
3.   The result of joining any two terms by "+," "·," or "^" and enclosing the result in parentheses is a *term*.
4.   The result of joining any two terms by "=" is a *wff*.

So "//" and "///////" are numerals (like our "2" and "7"); "///////," "(// · //)," and "((/ + /) ^ //)" are terms (like our "7," "2 · 2," and "$(1+1)^2$"); and "/// = ///" and "(// + //) = ////" are wffs (like our "3=3" and "2+2=4"). Wffs make true or false claims about numbers.

NC uses this one axiom (where any numeral can substitute for "a"):

Axiom: a = a

Any instance of this, any self-identity using the same numeral on both sides, is an axiom: "/=/," "//=//," "///=///," and so on.

Our six inference rules let us substitute one string of symbols for another. We will use "↔" to say that we can substitute the symbols on either side for those on the other side. We have two rules for "plus" (where "a" and "b" in our inference rules stand for any numerals):

R1.   (a+/) ↔ a/
R2.   (a+/b) ↔ (a/+b)

For example, R1 lets us interchange "(///+/)" and "/////." R2 lets us interchange "(//+//)" and "(///+/)"—moving the "+" one "/" to the right. We will see R3 to R6 in a moment.

A **proof** is a vertical sequence of wffs, each of which is either an axiom or else follows from earlier members by one of the inference rules R1 to R6. A **theorem** is any wff of a proof. Using our axiom and rules R1 and R2, we can prove any true wff of NC that does not use "·" or "^." Here is a proof of "(//+//)=////" ["2+2=4"]:

1. $////=////$ {from the axiom}
2. $(///+/)=////$ {from 1 using R1}
3. $(//+//)=////$ {from 2 using R2}

We start with a self-identity. We get line 2 by substituting "$(///+/)$" for "$////$" (by R1). We get line 3 by further substituting "$(//+//)$" for "$(///+/)$" (by R2). So "$(//+//)=////$" is a theorem.

Here are our rules for "times" and "to the power of":

R3. $(a \cdot /) \leftrightarrow a$
R4. $(a \cdot /b) \leftrightarrow ((a \cdot b) + a)$
R5. $(a \wedge /) \leftrightarrow a$
R6. $(a \wedge /b) \leftrightarrow ((a \wedge b) \cdot a)$

Our NC is sound and complete; any wff of NC is true if and only if it is provable in NC. This is easy to show, but we will not do the proof here.

Suppose we take our number calculus; add symbols and proof methods for propositional, quantificational, and identity logic; maybe add a few more axioms and inference rules; and call the result the "arithmetic calculus" (AC). We could then symbolize any statement of arithmetic in AC. So we could symbolize these:

If xy=8 and x=2y, then x=4 and y=2.
= $(((x{\cdot}y)=//////// \cdot x=(//{\cdot}y)) \supset (x=//// \cdot y=//))$

x is even.
= For some number y, x = 2 times y.
= $(\exists y)x=(// \cdot y)$

x is prime.
= For every number y and z, if x = y times z, then y=1 or z=1.
= $(y)(z)(x=(y \cdot z) \supset (y=/ \vee z=/))$

And we could symbolize **Goldbach's conjecture**, which no one has yet proved or disproved (here letters with primes are distinct variables):

Every even number is the sum of two primes.
= $(x)((\exists y)x=(2 \cdot y) \supset (\exists x')(\exists x'')(x=(x'+x'') \cdot ((y)(z)(x'=(y \cdot z)$
$\supset (y=/ \vee z=/)) \cdot (y)(z)(x''=(y \cdot z) \supset (y=/ \vee z=/)))))$

Gödel's theorem shows that any such arithmetic calculus has a fatal flaw: either it *cannot* prove some arithmetic truths, or it *can* prove some arithmetic falsehoods. This flaw comes not from an avoidable error in the choice of axioms and inference rules but rather from the fact that any such system can encode messages about itself.

To show how the reasoning for this works, it is helpful to use a minimal-vocabulary version of AC. So far we have used these symbols:

$$/ \quad + \quad \cdot \quad \wedge \quad (\quad) \quad = \quad \sim \quad \vee \quad \supset \quad \exists \quad x, y, z, x', \ldots$$

We will now economize. Instead of writing "∧" ("to the power of"), we will write "‥." We will drop "∨" and "⊃" and express the same ideas using "∼" and "·" (*see* **truth functions**). We will use "n," "nn," "nnn," "nnnn," . . . for our variables (instead of "x," "y," "z," "x'," . . .). We will drop "∃," and write "∼(n)∼" instead of "(∃n)." Our minimal-vocabulary version of AC uses only eight symbols:

$$/ \quad + \quad \cdot \quad (\quad) \quad = \quad \sim \quad n$$

Any statement of arithmetic can be symbolized using these symbols.

Our strategy for proving Gödel's theorem goes as follows. First we give ID numbers to AC formulas. Then we show how some AC formulas can encode messages about other AC formulas. Then we construct a special formula, called the Gödel formula G, that encodes this message about itself: "G is not provable." G asserts its own unprovability; this is the key to Gödel's theorem.

It is easy to give ID numbers to AC formulas. Let us assign to each of the eight symbols a digit from 1 to 8:

Symbol:	/	+	·	(	)	=	∼	n
ID Number:	1	2	3	4	5	6	7	8

Thus "/" has ID # 1 and "+" has ID # 2. To get the ID number for a formula, replace each symbol by its one-digit ID number:

The ID # for:	/=/	The ID # for:	(//+//)
is:	161	is:	4112115

The ID numbers follow patterns. For example, each numeral has an ID number consisting of all 1s:

Numeral:	/	/ /	/ / /	/ / / /
ID Number:	1	11	111	1111

So we can say,

Formula # n is a numeral	if and only if	n consists of all 1s.

We can formulate the right-hand box as the equation "(nine-times-n plus one) equals some power of ten," or "$(\exists x)9n+1=10^x$," which can be symbolized in a long AC formula. This AC formula is true of any number n if and only if formula # n is a numeral. This is how system AC encodes messages about itself.

An AC theorem is any formula provable in AC. The ID numbers for theorems follow definite but complex patterns. It is possible to find an equation that is true of any number n if and only if formula # n is a theorem. If we let "n is . . ." represent this equation, we can say,

Formula # n is a theorem	if and only if	n is . . . .

The equation in the right-hand box would be very complicated.

To help our intuition, let us pretend that all and only theorems have *odd* ID numbers. Then "n is odd" encodes "Formula # n is a theorem":

Formula # n is a theorem	if and only if	n is odd.

For example, "161 is odd" would encode the message that formula # 161 (which is "/=/") is a theorem:

Formula # 161 is a theorem	if and only if	161 is odd.

Then "n is even" would encode that formula # n is a non-theorem:

Formula # n is a non-theorem	if and only if	n is even.

Imagine that "485 . . ." is some specific large number; then:

| Formula 485 . . . is a non-theorem | if and only if | 485 . . . is even. |

Now imagine that this formula on the right, when expressed in AC's basic notation, itself happens to have ID number 485 . . . . Then the formula would be talking about itself, declaring that it itself is a non-theorem. This is what the Gödel formula G does. G, which itself has a certain ID number, encodes the message that the formula with this ID number is a non-theorem. G in effect says this:

| G | G is not a theorem. |

So G encodes the message "G is not a theorem." But this means that G is true if and only if it is not a theorem.

So G is true if and only if it is not provable. Now G, as a definite formula of arithmetic, is presumably either true or false. Is G true? Then it is not provable—and our system contains unprovable truths. Or maybe G is false? Then it is provable—and our system contains provable falsehoods. In either case, system AC is flawed.

We cannot remove the flaw by adding further axioms or inference rules. No matter what we add to the arithmetic calculus, we can use Gödel's technique to find a formula of the system that is true-but-unprovable or false-but-provable. Hence arithmetic cannot be reduced to any sound and complete formal system.

We have left out some details. We have not said how to construct the formula that encodes the message "formula # n is a theorem." And we have not said how to construct the Gödel formula G. These details can be worked out, but they are too complex to work through here.

So Gödel showed that arithmetic cannot be reduced to a formal system. Sometimes his theorem is worded differently, and often as two theorems that can be expressed roughly as follows:

1.   Every consistent axiomatization of arithmetic contains an undecidable formula.

2.   If an axiomatization of arithmetic is consistent, then the arithme-

tic statement that encodes the message "The system is consistent" is not itself provable in the system.

In theorem 1, an *undecidable formula* is one that cannot be proved or disproved in the system. But either it or its denial has to be true (since both make definite claims about numbers); so it follows that every consistent axiomatization of arithmetic contains truths not provable within the system. Theorem 2 shows that the consistency of arithmetic cannot be shown except perhaps by assuming principles more powerful, and thus more questionable, than arithmetic itself.

Gödel's theorem was a major blow to two views about the foundations of arithmetic: the *logicism* of **Gottlob Frege** and **Bertrand Russell** (which held that arithmetic could be axiomatized using a small group of logical axioms) and the *formalism* of David Hilbert (which held that truth in arithmetic was equivalent to provability—while Gödel's formula G is true if and only if it is not provable).

Two groups of people are especially happy about Gödel's theorem: logicians (who love the proof's intricate reasoning) and those who oppose the idea that everything in life can be reduced to a tight system (since Gödel showed that not even arithmetic can be reduced to a tight system). Sometimes it is said that Gödel showed that *nothing* can be reduced to a tight system; but this is incorrect—since Gödel himself was the first to prove the soundness and completeness of **classical symbolic logic**, which shows that the latter *can* be reduced to tight systems. The moral of Gödel's story, rather, is that some things can be tightly systematized while other things (including arithmetic) cannot.

GOLDBACH'S CONJECTURE. The idea suggested by Christian Goldbach (1690–1764), as yet neither proved nor disproved, that every even number is the sum of two primes. *See also* GÖDEL; INTUITIONIST LOGIC.

GRICE, PAUL (1913–1988). An English philosopher who introduced the idea of *conversational implication*. He suggested, as a general rule of communication, that we should not make a weaker claim rather than a stronger one unless we have a special reason to do so. Suppose our report says, "John published *an article*" when in fact he published *three articles*. While we have asserted nothing false, we have insinuated or "conversationally implied" something false. Following Grice's rule, we would not normally say "an article" while the stronger "three articles" was true, unless we were ignorant or trying to mislead.

Grice's rule sheds light on some puzzling aspects of logic. For example, while "some" is logically compatible with "all," many are inclined to see the two as incompatible. Grice would point out that it is normally misleading to say "some" when we could have said "all." So "some" conversationally implies "not all," even though "some" is logically compatible with "all." *See also* CONDITIONAL, KINDS OF; RELEVANCE LOGIC.

– **H** –

HASTY GENERALIZATION. The **fallacy** of arguing for a generalization from cases that are too few or not random enough (*see* "sample-projection syllogism" under **inductive logic**). For example, suppose that Jim, who is fat and never exercises, sees that you, who are slim and exercise regularly, are suffering from the flu. Jim says, "Your flu proves that staying slim and exercising does nothing for one's health." But your flu does not prove this; in fact, the larger evidence goes in the other direction.

HIGHER-ORDER LOGIC. *See* SECOND-ORDER LOGIC.

HINTIKKA, JAAKKO (1929–). A Finnish logician who now teaches in the United States. Hintikka has done important work in many areas, including **epistemic logic**, game theory, and **possible-world** semantics. His 1962 *Knowledge and Belief* was a pioneering work in the logic of **belief** and knowledge. Symbolizing "s believes that P" as "B_sP" and "s knows that P" as "K_sP," his book systematizes what our belief systems would be like if we were rational in certain ways. It also tries to analyze the difference between *de dicto* knowledge ("I know that someone robbed the bank," symbolized as "$K_i(\exists x)Rx$") and *de re* knowledge ("I know who robbed the bank" or "For some person x, I know that x robbed the bank," symbolized as "$(\exists x)K_iRx$").

Hintikka uses *model sets*, which are partial descriptions of possible worlds, instead of conventional **semantic** or **proof** methods. A *model set* for **propositional logic** is a set of **wffs** satisfying these conditions:

- If a **negation** is in the set, then the negated part is not in it.
- If a **conjunction** is in the set, then both conjuncts are in it.

- If a **disjunction** is in the set, then at least one disjunct is in it.
- If a **conditional** is in the set, then either the consequent is in it or the denial of the antecedent is in it.
- If a **biconditional** is in the set, then either both parts are in it or the denial of each part is in it.

A set of wffs is *consistent* if and only if some model set contains them all. A conclusion *follows from a set of premises* if and only if no model set contains the premises and also the negation of the conclusion.

HUME, DAVID (1711–1776). A Scottish philosopher known for his empirical and skeptical tendencies. Hume raised the question of how to justify **inductive reasoning**; his answer seems to be that we cannot justify such reasoning intellectually—but that does not matter, since inductive reasoning is built into our habits and feelings, and these are what move us. He thought all our ideas either are copies of impressions (experiences) or are compounded from these. *Hume's fork* says that genuine beliefs are either relations of ideas (**analytic** truths) or matters of fact (**empirical**); everything else is sophistry. *Humean rationality* sees the only rational constraint on actions and ethical beliefs as consistency between ends and means (*see* **formal ethics**). *Hume's law* says we cannot deduce an "ought" from an "is" (*see* **deontic logic**).

Hume's ideas are highly provocative (even though some interpreters try to tone them down) and have led **logical positivists** to follow him and **Kantians** to develop strongly opposing theories.

HUSSERL, EDMUND (1859–1938). The founder of phenomenology. He wrote much on logic. His early *Philosophy of Arithmetic* argued for the **psychologistic** view, then popular in Germany, that logic describes human thinking. Sometime after **Gottlob Frege** wrote a scathing criticism of this, Husserl became an opponent of psychologism; this came out clearly in his *Logical Investigations*.

HYPOTHETICALS. *See* CONDITIONAL.

– I –

IDENTITY LOGIC. A branch of logic that studies arguments whose validity depends on "=" ("equals"). Identity logic generally builds on

quantificational logic, to which it adds the "=" symbol and some new axioms or inference rules. The result is called "quantificational logic with identity" and is a common extension of quantificational logic.

Identity logic adds one new way to form **wffs:**

> The result of writing a small letter and then "=" and then a small letter is a wff.

This rule lets us construct wffs like these:

$$x=y \quad = \quad \text{x equals y.}$$
$$r=l \quad = \quad \text{Romeo is the lover of Juliet.}$$

One use of "=" is to translate an "is" that goes between singular terms. The difference between general and singular terms is crucial:

General terms	Singular terms
General terms *describe* or put in a *category*— and are symbolized by capital letters:	**Singular terms** pick out a *specific* person or thing—and are symbolized by small letters:
L = a lover C = charming R = drives a Rolls	l = the lover of Juliet c = this child r = Romeo

Compare these two forms:

Predication	*Identity*
Lr	r=l
Romeo is a lover.	Romeo is the lover of Juliet.

Use "=" for "is" if both sides are singular terms, and thus represented by small letters. The "is" of identity can be replaced with "is identical to" or "is the same entity as," and can be reversed (so if x=y then y=x).

We can use "=" to translate some numerical notions. For example:

$$(\exists x)(\exists y)(\sim x=y \cdot (Rx \cdot Ry)) \quad \begin{array}{l} = \quad \textit{At least two} \text{ are rich.} \\ = \quad \text{For some x and some y, x}\neq\text{y,} \\ \qquad \text{x is rich, and y is rich.} \end{array}$$

The pair of quantifiers "($\exists$x)($\exists$y)" ("for some x and some y") does not say whether x and y are identical; so we need "~x=y" to say they are not. Here is how we translate "exactly one" and "exactly two":

($\exists$x)(Rx · ~($\exists$y)(~y=x · Ry)) = *Exactly one* thing is rich.
= For some x, x is rich and there is no y such that y≠x and y is rich.

($\exists$x)($\exists$y)(((Rx · Ry) · ~x=y) = *Exactly two* things are rich.
· ~($\exists$z)((~z=x · ~z=y) · Rz)) = For some x and some y, x is rich, y is rich, x≠y, and there is no z such that z≠x and z≠y and z is rich.

Our notation can express "There are exactly n Fs" for any specific whole number n.

We can also express, but awkwardly, claims about the addition of specific whole numbers. Here is an English paraphrase of "1+1=2" and the corresponding formula:

If exactly one thing is F,   ((((∃x)(Fx · ~(∃y)(~y=x · Fy))
exactly one thing is G,   · (∃x)(Gx · ~(∃y)(~y=x · Gy)))
and nothing is F-and-G,   · ~(∃x)(Fx · Gx)) ⊃
then exactly two   (∃x)(∃y)(((Fx ∨ Gx) · (Fy ∨ Gy)) · (~x=y
things are F-or-G.   · ~(∃z)((~z=x · ~z=y) · (Fz ∨ Gz)))))

We could prove our "1+1=2" formula (or longer additions) by assuming its denial and deriving a contradiction.

Identity logic adds one new axiom and one new inference rule. This self-identity axiom holds regardless of what constant replaces "a":

Self-identity | a=a |

By this axiom, we can assert a self-identity as a step anywhere in a proof, no matter what the earlier lines are. We also need the following equals-may-substitute-for-equals inference rule, which is based on the idea that identicals are interchangeable: if a=b, then whatever is true of a is also true of b, and vice versa. This rule holds regardless of what

constants replace "a" and "b" and what wffs replace "Fa" and "Fb"—
provided that the two wffs are alike except that the constants are interchanged in one or more occurrences:

Substitute
Equals

$$Fa, a=b \rightarrow Fb$$

Here is an easy identity proof (read the argument as "I weigh 180
pounds; my mind does not weigh 180 pounds; therefore I am not identical to my mind"):

```
1      Wi
2      ~Wm
  [ ∴ ~i=m
3    ┌ asm: i=m
4    └ ∴ Wm    {from 1 and 3}
5    ∴ ~i=m    {from 3; 2 contradicts 4}
```

Line 4 follows by substituting equals; if i and m are identical, then
whatever is true of one is true of the other.

Our substitute-equals rule seems to hold universally in arguments
about matter or mathematics. But the rule can fail with mental phenomena. Consider this argument (where "Bx" stands for "Jones believes
that x is on the penny"):

Jones believes that Lincoln is on the penny.   Bl
Lincoln is the first Republican U.S. president.   l=r
∴ Jones believes that the first Republican U.S.   ∴ Br
   president is on the penny.

If Jones is unaware that Lincoln was the first Republican president, the
premises could be true while the conclusion is false. So the argument is
invalid. But yet we can derive the conclusion from the premises using
our substitute-equals rule. So something is wrong here.

To avoid the problem, we can disallow translating into quantificational logic any predicates or relations that violate the substitute-equals
rule; so we will not let "Bx" stand for "Jones believes that x is on the
penny." (Some say that "Bx" expresses a property about Jones, not
about x, and so is illegitimate on this basis.) Statements about beliefs

and other mental phenomena often violate the substitute-equals rule; so we have to be careful translating such statements into quantificational logic. And if we add **modal** or **belief logic** to identity logic, then we may want to qualify the substitute-equal rules so that it does not apply to modal or belief formulas.

So the mental seems to follow different logical patterns from the physical. Does this refute the materialist project of reducing the mental to the physical? Philosophers dispute this question. *See also* IDENTITY OVER TIME; IDENTITY, RELATIVE; LEIBNIZ; MIND, PHILOSOPHY OF; SECOND-ORDER LOGIC.

IDENTITY OVER POSSIBLE WORLDS. Under what conditions is *Bill Gates* in the actual world identical to *Bill Gates* in another possible world, for example one described by a story? Is this a matter of stipulation, so the storyteller stipulates that the story is about Bill Gates (making the name into a "rigid designator")? Or do we have to argue for the identity on the basis of similarities between the two people called "Bill Gates"? Or perhaps is it that, while entities in different possible worlds cannot be strictly identical, one can be a "counterpart" of another if the two are similar? *See also* KRIPKE; QUANTIFIED MODAL LOGIC.

IDENTITY OVER TIME. Under what conditions is an object x at one time identical with an object y at another time? If your car (or dog) keeps getting new parts, at what point, if any, does it cease to be the same entity you started with? Suppose your car, through the replacement of parts, becomes the basis for two new objects; are both of these identical with the entity you started with?

Standard **identity logic** might seem to give a simple answer to these questions, since it holds that if x has property F and y does not, then x is not identical to y. Does it not then follow that if your dog today has different properties from your dog yesterday, then the two are not identical—and so we must be talking about two different dogs? We could avoid this conclusion by building temporal qualifiers into the properties; so the same dog can have the properties of *being fat at t_1* and *being skinny at t_2*. Perhaps **temporal logic** can help here.

IDENTITY, RELATIVE. Those who think identity is relative see "Is x identical to y?" as incomplete; we should instead ask "Is x *the same F* as y?" Consider your dog at age 1 and at age 3; they might be the same dog but different material objects. On this view, the "=" of **identity**

logic should be replaced by something like "$=_F$," which would mean "is the same F as." Some critics see this as a needless complication, since we can keep identity as it is and just specify the identical objects better; so we can say "The dog at age 1 is identical to the dog at age 2—but the first material object is distinct from the second material object."

IDIOMS. English and other natural languages have various idiomatic ways to express central logical concepts like "all" and "if-then." When we symbolize English **arguments**, we need to know how to translate these idioms into the more central logical concepts.

Here are some common ways to express "all," "no," and "some" (*see* **syllogistic logic**, **quantificational logic**, and **any/all**):

	=	Every (any, each) A is B.
	=	Whatever is A is B.
All A is B.	=	Nothing is an A unless it is a B.
	=	Only Bs are As.
	=	None but Bs are As.
No A is B.	=	Not any A is B.
	=	It is false that there is an A that is a B.
Some A is B.	=	One or more As are Bs.
	=	It is false that no A is B.
Some A is not B.	=	One or more As are not Bs.
	=	Not all As are Bs.

Logicians usually take "As are Bs" to mean "All As are Bs"—even though this form in English sometimes means "Some As are Bs" or "Most As are Bs." Similarly, "As are not Bs" is taken as "No A is B."

Here are some common ways to express "and," "or," "if-then," and "if and only if" (the order of the letters only matters with "if-then"):

A and B	=	A but (yet, however, although) B
A or B	=	A unless B

	=	B if A
	=	Provided that A, B
	=	B, provided that A
If A then B	=	A is a sufficient condition for B
	=	B is a necessary condition for A
	=	A only if B
	=	Only if B, A
	=	A iff B
A if and only if B	=	A just if B
	=	A is necessary and sufficient for B

Translating "but" as "and" loses the rhetorical force of "but" to express surprise or shock. Logic is not concerned with this but only with information content.

IF. *See* CONDITIONAL; CONDITIONAL, KINDS OF; ENTAILMENT; RELEVANCE LOGIC.

IFF. Short form of "if and only if," which is used for **biconditionals**.

IMPERATIVE LOGIC. A branch of logic that studies arguments using imperatives, like "Do this." This branch is less established, with much diversity. Here we will sketch the approach of Hector-Neri Castañeda, which builds on **propositional** and **quantificational logic** and adds underlining for imperatives. Imperative logic is related to **deontic logic**, which studies arguments involving "ought" and "permissible."

Castañeda's imperative logic adds two ways to form **wffs**:

1. Any underlined capital letter is a wff.
2. The result of writing a capital letter and then one or more small letters, one small letter of which is underlined, is a wff.

Underlining turns indicatives into imperatives. So if "A" means "You are doing A" (indicative) then "<u>A</u>" means "Do A" (imperative). Here are some further translations:

$$\sim\underline{A} \quad = \quad \text{Do not do A.}$$
$$(\underline{A} \cdot \underline{B}) \quad = \quad \text{Do A and B.}$$
$$\sim(\underline{A} \cdot \underline{B}) \quad = \quad \text{Do not combine doing A with doing B.}$$
$$(\underline{A} \vee \underline{B}) \quad = \quad \text{Do A or B.}$$

Underline imperative parts but not factual ones:

$$(A \cdot B) \quad = \quad \text{You are doing A and you are doing B.}$$
$$(A \cdot \underline{B}) \quad = \quad \text{You are doing A, but do B.}$$
$$(\underline{A} \cdot \underline{B}) \quad = \quad \text{Do A and B.}$$

$$(A \supset B) \quad = \quad \text{If you are doing A, then you are doing B.}$$
$$(A \supset \underline{B}) \quad = \quad \text{If you (in fact) are doing A, then do B.}$$
$$(\underline{A} \supset B) \quad = \quad \text{Do A, only if you (in fact) are doing B.}$$

Since English cannot put an imperative after "if," we cannot read "$(\underline{A} \supset B)$" as "If do A, then you are doing B." But we can read it as the equivalent "Do A, only if you are doing B." This means the same as "$(\sim B \supset \sim\underline{A})$"—"If you are not doing B, then do not do A."

These examples, which build on quantificational logic, underline the letter for the agent:

$$A\underline{x} \quad = \quad \text{X, do (or be) A.}$$
$$A\underline{xy} \quad = \quad \text{X, do A to Y.}$$

$$(x)Ax \quad = \quad \text{Everyone does A.}$$
$$(x)A\underline{x} \quad = \quad \text{Let everyone do A.}$$

$$(x)(Ax \supset B\underline{x}) \quad = \quad \text{Let everyone who (in fact) is doing A do B.}$$
$$(\exists x)(Ax \cdot B\underline{x}) \quad = \quad \text{Let someone who (in fact) is doing A do B.}$$
$$(\exists x)(A\underline{x} \cdot B\underline{x}) \quad = \quad \text{Let someone both do A and do B.}$$

Imperative **proofs** work much like indicative ones and require no new inference rules. But we must treat "A" and "$\underline{A}$" as different wffs; "A" and "$\sim\underline{A}$" are not contradictories, since it is consistent to say "You are doing this but do not do it." Here is an imperative *modus ponens*:

If the cocoa is about to boil, then remove it from the heat.	$(B \supset \underline{R})$
The cocoa is about to boil.	B
∴ Remove it from the heat.	∴ $\underline{R}$

While this seems intuitively **valid**, there is a problem with calling it "valid," since this term is normally defined using "true" and "false": an **argument** is *valid* if it would be contradictory to have the premises all *true* and conclusion *false*. "Remove it from the heat" and other imperatives are not true or false. So how can the valid/invalid distinction apply to imperative arguments?

We need a broader notion of "valid" that applies to both indicative and imperative arguments. This one (which avoids "true" and "false") does the job: an argument is *valid* if the conjunction of its premises with the contradictory of its conclusion is inconsistent. On this definition, saying that our argument is *valid* means that this combination is inconsistent: "If the cocoa is about to boil, then remove it from the heat; the cocoa is about to boil; do not remove it from the heat." Since this *is* inconsistent, our argument is *valid* in this extended sense.

Here is an example of an *invalid* imperative argument:

If you get 100 percent, then celebrate.	$(G \supset \underline{C})$
Get 100 percent.	$\underline{G}$
∴ Celebrate.	∴ $\underline{C}$

Do not celebrate yet—maybe you will flunk. To derive the conclusion, we need not an imperative second premise, but rather a factual one saying that you did get 100 percent. The argument is invalid because it is consistent to say the following:

> If you get 100 percent, then celebrate. If you flunk, then do not celebrate. Get 100 percent. Oops, you did not get 100 percent; instead you flunked. So do not celebrate.

If we try to construct a proof for this argument, we instead get something resembling a refutation:

1	$(G \supset \underline{C})$	Invalid
2	$\underline{G}$	
[∴ $\underline{C}$		$\boxed{\sim G,\ \underline{G},\ \sim \underline{C}}$
3	asm: $\sim \underline{C}$	
4	∴ $\sim G$ {from 1 and 3}	

We get no contradiction. "$\sim \underline{G}$" of line 4 does not contradict "$\underline{G}$" of line 2, since the two differ in underlining; it is consistent to say "You are not doing this, but do it." The refutation works formally: if we assign F to "G," T to "G̲," and F to "C̲" then the premises come out all T and the conclusion F. But there is a problem. Our refutation assigns false to the imperative "Celebrate"—even though imperatives are not true or false. So what does "C̲ = F" mean?

We can here read "T" as "correct" and "F" as "incorrect." Applied to indicatives, these mean "true" or "false." Applied to imperatives, these mean that the action is "correct" or "incorrect" relative to some standard that divides actions prescribed by the imperative letters into *correct* and *incorrect* actions. The standard could be of different sorts, based on things like morality, law, or consistent personal goals.

Suppose we have an argument with just indicative and imperative letters and propositional connectives. The argument is *valid* if and only if, relative to every assignment of "T" or "F" to the indicative and imperative letters, if the premises are "T," then so is the conclusion. Equivalently, the argument is *valid* if and only if, relative to any possible facts and any possible standards for correct actions, if all the premises are correct, then so is the conclusion.

Our invalid "celebrate" argument shows the error of two older views about the validity of an imperative argument. The *obedience view* says that an imperative argument is valid if doing what the premises prescribe necessarily involves doing what the conclusion prescribes. This is fulfilled in the present case; if you do what both premises say, you will get 100 percent and celebrate. So the obedience view says our argument is valid. So the obedience view is wrong.

The *threat view* analyzes "Do A" as "Either you will do A or else S will happen"—where sanction "S" is an unspecified bad thing. So "A̲" is taken to mean "(A ∨ S)." But if we replace "C̲" with "(C ∨ S)" and "G̲" with "(G ∨ S)," then our argument becomes valid. So the threat view says our argument is valid. So the threat view is wrong.

INDIAN LOGIC. *See* BUDDHIST LOGIC.

INDIFFERENCE, PRINCIPLE OF. The idea that possibilities are to be assumed to be equally **probable** unless we have reason to think otherwise. So if we are picking a card randomly from a deck, each card is normally assumed to have an equal chance to be picked. This principle must be used with care, since drawing-an-ace and drawing-a-non-ace are both possibilities but are not equally probable.

INDIRECT PROOF. A form of proof in which one derives a conclusion C by first assuming not-C and then deriving a contradiction. *See also* PROOF; RAA.

INDUCTIVE/DEDUCTIVE. The distinction between *deductive* and *inductive* can be applied to kinds of argument or to kinds of validity. Consider these two arguments:

Deductively valid	*Inductively valid*
All who live in Iowa	Most who live in Iowa were born in Iowa.
live in the U.S.	Jones lives in Iowa.
Jones lives in Iowa.	This is all we know about the matter.
∴ Jones lives in the U.S.	∴ Jones was born in Iowa (probably).

The first argument has a tight connection between premises and conclusion; it would be impossible for the premises to be all true but the conclusion false. The second argument has a looser connection between premises and conclusion; here, relative to the premises, the conclusion is only a good guess—it is likely true but could be false.

A *deductive argument* is one in which the conclusion is claimed to follow with necessity; this claim is either implicit or expressed by terms like "must" ("So Jones *must* live in the U.S."). An *inductive argument* is one in which the conclusion is claimed to follow not with necessity but with probability; this claim again is either implicit or else expressed by terms like "probably" ("So Jones *probably* was born in Iowa").

Some logicians dislike this idea of an "implicit claim" and prefer to talk about two kinds of validity. An argument is *deductively valid* if it would be impossible to have the premises all true and conclusion false; it is *inductively valid* if the truth of the premises would make the conclusion only probably true. *See also* INDUCTIVE LOGIC.

INDUCTIVE LOGIC. The study of inductive reasoning, where we extrapolate from observed patterns to conclude that a given conclusion is *probably* true (*see* **inductive/deductive**).

The statistical syllogism is a common form of inductive reasoning. It has two main forms, depending on whether percentages are specified:

Statistical Syllogism

Most As are Bs.	N percent of As are Bs.
X is an A.	X is an A.
This is all we know about the matter.	This is all we know about the matter.
∴ X is probably a B.	∴ It is N percent probable that X is a B.

Suppose we are hiking the Appalachian Trail (AT) and plan to spend the night at Rocky Ridge Shelter. We would like to know beforehand whether there is water (a spring or stream) close by. If all we know about the matter is that Rocky Ridge is an AT shelter and that roughly 90 percent of the AT shelters have water, we could reason inductively:

> 90 percent of AT shelters have water.
> Rocky Ridge is an AT shelter.
> This is all we know about the matter.
> ∴ Probably Rocky Ridge has water.

This is a strong inductive argument. Relative to the premises, the conclusion is a good bet. But it is partially a guess; it could turn out false, even though the premises are all true.

"This is all we know about the matter" means "We have no further information that influences the probability of the conclusion." Suppose we just met a thirsty backpacker complaining that the water at Rocky Ridge had dried up; that would change the probability of the conclusion. The premise claims that we have no such further information.

Two features set inductive arguments apart from deductive ones. (1) Inductive arguments vary in how strongly the premises support the conclusion. The premise "99 percent of the AT shelters have water" supports the conclusion more strongly than does "60 percent of the AT shelters have water." We have shades of gray here—not the black and white of deductive validity/invalidity. (2) Even a strong inductive argument has only a loose connection between premises and conclusion. The premises make the conclusion at most only highly probable; the premises might be true while the conclusion is false. Inductive reasoning is a form of guessing based on recognizing and extending known patterns and resemblances.

Statistical syllogisms apply more cleanly if we have less knowledge. Suppose we know these *two* things about Rocky Ridge:

- Rocky Ridge is an AT shelter—and 90 percent of all AT shelters have water.
- Rocky Ridge is a ridge campsite—and 20 percent of all ridge campsites have water.

Relative to 1, Rocky Ridge probably has water. Relative to 2, it probably does not. It is unclear what to conclude relative to 1-and-2. Which group of shelters provides the better *reference class* for figuring out whether Rocky Ridge has water? Things gets worse if we add further facts about Rocky Ridge, some of which lead toward the "water" conclusion and some toward "no water." Each fact by itself may lead to a clear conclusion; but the combination muddies the issue. Too much information can confuse us when we apply statistical syllogisms.

Here is another common form of inductive reasoning:

Sample-Projection Syllogism

Most (or N percent of) examined As are Bs.
A large and varied group of As has been examined.
∴ Probably most (or roughly N percent of) As are Bs.

We might base our earlier premise, "90 percent of the AT shelters have water," on sample-projection reasoning. Maybe we have visited many AT shelters and noted that about 90 percent of them have water. We conclude that probably roughly 90 percent of *all* the shelters (including those we have not visited) have water:

90 percent of examined AT shelters have water.
A large and varied group of AT shelters has been examined.
∴ Probably roughly 90 percent of all AT shelters have water.

Three factors determine the strength of such an argument: (1) *size* of sample, (2) *variety* of sample, and (3) *cautiousness* of conclusion.

1. Other things being equal, a larger sample gives a stronger argument. So we would have a stronger case if we observed 100 shelters instead of only 10.

2. Other things being equal, a more varied sample gives a stronger argument. A sample is *varied* to the extent that it proportionally represents the diversity of the whole. AT shelters differ. Some are on high ridges, while others are in valleys. Some are on the main trail, while others are on blue-blazed side trails. Some are in the wilderness of New England and the South, while others are in rural areas of the middle states. Our sample is varied to the extent that it reflects this diversity.

3. Other things being equal, we get a stronger argument if we have a more cautious conclusion. We have stronger reason for thinking the proportion of shelters with water is "between 80 and 95 percent" than for thinking it is "between 89 and 91 percent."

Suppose our sample-projection argument is strong and has true premises. Then it is likely that roughly 90 percent of the shelters have water. But the conclusion is only a rational guess; it could be far off. It even could happen that every shelter we did not check is bone dry. Inductive reasoning involves risk.

Other common forms of inductive reasoning, besides the two considered above, include **analogical reasoning**, **inference to the best explanation**, and **Mill's** methods. All are subject to two perplexing problems: about formulations and about justification.

The first problem is about how to *formulate* the inductive principles. The wordings that we gave can lead to absurdities if taken literally. Consider this instance of our statistical-syllogism formulation:

> 80 percent of all Cleveland voters are Democrats.
> This non-Democrat is a Cleveland voter.
> This is all we know about the matter.
> ∴ It is 80 percent probable that this non-Democrat is a Democrat.

Actually, the conclusion is 0 percent probable, since it is self-contradictory. So our statistical syllogism principle is not entirely correct.

Our sample-projection syllogism suffers from a problem raised by Nelson Goodman. Consider this argument:

> All examined diamonds are hard.
> A large and varied group of diamonds has been examined.
> ∴ Probably all diamonds are hard.

Let us suppose the premises are true; then the argument would seem to be a good one. Now this second argument has the same form but a more complex phrase instead of "hard":

> All examined diamonds are such that they are hard-if-examined-before-2222-but-soft-otherwise.
> A large and varied group of diamonds has been examined.
> ∴ Probably all diamonds are such that they are hard-if-examined-before-2222-but-soft-otherwise.

Since it is not yet the year 2222, all our observations about diamonds fit the first premise of this argument as well as they fit the first premise of the previous argument. In this respect, the two arguments are equally good. But consider a diamond X that will be first examined after 2222. By our first argument, diamond X probably *is* hard; by the second, it probably *is not* hard. So our sample-projection formula leads to conflicting conclusions.

Philosophers have discussed this problem for decades. Some suggest that we qualify the sample-projection syllogism form to outlaw the second argument; but it is not clear how to eliminate the bad apples without also eliminating the good ones. As yet, there is no agreement on how to solve the problem.

Goodman's problem is like one that arises in **scientific reasoning**. No matter what observations we make, there is more than one theory that could explain these observations. No matter how many dots we put on a chart (representing test results), we could draw an unlimited number of lines that go through all these dots but diverge in further cases. In practice, we often use simplicity to choose between such rival views: other things being equal, we ought to prefer the simplest theory that adequately explains the data. So if two theories explain all the same test results, the simpler theory is to be preferred. While "simpler" here is vague and difficult to explain, we seem to need some such *simplicity criterion* (*see* **Ockham**'s razor) to justify any scientific theory.

Simplicity is important in our diamond case, since our first conclusion uses the simpler "hard" rather than the complex "hard-if-examined-before-2222-but-soft-otherwise." By our simplicity criterion, we ought to prefer the first conclusion to the second, even if both have equally strong inductive backing. So the sample-projection syllogism needs a simplicity qualification; but it is not clear how to formulate it.

So it is difficult to formulate clear inductive principles that do not lead to absurdities. Induction is less neat and tidy than deduction.

Our second problem is how to justify inductive principles. Let us pretend that we have clear inductive principles that accord with our practice and do not lead to absurdities. Why follow these principles?

Consider this argument (which says roughly that the sun will probably come up tomorrow, since it has come up every day in the past):

> All examined days are days on which the sun comes up.
> A large and varied group of days has been examined.
> Tomorrow is a day.
> ∴ Probably tomorrow is a day on which the sun comes up.

Even though the sun has come up every day in the past, it still might not do so tomorrow. Why think that the premises give good evidence for the conclusion? Why accept this or any inductive argument?

David Hume several centuries ago raised this problem about the justification of induction. We will discuss five responses.

1. Some suggest that to justify induction we need to presume that nature is uniform. If nature works in regular patterns, then the cases we have not examined will likely follow the same patterns as the ones we have examined. But there are two problems with this suggestion. First, "Nature is uniform" is vague and doubtful; is not the universe marvelously diverse in so many ways? Second, the truth of "Nature is uniform" could only be decided by experience, using inductive reasoning; but this makes the justification **circular**.

2. Some suggest that we justify induction by its success. Inductive methods work. Using inductive reasoning, we know what to do for a toothache and how to fix cars. We use such reasoning continuously and successfully in our lives. What better justification for inductive reasoning could we have than this? But this is circular too. "Induction has worked in the past, so it probably will work in the future" is an inductive argument; why accept it?

3. Some suggest that it is part of the meaning of "reasonable" that beliefs based on inductive reasoning are *reasonable*. "Reasonable belief" just means "belief based on experience and inductive reasoning." So it is true by definition that beliefs based on experience and inductive reasoning are reasonable. But there are two problems with this. First, the definition is wrong, since we can disagree about this standard of reasonableness (as mystics and skeptics do) without contradicting ourselves. Second, we would have the same problem even if the definition were correct; for verbal usage gives us no reason for following what is "reasonable" in the sense of "based on experience and inductive rea-

soning" instead of going with mystics or skeptics.

4. Karl Popper suggests that we avoid inductive reasoning. But we seem to need such reasoning in our lives; without it, we have no basis for believing that bread nourishes and arsenic kills. And suggested substitutes for inductive reasoning do not seem adequate.

5. Some suggest that we approach justification in inductive logic in the same way we approach it in deductive logic. How can we justify the validity of a deductive principle like **modus ponens** ("If A then B, A ∴ B")? There seems to be no way to prove it. Suppose we attempt to prove it by an argument of this form: "If P, then *modus ponens* is valid; but P; therefore *modus ponens* is valid." Any such argument simply assumes either *modus ponens* or else some less obvious inference rule.

Aristotle long ago showed that every proof must eventually rest on something unproved; otherwise, we would need either an infinite chain of proofs or circular arguments—and neither is acceptable. So why not just accept the validity of *modus ponens* as a self-evident truth—a truth that is evident but cannot be based on anything more evident? If we have to accept some things as evident without proof, why not accept *modus ponens* as evident without proof?

If we accept this approach, we should not think that picking logical principles is purely a matter of following untrained "logical intuitions." Many people before they study logic cannot distinguish valid forms like *modus ponens* from invalid forms like *affirming the consequent* (*see* **conditionals**). But they can distinguish the two after some training in logic; this training consists largely in examining concrete instances in which the validity or invalidity of forms is more obvious (*see* the introduction).

If we approach inductive principles in the same way, we would not demand a proof for such principles. Rather, we would look for clear formal inductive principles that lead to intuitively correct results in concrete cases without leading to absurdities. Once we reach these, we would be content with them and not look for further justification.

But deductive and inductive principles, as mentioned above, differ in an important way. Deductive principles can be formulated in clear-cut ways; inductive principles cannot. We seem unable to formulate clear *inductive* principles that do not sometimes lead to absurdities. This is what makes the current state of inductive logic intellectually unsatisfying. Even though inductive reasoning is useful and even essential to human life, its intellectual basis is somewhat shaky. *See also* BAYES; LOGICAL PRINCIPLES, STATUS AND JUSTIFICATION

OF; NON-MONOTONIC; PROBABILITY.

INFERENCE. The mental act of drawing a conclusion from premises. **Arguments** are the verbal expression of such acts of **reasoning**. While logic focuses on the validity of arguments, it is indirectly about the validity of inferences.

INFERENCE RULE. Rule stating that certain formulas can be derived or deduced from certain other formulas. *See also* AXIOM; PROOF.

INFERENCE TO THE BEST EXPLANATION. Argument that says that a given view is reasonable to accept since it provides the best explanation for the data. The best argument for the theory of evolution probably has this form:

> It is reasonable to accept the best explanation for the wide range of empirical facts about biological species.
> The best explanation for the wide range of empirical facts about biological species is evolution.
> ∴ It is reasonable to accept evolution.

A fuller formulation would elaborate on what these empirical facts are, what alternative ways there are to explain them, and why evolution provides a better explanation than its rivals.

Some propose that most of our core beliefs about things, including the existence of material objects, other minds, and perhaps even **God**, are to be justified as inferences to the best explanation.

INFINITARY LOGIC. Any system of logic that allows infinitely long formulas. While we finite humans cannot write down such formulas, we can cheat a little and use an ellipsis ("..."). With infinitary logic, we can replace "Some positive integer is F" with "1 is F, or 2 is F, or 3 is F, or ...." And we can replace "All positive integers are F" with "1 is F, and 2 is F, and 3 is F, and ...." In infinitary logic, if we are given an infinite set of statements, then we can define another statement that is the infinite **conjunction** or **disjunction** of them all.

INFINITE-REGRESS ARGUMENT. One that says that a specific series must have a first member, since otherwise the series would be infinite, which is impossible. We can construe this as a *modus tollens*:

If this series has no first member, then this series is infinite.
This series is not infinite.
∴ This series has a first member.

Thinkers have thus argued that there must be a first cause, or a known but unproved premise, or something good in itself—because otherwise there would have to be an infinite series of causes, or arguments, or things being good because they promote other things—which is impossible. Critics tend to dispute one of the premises; their grounds for doing this vary with the particular argument in question.

INFORMAL/FORMAL LOGIC. *Formal logic* focuses on systems of deductive logic (*see* **logic: deductive systems**) and on determining whether a conclusion follows **validly** from a set of premises. *Informal logic* includes other skills that relate to the appraisal of **arguments**— including things like locating the premises and conclusion in a passage that contains reasoning, supplying implicit premises, symbolizing English arguments, appraising the plausibility of premises, clarifying the **meaning** of a statement, and recognizing informal **fallacies**. While the "critical thinking" model for teaching logic stresses informal over formal logic, many or perhaps most introductory **logic courses** try to cover both areas.

INFORMAL FALLACY. *See* FALLACY.

INTENSION. *See* SENSE/REFERENCE.

INTENSIONAL. *See* EXTENSIONAL/INTENSIONAL.

INTERPRETATION. A way of assigning meaning to a string of symbols or words that makes it express a true or false claim. A *model* is an interpretation that makes the string express something that is **true**. *See also* SEMANTICS/SYNTAX.

INTUITIONIST LOGIC. An alternative to standard **propositional logic** that drops the **law of excluded middle** and double **negation**:

- *Excluded middle:* $(A \lor \sim A)$ ("A or not-A")
- *Double negation:* $(\sim \sim A \supset A)$ ("If not-not-A, then A")

Intuitionists believe that "A" and "~A" are sometimes both false in mathematical cases involving infinite sets. To emphasize these differences, intuitionists use "¬" for negation instead of "~."

The Dutch mathematicians Luitzen E. J. Brouwer (1881–1966) and Arend Heyting (1898–1980) proposed intuitionist logic in order to block some mathematical **proofs**. The suspicious proofs argue for a result "A" by first showing that "~A" is false (because it leads to a contradiction) and then concluding "A" from the falsity of "~A":

"~A" is false.   ~~A
∴ "A" is true.   ∴ A

Intuitionists contend that this is invalid in some cases involving infinite sets—because if "~A" is false then "A," too, might be false.

Intuitionist mathematicians see the natural numbers (0, 1, 2, . . .) as grounded in our experience of counting. Mathematical **truths** are constructions of the human mind; mathematical formulas should not be considered "true" unless somehow the mind can prove their truth. Consider **Goldbach's conjecture**: "Every even number is the sum of two primes." If you start going through the even numbers, this seems to work for every one you pick (since 2=1+1, 4=3+1, 6=5+1, 8=7+1, 10=7+3, and so on). But no one has ever proved or disproved that it holds for *all* even numbers. Some think Goldbach's conjecture must be true or false objectively, even though we may never prove which it is. Intuitionists disagree. They say that truth in mathematics is provability; if we assume that neither Goldbach's conjecture nor its negation is provable, we must conclude that neither it nor its negation is true. This is why intuitionists think that in some cases involving infinite sets (like the set of even numbers), neither "A" nor "~A" is true—and so both are false. The law of excluded middle does apply if we limit ourselves to finite sets; so "Every even number under 1,000,000,000 is the sum of two primes" is either true or false, and we could write a computer program that could in principle eventually tell us which it is.

Intuitionists were influenced by the philosophy of **Immanuel Kant**. They see mathematical ideas as based on the temporal structure that the human mind gives to sense experience. When we try to apply these mathematical ideas beyond possible experience—to things in themselves—we get lost in sophistry and contradictions.

Intuitionists insist on *constructive methods* of proof, whereby a genuine mathematical entity has to be actually constructed; it does not suffice to show that it could in principle be constructed or that its non-existence would lead to a contradiction. Some insist on *finitist methods* that reject infinite sets altogether. Much of classical mathematics (*see* **Georg Cantor**) is lost if we restrict proofs in these ways; but intuition-ists do not see this as a great loss. *See also* MANY-VALUED LOGIC.

INVALID ARGUMENT. Argument in which the premises could be all true while the conclusion is false; equivalently, an argument that is not **valid**, or one whose conclusion does not follow from its premises.

IS. Modern logic recognizes three main senses of the verb "is"; these are translated in distinct ways and follow distinct logical patterns:

1. *Existence:* "There are dogs" ("Dogs exist") is "$(\exists x)Dx$" ("For some x, x is a dog" or "One or more entities are dogs"). This claims that a category (dogs) has one or more existing instances; it follows logically from a claim like "Df" ("Fido is a dog"). (But *see* **free logic**.)

2. *Predication (essence):* "Fido is a dog" ("The individual entity Fido has the property of being a dog") is "Df."

3. *Identity:* "Fido is my dog" ("Fido is identical to my dog") is "f=m." Here "Fido" and "my dog" are singular terms referring to the same individual entity; the "is" of identity (*see* **identity logic**) can be replaced with "is identical to" or "is the same entity as," can be reversed (so if x=y then y=x), and entails that what is true of one entity must also be true of the other (since they are the same entity).

– K –

KANT, IMMANUEL (1724–1804). German philosopher who had a huge impact on philosophy. Despite his innovations in other areas, Kant followed **traditional logic**. He echoed the thinking of his time when he stated that **Aristotle** not only was the first to conceive of logic but also had substantially brought the subject to its completion.

Kant thought that opposite conclusions about many big questions (like whether the world has a beginning in time, is infinitely divisible, is infinite spatially, or depends on a necessary being) could seemingly be proved equally well. His critical solution to this problem says that the categories that we use to describe the world (like substance/pro-perty and cause/effect) are imposed by our own minds and apply only to objects of possible experience; when applied further, they lead to sophistry and illusion. From his table of the 12 logical functions he derived his 12 categories of the understanding:

Logical Functions	*Categories*
Quantity: universal, particular, singular	Quantity: unity, plurality, totality
Quality: affirmative, negative, indefinite	Quality: reality, negation, limitation
Relation: categorical, hypothetical, disjunctive	Relation: substance/property, cause/effect, reciprocity
Modality: problematic, assertoric, apodeictic	Modality: possibility, existence, necessity

Some aspects of his tables are mysterious. For example, it is unclear why "indefinite" is a separate logical function while "conjunction" is not. And many of the categories of the understanding are hard to distinguish from the corresponding logical functions.

Kant was important also for his **analytic/synthetic** and *a priori/a posteriori* distinctions, for his interesting explanation of how **geometry** can be *a priori* and also tell us about the physical world, and for how he applied consistency to **ethics** (*see* **deontic logic** and **formal ethics**).

KRIPKE, SAUL (1940–). An influential American logician. At age 19, he published a **semantics** for **quantified modal logic** based on **possible worlds**; this came from earlier work done as a high school student. Before this time, **modal logic** had struggled for respectability; **Willard Van Orman Quine** attacked it, while **Ruth Barcan Marcus** tried to defend it. Kripke's possible worlds made more sense of modal logic and gave it new respect among logicians. His possible-worlds techniques have since proved useful in many other areas of logic— including, for example, **deontic**, **belief**, and **relevance logic**.

Kripke rejected the idea, proposed by **Gottlob Frege** and **Bertrand Russell**, that **names** are short for **definite descriptions**. The name "Bertrand Russell" does not mean, for example, "the author of the 1905 article on denoting," since we can imagine a possible world where Russell did not write this article. A name refers to a specific individual more directly. We give a reference to a name by a kind of stipulation or baptism ("We hereby name you 'Bertrand Russell'"); henceforth we use the name as a *rigid designator*, to refer to the same individual in every possible world where that individual exists (even in worlds where people there refer to the individual using a different name). In contrast, "the author of the 1905 article on denoting" is only a *weak designator*; it can refer to different individuals in different possible worlds—since we might imagine a possible world in which Frege wrote the article.

The planet Venus in ancient times had two names. In the evening, people pointed to a bright celestial object and called it *Vesperus*; in the morning, they pointed to a bright object and called it *Phosphorus*. Later they discovered empirically that they pointed to the same object: Vesperus = Phosphorus. Since a name refers to the same object in every possible world, the "Vesperus = Phosphorus" identity statement is true in all possible worlds, and hence necessary—even though it was discovered empirically. So, Kripke argues, it is wrong to think that only *a priori* statements can be necessary.

The kind of necessity that "Vesperus = Phosphorus" has, according to Kripke, is *metaphysical necessity*. This is not based on language conventions. So Kripke rejected the then-popular view that all necessity comes from language conventions. He also supported a **metaphysics** that distinguishes necessary from contingent properties (*see* **Aristotelian essentialism** and **Alvin Plantinga**). Kripke's work is sometimes said to have made metaphysics respectable again.

– **L** –

LADD-FRANKLIN, CHRISTINE (1847–1930). American logician and psychologist. She introduced *antilogisms*, which are sets of three categorical statements of **syllogistic logic**, using three terms, each occurring in exactly two of the statements. Every syllogism has a corresponding antilogism that replaces its conclusion with its contradictory:

Syllogism	Antilogism
all M is P	all M is P
all S is M	all S is M
∴ all S is P	some S is not P

A syllogism is *valid* just if the corresponding antilogism is inconsistent.

Ladd-Franklin was delayed for 40 years in getting her Ph.D. from Johns Hopkins, even though she fulfilled all the requirements, because the school did not grant this degree to women.

LANGUAGE, PHILOSOPHY OF. The study of philosophical questions raised by language. To some extent, philosophy of language overlaps with **philosophy of logic**, since both are concerned with symbols (**syntax**), their relationship to reality (**semantics**), and their use by people (*pragmatics*). So both, for example, deal with topics like **abstract entities** and **ontology**, the **analytic/synthetic** distinction, **ambiguity**, **definite descriptions**, **definitions**, **general/singular** terms, **meaning**, **truth** and **paradoxes** about truth, and the **sense/reference** distinction. But philosophy of language deals also with more specific questions about *language* (such as its definition, how we learn language, its similarity or variability across cultures, whether we could think without it, and whether there could be a private language)—while philosophy of logic deals also with more specific questions about *logic* (such as its definition and scope, the analysis of its key concepts, the relationship between logic and ordinary language, the possibility of deviant logics, and the status and justification of basic logical principles).

LANGUAGE, USES OF. Grammarians traditionally distinguish four sentence types, which broadly reflect four important uses of language:

- Declarative (making assertions): "Michigan beat Ohio State."
- Interrogatory (asking questions): "Did Michigan win?"
- Imperative (telling what to do): "Beat Ohio State."
- Exclamatory (expressing feelings): "Hurrah for Michigan!"

Sentences can do various jobs at the same time. Besides making assertions, "I wonder whether Michigan won" can serve as a question, "I want you to throw the ball" can tell what to do, and "Michigan won!" can express positive or negative feelings.

Arguments too can exemplify different uses of language. Suppose someone argues this way about the Cleveland river that used to catch on fire: "You can see that the Cuyahoga River is polluted from the fact that it even catches on fire!" We can recast this as follows:

> No pure water is burnable.
> Some Cuyahoga River water is burnable.
> ∴ Some Cuyahoga River water is not pure water.

One who argues thusly might also be implicitly raising a question ("What can we do to clean up this polluted river?"), directing people to do something ("Let us all resolve to take action on this problem"), or expressing feelings ("How disgusting is this polluted river!"). Arguments have a wider human context and purpose. We should remember this when we study detached specimens of argumentation.

When we do logic, our focus narrows and we concentrate on assertions and reasoning. For this purpose, detached specimens of argumentation are better. Expressing an argument in a clear, direct, emotionless way can make it easier to appraise the **truth** of the premises and the validity of the **inference**.

It is important to avoid emotional language when we reason. Of course, there is nothing wrong with **feelings** or emotional language. Reason and feeling are both important parts of life; we need not choose between the two. But we often need to focus on one or the other for a given purpose. At times, expressing our feelings is the important thing and argumentation only gets in the way. At other times, we need to reason things out in a cool-headed manner.

Emotional language can discourage clear reasoning. When reasoning about abortion, for example, it is wise to avoid slanted phrases like "the atrocious, murderous crime of abortion" or "Neanderthals who oppose the rights of women." **Bertrand Russell** gave this example of how we slant language: "I am *firm*; you are *obstinate*; he is *pig-headed*." Slanted phrases can mislead us into thinking we have defended our view by an argument (premises and conclusion), when in fact we have only expressed our feelings. Careful thinkers try to avoid highly emotional terms when constructing their arguments.

LAW OF EXCLUDED MIDDLE. The claim, symbolized in **propositional logic** as "(P ∨ ~P)," that "P or not-P" is always true for every statement P.

The **truth table** for "(P ∨ ~P)" shows that the formula is true in all cases; this makes it a *truth-table tautology*, and thus an **analytic truth**:

P	(P ∨ ~P)
F	T
T	T

"I went to Paris or I did not go to Paris."

This law holds in propositional logic, which normally stipulates that capital letters stand for true-or-false statements.

Matters are messier in English, where we sometimes have sentences that are meaningless (like "Glurklies glurkle") or vague ("Her shirt is white," when it is somewhere between white and gray); these seem to be too obscure to be either true or false. So "(P ∨ ~P)" is an idealization when applied to some English sentences. We could still accept it as a universal logical truth by stipulating that our propositional letters only cover claims that are definitely true or false. Or we might let propositional letters cover claims too vague to be true or false, but instead understand a "logical truth" to be one which is true on any "supervaluation" of the letters, any way of resolving the vagueness to make the letters represent statements that are definitely true or false.

The law of excluded middle is sometimes expressed as "Every statement is either true or false," which seems to assert the **bivalence** view, against **many-valued logic**, that there are only two truth values. But one could hold that there is a third truth value, half-true, and that *half-true or half-true* is true: $(\frac{1}{2} \vee \frac{1}{2}) = 1$; such a view would accept the law of excluded middle but reject bivalence. So it is better to formulate the law of excluded middle as claiming the universal truth of "P or not-P." *See also* ARISTOTLE; INTUITIONIST LOGIC.

LAW OF NON-CONTRADICTION. The claim, symbolized in **propositional logic** as "~(P · ~P)," that "P and not-P" is always false for any statement P. This law is subject to the proviso that we have to take the statement P in the same sense in both its occurrences.

The **truth table** for "(P · ~P)" is false in all cases—which makes this formula a **self-contradiction**:

P	(P · ~P)
F	F
T	F

"I went to Paris and I did not go to Paris."

"P and not-P" is always false in propositional logic, which presupposes that "P" stands for the same statement throughout. English is looser and lets us shift the meaning of a phrase in the middle of a sentence. "I went to Paris and I did not go to Paris" may express a truth if it means "I went to Paris (in that I landed once at the Paris airport)—but I did not really go there (in that I saw almost nothing of the city)." Because of the shift in meaning, this would better be translated as "(P · ~Q)."

Aristotle formulated and defended the law of non-contradiction; he saw adherence to this law as a necessary condition of coherent thought and he criticized earlier thinkers whom he took to violate this law (*see* **ancient logic before Aristotle**). While **classical symbolic logic** agrees with Aristotle on this, contemporary **dialethists** insist that a few self-contradictions are true; they propose a **paraconsistent logic** to help make sense of this contention.

Is the law of non-contradiction a claim about the world, or is it a language convention? Some people could have a convention that vague statements (like "This shirt is white") in borderline circumstances are *both true and false* (instead of being *neither true nor false*). We could choose to speak this way; and we could easily translate between this and normal speech. If so, then perhaps a strict adherence to the law of non-contradiction is at least partly conventional.

LEIBNIZ, GOTTFRIED WILHELM (1646–1716). A German philosopher and mathematician who coinvented calculus. His work hinted at the changes in logic that were to come later. He proposed the idea of an artificial language that would reduce reasoning to **arithmetic** calculation; if controversies arose, the parties could take up their pencils and say to each other, "Let us calculate." He created a logical notation much like that of **George Boole**, but his work on this was not published until 1903. He invented an *arithmetic machine* that could multiply and divide. And he spoke of **"possible worlds"**; these would much later, through **Saul Kripke** and others, become important in **modal logic**.

Two principles about identity are sometimes called "Leibniz's law." (1) *Indiscernability of identicals*: if x=y, then whatever is true of x is true of y, and whatever is true of y is true of x. (2) *Identity of indiscernables*: if whatever is true of x is true of y (and vice versa), then x=y. The second is trivially true if the antecedent covers truths about identities (*see* **identity logic** and **second-order logic**).

LEŚNIEWSKI, STANISŁAW (1886–1939). A Polish logician who presented three related systems: protothetic, **ontology**, and **mereology**. *Protothetic* encompasses **propositional logic** but is more powerful because it allows things like propositional **quantifiers**. *Ontology* is about names and is somewhat like **quantificational** and **identity logic**. **Mereology** is about parts and wholes and is somewhat like **set theory**, but without bringing in **abstract entities**. Leśniewski was much concerned with the foundations of **arithmetic** and how to avoid problems associated with **Russell**'s paradox. While he was very creative, the unconventionality of his approach discouraged its wider influence.

LEWIS, CLARENCE IRVING (1883–1964). An American philosopher and logician. His 1932 *Symbolic Logic* (coauthored with Cooper Harold Langford) started the contemporary interest in **modal logic**; since he was unsure about some modal principles, he sketched a variety of modal systems, which he called "S1" to "S9." Lewis got interested in modal logic because he was unhappy with *material implication* as an analysis of **conditionals**; he saw **entailment** as better capturing the logic of "if-then" (*see* **relevance logic**). Lewis's 1955 *The Ground and Nature of the Right* was important for its application of logic to ethics.

LIAR PARADOX. A statement that asserts its own falsity and thus, **paradoxically**, appears to be both **true** and **false**. Consider claim P:

(P) P is false.

Is P true? Then things must be as P says they are, and thus P has to be false. Is P false? Then things are as P says they are, and thus P has to be true. So if P is either true or false, then it has to be both true and false.

Graham Priest and other **dialethists** say that P is *both true and false*; this requires rejecting **Aristotle**'s venerable **law of non-contradiction**. Most other philosophers say that P is *neither true nor false*; this requires rejecting, or at least qualifying, Aristotle's **law of excluded middle**. But why is P neither true nor false?

Bertrand Russell, to deal with such paradoxes, proposed a *theory of types* that outlaws certain forms of self-reference. Very roughly, there are ordinary objects (type 0), properties of these (type 1), properties of these properties (type 2), and so on. Any meaningful statement can talk only about objects of a lower type; so no speech can talk meaningfully about itself. P violates this condition, and so is meaningless—and thus neither true nor false.

However, Russell's theory seems to refute itself. "Any meaningful statement can talk only about objects of a lower type," to be useful, has to restrict *all statements*, of *every type*; but then it violates its own rule and declares itself meaningless. So the paradox reappears.

Alfred Tarski, to deal with the liar paradox, proposed that no language can contain its own truth predicate; to ascribe truth or falsity to a statement in a given language, we must ascend to a higher-level language, called the **metalanguage**. P violates this condition, and so is meaningless—and thus neither true nor false.

Opponents say Tarski's view is too restrictive. English and other languages *do* contain their own truth predicates, and they need to for many purposes. So it would be better to have a less sweeping restriction to take care of the liar paradox. But there is little agreement about what this restriction should be.

The Cretian Epimenides in the sixth century BC proposed the liar paradox, and St. Paul mentioned it in his letter to Titus (1:12). It has been widely discussed ever since. While most think that a theory of truth must deal with the paradox, how best to do this is still unclear.

LOCUTIONARY/ILLOCUTIONARY/PERLOCUTIONARY. A distinction made by John L. Austin (1911–1960) among three different kinds of speech act. I can utter certain words (*locutionary act*); and thereby do something further, like state or apologize (*illocutionary act*); and by doing this cause some further effect, like convince you or cause alarm (*perlocutionary act*). Austin also spoke much about *performatives*, which are speech acts, like "I thereby name this ship the Carol Ann," that bring about something but are not true or false.

LOGIC. "Logic" is often defined in ways like "the analysis and appraisal of arguments," "the study of valid reasoning," "the science of the principles governing the validity of inference," and "the art and science of right reasoning." These definitions work out much the same in practice.

Actually, the meaning of the term "logic" is more complicated than this, because the term can be used in a narrow and a broad sense. *Logic in the narrow sense* is the study of deductive reasoning, which is about what logically follows from what. *Logic in the broad sense* includes also various other studies that relate to the analysis and appraisal of arguments; these include areas like **informal logic, inductive logic, metalogic**, and **philosophy of logic**.

Sometimes "logic" is used to mean "a system of logic"; so then "deviant logics" would mean "systems of deviant logic." Sometime "logic" covers what a simple-minded application of a particular view or attitude would lead to; so we speak of "the logic of terrorism" or "the logic of revenge." Sometimes the term simply means "reasoning"; so we speak of "the logic behind her decision."

The four general entries in this dictionary that start with "logic:" serve mainly to point to more specific entries, and these in turn often point to further related topics. So we have here a hierarchy of topics. Here are the four "logic:" entries:

- **logic: deductive systems** points to entries like **propositional logic, modal logic, deontic logic, temporal logic, set theory, many-valued logic, mereology,** and **paraconsistent logic.**
- **logic: history of** is about historical periods and figures and includes entries like **medieval logic, Buddhist logic, twentieth-century logic, Aristotle, Ockham, Boole, Frege,** and **Quine.**
- **logic: and other areas** relates logic in an interdisciplinary way to other areas and includes entries like **biology, computers, ethics, gender, God,** and **psychology.**
- **logic: miscellaneous** is about everything else and includes entries like **abstract entities, algorithm,** *ad hominem,* **inductive logic, informal/formal logic, liar paradox, metalogic, philosophy of logic,** and **software for learning logic.**

See also LOGIC, SCOPE OF.

LOGIC CHOPPING. The **fallacy** of using the technical tools of logic in an unhelpful and pedantic manner by focusing on trivial details instead of directly addressing the main issue in a dispute. That this is done is truly a shame, because logic can be so useful, if we know how to use it, for clarifying our reasoning on important issues.

LOGIC COURSES. Popular at universities, since students and teachers see them as useful in promoting disciplined, logical thinking. Logic is usually required of all philosophy majors. It is often highly recommended for students interested in other disciplines that involve close reasoning and argumentation—like law, mathematics, and computer programming; it can be useful in any area where good reasoning is important, including politics, religion, medicine, science, journalism, business, and education. Introductory "baby logic" courses tend to focus on **propositional logic,** including **truth tables** and **proofs;** they may include also further topics like **syllogistic** or **quantificational logic, fallacies,** and **definitions.** An alternative "critical thinking" model stresses informal over formal areas; many teachers try to combine both models and show how formal and informal tools work together in appraising actual arguments. Logic is taught also in math and computer science departments, but with a different emphasis. One practical problem with logic courses is that students vary greatly in their natural aptitude for logic, with many picking up the material very easily and a few struggling with the easiest ideas.

LOGIC GATE. An electrical or electronic device whose input-output function mirrors the **truth table** of a **propositional logic** formula. The idea that logic gates were possible, first discovered by **Charles Sanders Peirce** in 1867 and later rediscovered by Claude Shannon in 1937, was the key insight that led to the creation of modern digital **computers**.

Computers represent "1" and "0" (which are often used for "true" and "false") by different physical states; "1" might be a positive voltage and "0" a zero voltage. An AND-GATE would then be a physical device with two inputs and one output, where the output has a positive voltage if and only if *both* inputs have positive voltages:

An OR-GATE would be similar, except that the output has a positive voltage if and only if *at least one* input has a positive voltage. For any formula, we can construct an input-output device, a *logic gate*, that mimics that formula.

A computer basically converts input information into 1s and 0s, manipulates these 1s and 0s by logic gates and memory devices, and then converts the resulting 1s and 0s back into a useful output. Logic gates use propositional logic, which is thus central to the operation of computers. The logicians **John von Neumann** and Arthur Burks were the key logicians on the team in the 1940s that produced the ENIAC— the first large-scale electronic computer. So logic had an important role in moving us into the computer age.

LOGIC PUZZLE. You are shipwrecked on an island and learn that the natives form two groups: the *truth-tellers* always tell the truth and the *liars* always lie. You meet two natives, A and B, and know that one is a truth-teller and the other a liar, but you do not know which is which. You need to find out which road, right or left, goes to the airport. What do you ask them?

This is a logic puzzle. You can get the answer, if you are clever enough, just by thinking; you need not appeal to additional information. Some find logic puzzles recreational; others use them to prepare for the Law School Admissions Test, which has many very difficult ones.

By the way, you could ask A this: "How would B answer, yes or no, if I asked B whether the left road goes to the airport?" The truth is then the opposite of what A tells you.

LOGIC SOFTWARE. *See* SOFTWARE FOR LEARNING LOGIC.

LOGIC, PHILOSOPHY OF. *See* PHILOSOPHY OF LOGIC.

LOGIC, SCOPE OF. This book uses the term "logic" in two senses, narrow and broad. *Logic in the narrow sense* is the study of deductive reasoning, which is about what logically follows from what. *Logic in the broad sense* includes various other studies that relate to the analysis and appraisal of arguments—for example, informal logic, inductive logic, metalogic, and philosophy of logic.

Even if we take "logic" in this narrow deductive sense, there still is some unclarity on what it includes. Suppose you say, "I have $30; therefore I have more than $20." Is this part of logic, part of mathematics, or both?

Willard Van Orman Quine suggested that we limit "logic" to **classical symbolic logic**, which he saw as fairly uncontroversial and as focusing on *topic-neutral* terms like "and" and "not" that arise in every area of study. *Philosophical extensions* to this, like **modal** and **deontic logic**, if legitimate at all, are part of philosophy in general. *Mathematical extensions*, like **set theory** and axiomatizations of **arithmetic**, are part of **mathematics**. And **deviant logics** are illegitimate.

Most logicians today tend to use "(deductive) logic" in a broader way that is hard to pin down (*see* **logic: deductive systems**). Deductive logic is commonly taken to include, besides classical symbolic logic and traditional **syllogistic logic**, philosophical extensions (like modal and deontic logic), deviant logics, and sometimes even mathematical extensions (like **set theory**). Logic is seen as part of at least three disciplines—philosophy, mathematics, and **computer** science—which approach it from different angles. Any attempt to give sharp and final boundaries to the term "logic" would be artificial.

LOGIC: AND OTHER AREAS. Many entries in this dictionary are interdisciplinary and relate logic to other areas—like **aesthetics,**

anthropology, arithmetic, biology, children, cognitive science, computers, epistemology, ethics, gender, geometry, God, language, machines, mathematics, metaphysics, ontology, philosophy, physics, psychology, rhetoric, and scientific reasoning.

LOGIC: DEDUCTIVE SYSTEMS. It is useful to divide systems of deductive logic into five rough groupings:

Classical symbolic logic covers standard **propositional** and **quantificational logic**, and sometimes **identity logic**. "Classical" or "standard" approaches agree with the systems of **Gottlob Frege** and **Bertrand Russell** on which arguments are valid, despite differences in symbolization and **proof** techniques. Sometimes the next three groupings are collectively called "non-classical" logics.

Philosophical extensions are systems of philosophical interest that build on classical logic without contradicting it; these supplements to classical logic include **axiological, belief, deontic, epistemic, imperative, infinitary, mereological, modal, question, second-order**, and **temporal logic**.

Mathematical extensions are systems of mathematical interest that build on classical logic without contradicting it. These include axiomatizations of **arithmetic, geometry**, and **set theory**.

Deviant logic includes systems that conflict with classical logic on which arguments are valid; these suggested replacements for parts of classical logic include **free, fuzzy, intuitionist, many-valued, paraconsistent, relative identity**, and **relevance logic**.

Traditional logic covers **syllogistic logic** (from **Aristotle** but with later refinements), some additions from the Stoics (*see* **ancient logic since Aristotle**), **fallacies**, and **definitions**.

LOGIC: HISTORY OF. While logic proper started with **Aristotle**, there were precursors of **ancient logic before Aristotle** and significant **ancient logic since Aristotle**. In the West, **medieval logic** and **Renaissance-to-nineteenth-century logic** developed further the same Aristotelian tradition; in the East, there developed a parallel tradition of **Buddhist logic**. **Gottlob Frege** and **Bertrand Russell** created **classical symbolic logic**; and **twentieth-century logic** has further developed classical and non-classical logics. Individuals with their own entries include **Aristotle, Arrow, Bayes, Boole, Cantor, Carroll, Church, Craig, De Morgan, Euler, Frege, Gödel, Grice, Hintikka, Hume, Husserl, Kant, Kripke, Ladd-Franklin, Leibniz, Leśniewski,**

Lewis, Löwenheim, Łukasiewicz, Marcus, Mill, Ockham, Peano, Peirce, Plantinga, Priest, Quine, Russell, Tarski, Turing, Venn, von Neumann, and **Wittgenstein.** *See also* the chronology.

LOGIC: MISCELLANEOUS. This covers everything not covered by the previous three entries, including broad areas like **informal logic, inductive logic, metalogic,** and **philosophy of logic**; more specific topics, like individual **fallacies, Bayes's** theorem, **semantics,** and **paradoxes**; and various technical terms, like **bivalence, biting the bullet, complete system,** and *de re/de dicto.*

LOGICAL ATOMISM. An approach to **metaphysics** that asks, "What language structure would suffice to describe reality completely?"—and answers that we would need the framework of **classical symbolic logic** plus terms that refer to the ultimately simple elements of reality. *See also* RUSSELL; WITTGENSTEIN.

LOGICAL CONSTRUCT. Sometimes a noun phrase does not refer to real entities, but yet sentences using that phrase make genuine assertions. Suppose you say, "The average American has 2.4 children." While "the average American" does not refer to any actual entity, the sentence as a whole is meaningful; the sentence claims that the average number of children that Americans have is 2.4. So "the average American" is a *logical construct.* **Bertrand Russell** introduced the notion and asked whether things like sets, numbers, material objects, persons, electrons, and experiences were real entities or logical constructs. *See also* NOTHING; ONTOLOGY; QUINE.

LOGICAL FORM. *See* FORM.

LOGICAL POSITIVISM. A philosophical movement that was strong in the first half of the 20th century. It started with the *Vienna Circle*, a group of Austrian scientists and philosophers who met in the 1920s to develop a scientific approach to philosophy. Important logical positivists included Rudolf Carnap and Alfred Jules Ayer. They saw logic and **mathematics** as true by virtue of language conventions, and hence **analytic.** Their central doctrine was the *verifiability criterion of meaning*, which claimed that the **meaning** of a non-analytic statement is determined by what conceivable observable tests would settle whether the statement is true. Any non-analytic statement whose truth could not, at least in principle, be tested empirically was meaningless (in the sense of being neither true nor false). On this basis, logical positivists declared that **ethics**, theology, and **metaphysics** were meaningless.

Logical positivism has now been almost universally abandoned, since (1) the view is **self-refuting** (its central doctrine is neither analytic nor empirically testable, and so is meaningless on its own terms); (2) the difference between what is or is not empirically testable proved impossible to draw in a clear way; and (3) many of the things that positivists declared meaningless did not seem to be meaningless at all.

LOGICAL PRINCIPLES, STATUS AND JUSTIFICATION OF. Take your favorite principle of logic, one that you think is clearly correct, perhaps *modus ponens* or **Barbara**. Why is it correct and how do we know that it is correct?

Most philosophers have thought that logical and mathematical principles are *a priori*. **John Stuart Mill** said they are empirical, and **Willard Van Orman Quine** said that the *a priori*/empirical distinction is unclear, that logic is a bit of both, and that we need to pick logical principles on pragmatic grounds. But the *a priori* view is the most common one; indeed, we seem to come to the correctness of a principle of logic by just thinking, and not by some sort of sense experience.

For those who hold the *a priori* view, some say that logical principles are based on convention, on what we mean by terms like "if-then" and "all" (*see* **analytic**). Others object that this makes logic too arbitrary and that it confuses the logical principles themselves (which are necessary truths) with how we express them (which depends on conventions about words like "if-then" and "all"). If we changed our language, the logical principles would still be true; but we would have to use different words to express them.

Others who hold the *a priori* view say that logic and **mathematics** are about independent, abstract **truths**, which are distinct from truths about matter or mind, and which can be somehow present to our minds. Others say that this very Platonic approach makes logic and mathematics too mysterious and that we have no good evidence that **abstract entities** even exist.

Some claim that many of the basic principles of logic and mathematics are clear and evident, even though we may have difficulty in explaining why they are true and how we know this. Most agree that we cannot prove a principle like *modus ponens* without circularity. If we say that we know that *modus ponens* is correct because, for example, it checks out as correct on the **truth-table** test, we are presuming a further *modus ponens* argument: "If modus ponens checks out on the truth-table test, then it is valid; but it checks out on the truth-table test; therefore, it is valid." Any justification of *modus ponens* assumes either *modus ponens* or else some similar inference rule.

Aristotle long ago showed that every **proof** must eventually rest on something unproved; otherwise, we would need either an infinite chain of proofs or circular arguments—and neither is acceptable. So why not just accept the validity of *modus ponens* as a self-evident truth—a truth that is evident but cannot be based on anything more evident? If we accept this, we should not think that picking logical principles is just a matter of following untrained "logical intuitions." Many people before they study logic cannot distinguish valid **conditional** forms like *modus ponens* from invalid ones like affirming the consequent. But they can distinguish the two after some training in logic; this training consists largely in examining concrete instances in which the validity or invalidity of forms is more obvious (*see* the introduction).

So are *modus ponens* and other logical principles clear and evident on this basis? Many defenders of **deviant logics** would say no; they have raised doubts about many of these principles, including *modus ponens* (*see* **relevance logic**). But are their doubts justified?

It should not surprise us that there are deep controversies about the basic principles of logic. Every area, including simple claims about material objects like "I see a chair," raises deep controversies if we push things far enough. *See also* INDUCTIVE LOGIC; LAW OF NON-CONTRADICTION; KRIPKE; PSYCHOLOGISM.

LOGICAL SYMBOLS. *See* the notation section.

LOGICAL TRUTH. *See* ANALYTIC/SYNTHETIC.

LOGICIANS, STEREOTYPES OF. Logicians are people with expertise in **logic**; that is what they have in common. Beyond that, logicians come in all sizes, shapes, and personality types.

Some people who do not know much about logic or logicians have **false stereotypes** of what logicians are like. Logicians are seen as people who are non-emotional, or who reject spontaneity in favor of carefully thinking everything out, or who will not accept what is not proved. In fact, however, many logicians are highly emotional; some have been known to jump up and down with enthusiasm while teaching logic. Many logicians are highly spontaneous and do not like to plan things out; one prominent logic professor was so against following a schedule that he did not even own a watch. And few if any logicians believe in rejecting everything that is not proved; **Aristotle**, the first logician, showed that this would lead to the rejection of all knowledge.

LOGICISM. *See* ARITHMETIC.

LOOP, ENDLESS. *See* ENDLESS LOOP.

LÖWENHEIM, LEOPOLD (1878–1957). A German mathematician best known for the *Löwenheim-Skolem theorem*, which says that any consistent set of first-order **quantificational** formulas all come out true under some **interpretation** in the realm of natural numbers. Hence, first-order quantificational logic cannot characterize structures involving nondenumerable infinities (*see* **Georg Cantor**).

ŁUKASIEWICZ, JAN (1878–1956). A Polish logician. He invented **Polish notation**, which is a way of writing logical formulas that avoids parentheses and results in shorter formulas. He put the **syllogistic logic** of **Aristotle** into a strict **formal system**, using Polish notation to symbolize the four categorical statements (using "Aab" for "all a is b," "Iab" for "some a is b," "Eab" for "no a is b," and "Oab" for "some a is not b"). Inspired by Aristotle's discussion of the sea battle, he invented three-valued logic (*see* **many-valued logic**).

– M –

MACHINES, LOGIC. The first mechanical computing device was probably the abacus, which the Babylonians developed before 1000 BC. The abacus was a common business tool in medieval Europe and was widely used in many cultures; the still-popular Chinese version dates from the Yuan Dynasty (c. 1300 AD). An abacus uses strings of beads that can be moved to the right or left to represent numbers. A simple abacus might represent 423 in this way:

A more complex version uses two columns, one with five beads (representing units) and one with two beads (representing fives). A Web search will reveal how to buy an abacus and how to use one to add, subtract, multiply, and divide.

The German Wilhelm Schickard in 1623 built a gear-based calculator; his *calculating clock* could add, subtract, and carry or borrow digits between columns. Leonardo da Vinci (1452–1519) had earlier designed but not built a similar device. Blaise Pascal in 1642 built a practical calculator, which **Gottfried Wilhelm Leibniz** improved in 1673. Mechanical calculators after this became increasingly popular. IBM began in 1896 as the Tabulating Machine Company; it built a sophisticated machine, which used paper cards with holes for storing information, to help process data from the 1900 U.S. census.

All these devices are *digital*: values are represented in discrete steps. In contrast, the slide rule, which the Englishman William Oughtred invented in 1632, was *analog*: values are represented by continuously variable physical states. Slide rules, which gave quick but approximate multiplications and divisions, were widely used until inexpensive electronic calculators became available in the 1970s.

The first non-numerical logic machine may have been a contraption built by the Spaniard Ramón Llull (c. 1235–1315), which used rotating disks with words written on them to combine basic concepts into complex statements. Llull's purpose was Christian apologetics, to demonstrate beliefs about God. In 1832 the Englishman Charles Babbage designed a steam-driven Analytical Engine, a computer that would (if actually built) accept punch-card programs. Augusta Ada King, the Countess of Lovelace, wrote an essay explaining Babbage's machine and showing how it related matter to abstract logical processes; she is widely hailed as the first **computer** programmer. These examples remind us that the connection of computing with electricity is just a matter of speed, power, and convenience; it is possible in principle to design non-electrical computers that use steam or hydraulic power.

Hobbyists have made various mechanical or electrical devices to mirror standard logical systems. For example, statements can be put on file cards, with holes punched out to represent key logical information (like which terms are distributed in a **syllogistic** formula or which **truth-table** cases are true for a **propositional** formula); then we can read off whether an argument using these statements is valid by whether the holes match up so we cannot see through them. And an electrical syllogistic logic machine can be built with switches whose positions represent various statement forms and figures, in such a way that a light will light if and only if the syllogism we select is valid. *See also* LOGIC GATE; TURING.

MANY-VALUED LOGIC. Any alternative to standard **propositional logic** that has more than two **truth values** (and thus rejects **bivalence**).

A three-valued logic might use "1" for true, "0" for false, and "½" for a third truth value, which we will call "half-true." This last category might apply to statements that are unknowable, or too vague to be true-or-false, or plausible but unproved, or meaningless, or about future events not yet decided either way (*see* **Aristotle**'s discussion of the future sea battle). A three-valued **truth table** for NOT looks like this:

P	~P	
0	1	If P is false, then ~P is true.
½	½	If P is half-true, then ~P is half-true.
1	0	If P is true, then ~P is false.

Here are tables for the other connectives:

P	Q	(P · Q)	(P ∨ Q)	(P ⊃ Q)	(P ≡ Q)
0	0	0	0	1	1
0	½	0	½	1	½
0	1	0	1	1	0
½	0	0	½	½	½
½	½	½	½	1	1
½	1	½	1	1	½
1	0	0	1	0	0
1	½	½	1	½	½
1	1	1	1	1	1

An AND takes the value of the lower conjunct, and an OR takes the value of the higher disjunct. An IF-THEN is true if the consequent is at least as true as the antecedent and is half-true if the consequent is a little less true than the antecedent. An IF-AND-ONLY-IF is true if both parts have the same truth value and is half-true if they differ a little.

Then "(P ∨ ~P)" (**law of excluded middle**) and "~(P · ~P)" (**law of non-contradiction**) are sometimes only half-true but are never false. "(P ⊃ Q)" is no longer equivalent to "~(P · ~Q)" or to "(~P ∨ Q)"; to preserve these equivalences, we would need to make "(½ ⊃ ½)" half-true, which would then make "(P ⊃ P)" sometimes only half-true. Or we could make "(½ ∨ ½)" true and "(½ · ½)" false, which would keep the standard equivalences and the laws of excluded middle and non-contradiction as true; but then "P" would not be logically equivalent to "(P ∨ P)" or to "(P · P)," which goes against our intuitions.

We might have more than three truth values—perhaps six, or 10, or an indefinite number n, or an infinite number; **fuzzy logic** often takes any real number between 0.00 and 1.00 to be a truth value. On the last scheme, we might define a "valid argument" as one in which, if the premises have at least a certain truth value (perhaps .9), then so does the conclusion; depending on how we set up the "⊃" truth table, *modus ponens* and other traditional logical principles may fail to hold (since if "A" and "(A ⊃ B)" are both .9, then "B" might be less than .9).

Opponents say many-valued logic is weird and arbitrary and has little or no application to real-life arguments. Even if this is so, still the many-valued approach has other sorts of application. It can be used, for example, for **computer** storage systems that have more than two states. And it can be used to show the independence of **axioms**; axiom A can be shown to be independent of the other axioms of a certain system if, for example, the other axioms (and theorems derived from these) always have a value of "7" on a certain multi-valued truth-table scheme, while axiom A sometimes has a value of "6."

Some **paraconsistent logics** use four truth values: "1" for "true and not false," "0" for "false and not true," "B" for "both true and false," and "N" for "neither true nor false." The truth table for NOT would then look like this:

P	~P	
0	1	If P is just-false, then ~P is just-true.
B	B	If P is true-&-false, then ~P is true-&-false.
1	0	If P is just-true, then ~P is just-false.
N	N	If P has no truth value, then ~P has no truth value.

On this scheme, a "valid argument" is defined as one in which, if the premises are true (either just-true "1" or both-true-and-false "B"), then so must be the conclusion; a "valid argument" might then have true premises and a conclusion that is both true and false. Another scheme uses just two truth values but allows "A" and "not-A" to have these independently of each other; so we then have four possibilities:

P	~P	
0	0	P and not-P are both false.
0	1	P is false and not-P is true.
1	0	P is true and not-P is false.
1	1	P and not-P are both true.

This approach rejects the usual understanding of "not," on which "not-A" has only one truth value, and the opposite truth value as "A." *See also* INTUITIONIST LOGIC.

MARCUS, RUTH BARCAN (1921–). American logician who was a pioneer in **quantified modal logic**. In the 1940s, she entered into a continuing debate with **Willard Van Orman Quine**, who strongly attacked **modal logic** in general and quantified modal logic in particular. **Saul Kripke** later built on Marcus's work but added a **semantics** for modal logic that involves **possible worlds**; since then, modal logic has become much more respectable among logicians.

MATERIAL IMPLICATION. Conditional using the simple "⊃" sense of "If A then B," which just denies that we have A true and B false.

MATHEMATICAL INDUCTION. Suppose something holds in the first case, and if it holds in the first n cases, then it holds in the n+1 case; then it holds in all cases. This is the rule of mathematical induction.

MATHEMATICAL LOGIC. Sometimes this term is used interchangeably with "**symbolic logic**" and sometimes it is used to refer to those areas of symbolic logic that are of greater interest to mathematicians than to philosophers (for example, highly technical results about **metalogic** or **set theory**). *See also* BOOLE.

MATHEMATICS. The founders of modern logic were very interested in mathematics. They wanted to see what logical principles were presupposed by mathematicians in their proofs. They speculated also about the nature of **arithmetic** and whether it could be reduced to logic. *See also* FREGE; GEOMETRY; GÖDEL; PEANO; RUSSELL; SET THEORY; TWENTIETH-CENTURY LOGIC.

MEANING. To appraise whether **premises** are true, we must first understand what they mean. So meaning is important for **arguments**.

We can often explain the meaning of a term by using a **definition**, which explains one term using other terms. But assuming that we must avoid circular sets of definitions, we cannot define all our terms; instead, we must leave some terms undefined. But how can we explain the undefined terms? One way is by examples.

To teach "red" to someone who understands no language that we speak, we could point to red objects and say "Red!" We would want to point to different sorts of red objects; if we pointed only to red shirts, the person might think that "red" meant "shirt." If the person understands "not," we could point also to non-red objects and say "Not red!" The person, unless color-blind, soon will catch our meaning. Such *ostensive definitions* are a basic, primitive way to teach language.

We sometimes point to examples through words. We might explain "plaid" to a child by saying, "It is a color pattern like that of your brother's shirt." We might explain "love" by mentioning examples: "Love is getting up to cook a sick person's breakfast instead of staying in bed, encouraging someone instead of complaining, listening to others instead of telling them how great you are, and similar things." It is often useful to combine a definition with examples, so the two can reinforce each other.

In abstract discussions, people sometimes use words so differently that they fail to communicate. Asking for definitions may then lead to the frustration of hearing one term you do not understand being defined using other terms you not understand. In such cases, it might be more helpful to ask for examples instead of definitions. The request for examples can bring a bewilderingly abstract discussion back down to earth and mutual understanding.

Theories about meaning are a central topic in philosophy of **language** and have a great influence on other issues. For example, **logical positivists** like A. J. Ayer proposed that we analyze the meaning of a non-**analytic** statement in terms of how we could empirically test its truth and that we regard a statement as meaningless if it has no empirical test; they concluded that **ethics**, theology, and **metaphysics** were meaningless. Other philosophers proposed that we analyze the meaning of a statement in terms of what practical differences its truth or falsity could make (William James's **pragmatism**), its use in language games and forms of life **(Ludwig Wittgenstein)**, or its contribution to certain sorts of speech acts (J. L. Austin, *see* **locutionary/illocutionary/perlocutionary**)—or that we give up the notion of "meaning" because it is so unclear **(Willard Van Orman Quine)**. *See also* LANGUAGE, USES OF; SENSE/REFERENCE.

MEDIEVAL LOGIC. Medieval logicians carried on the basic framework of **Aristotle** and the Stoics (*see* **ancient logic since Aristotle**) as logic became increasingly more important in higher education.

Boethius (480–524) was important for his works on logic (including commentaries) and his translations of Aristotle's logic into Latin. Most of these translated works were lost until the 12th century, except for *Categories* and *On Interpretation*, which became the main sources of logic for the next few centuries; the tradition based on these was later called the *logica vetus* (old logic). Except for Boethius, European Christians did little creative work on logic until the 11th century.

Boethius's *Consolations of Philosophy* gave a clear formulation of the modal **box-inside/box-outside ambiguity**. This is part of his criticism of an argument that tries to show that divine foreknowledge is incompatible with human freedom:

> God knew that you would do it.
> If God knew that you would do it, then it was
>   necessary that you would do it.
> If it was necessary that you would do it, then
>   you were not free.
> ∴ You were not free.

Boethius saw the second premise as ambiguous; it could express either "necessity of the *consequence*" (what is necessary is "if A then B": "□(A ⊃ B)" in **modal logic**) or the "necessity of the *consequent*" (where what is necessary is the second part: "(A ⊃ □B)"). The former makes the argument invalid while the latter makes the second premise doubtful; the argument fails either way. This is not the end of the story, since we can still ask how God knows future contingents; Boethius and others talked about this too, saying that God grasps simultaneously the whole past-present-future of history. In any case, Boethius's modal distinction became an important tool of medieval philosophy. (For a related argument, *see* **Aristotle**'s discussion of the sea battle.)

The Arab world dominated in logic from about 800–1200. Some Arab logicians were Christian, but most were Moslem; both groups saw logic as an important tool for theology and for areas like medicine. First they focused on translating Aristotle into Arabic; then they wrote commentaries, textbooks, and original works on logic and other areas. They worked on topics like modal logic, conditionals, universals, predication, existence, and categorical propositions. Baghdad and Moorish Spain were centers of logic studies; figures include Al-Farabi, Al-Tayyib, Avicenna, and Averroës.

The 11th and 12th centuries brought increased interest in logic in Christian Europe, first with the writings of Anselm and Péter Abelard and then later with more of Aristotle's works becoming available in Latin. The *logica vetus* (old logic) was based on Aristotle's *Categories* and *On Interpretation*; in contrast, the *logica nova* (new logic) was based on other books of his Organon that were coming into circulation: *Prior Analytics*, *Posterior Analytics*, *Topics*, and *Sophistical Refutations*. This new logic was also called "terminist," because of its emphasis on terms and how they signify (*see* **supposition**). There also was much interest in **fallacies** and **paradoxes**.

Peter of Spain (c. 1215–1277), who seems to have been a medical doctor and teacher and later Pope John XXI (although there is scholarly debate about these), wrote the comprehensive *Summulae Logicales*; this contained the **Barbara, Celarent** verse and became a popular logic textbook for several centuries. William of Sherwood (c. 1206–1268) wrote a logic book at about the same time, and it is difficult to decide who first came up with which ideas.

Thomas Aquinas (1224–1274), the most influential of the medieval philosophers, had little impact on the development of logic itself; but he made great use of logic in his writings. In light of the sheer bulk of his writings and his heavy stress on argumentation, it is likely that he produced a greater number of philosophical **arguments** than anyone else who has ever lived.

William of Ockham (c. 1285–1349) wrote an influential *Summa Logicae* ("Summary of Logic"). While best known for Ockham's razor and nominalist analyses of terms like "humanity," he also developed principles of modal logic; for example, he suggested that a disjunction is possible if either part is possible—and that whatever follows from something possible is itself possible. Modal ideas were important for Ockham's view of **God**, since he thought God could bring about whatever is possible; since God could have created the world in any possible way, he argued, the structure of the world has no necessity and thus cannot be known by abstract speculation apart from experience.

Jean Buridan (1300–1358) formulated the rules for valid **syllogisms** that have become standard. He raised questions also about choice, claiming that a dog placed at an equal distance between two bowls of food would choose one randomly. His disciple Albert of Saxony (c. 1316–1390) carried on Buridan's work, especially through his popular *Perutilis Logica* textbook.

There was much dispute among medieval logicians about universals, like redness or roundness. The majority *realist* view saw universals as having real existence ourside of our minds—existing either as independent entities (*Platonic realism*) or merely in the concrete things that are red or round (*Aristotelian realism*). But *conceptualist* or *nominalist* thinkers like Peter Abelard, Roscelin, and William of Ockham objected that only individual entities have real existence; universals are mental concepts or words that we use to apply to similar things. There are somewhat similar debates today about the existence of **abstract entities** (*see also* **Willard Van Orman Quine**).

Ramón Llull (c. 1235–1315) hinted at later developments. His *ars magna* (great art) tried to reduce knowledge to its 54 simplest concepts, from which other ideas could be constructed using a special logical notation. He built a logic **machine** based on these ideas. His goal was Christian apologetics, to demonstrate beliefs about God.

Logic was important in the Middle Ages—both in philosophical writings and in higher education. The world's first universities were then springing up in Catholic Europe, and these emphasized philosophy and logic. The "core curriculum" included the seven liberal arts: logic, grammar, and rhetoric (the *trivium*)—plus **arithmetic**, **geometry**, astronomy, and music (the *quadrivium*); these prepared students for graduate work in medicine, law, or theology. One sign of the continuing influence of medieval logic is the persistence of Latin terms (like **modus ponens** and *a priori*) even today.

MEREOLOGY. The logic of parts and wholes, first developed by **Stanisław Leśniewski**. Mereology raises issues like these: Is a whole the sum of its parts? Are entities with all the same parts thereby identical? Is part-hood transitive, so if X is a part of Y, and Y is a part of Z, then X must be a part of Z? Is part-hood reflexive, so everything is part of itself? (Clearly nothing can be a "proper part" of itself, where X is a *proper part* of Y just if X is a part of Y but Y is not a part of X.) If X is a part of Y and Y is a part of X, must X be identical to Y? Must there be smallest parts (atoms), or do parts break into smaller parts without limit? Are copper and a round shape (matter and form) parts of a penny? Given an entity X and an entity Y, is there automatically a further entity that has X and Y as parts? Is there an "empty object" (like the null set) that has no parts and is part of everything else? Is there an entity that has as parts all and only those things that are not parts of themselves? Is set-membership an instance of the part-whole relationship? Can mereology do much of the work of **set theory** but avoid **abstract entities** and problems associated with **Russell**'s paradox?

METALANGUAGE/OBJECT LANGUAGE. To avoid contradictions from statements that refer to their own **truth** or **falsity** (like the **liar paradox**, where a statement P says "P is false") **Alfred Tarski** and others suggest that no language can talk about the truth or falsity of its own statements. We must distinguish levels of language; when we talk about the truth or falsity of statements of a given language (the *object language*), we use a higher-order language (the *metalanguage*).

METALOGIC (metamathematics). The study of **formal systems** and the attempt to prove things about these systems. Metalogic focuses on the systems themselves, not on how to use them to test arguments.

Branches of logic, like **propositional logic**, are often presented in the guise of formal systems—as little languages that we invent to help us test **arguments**. Formal systems specify certain sequences of symbols to be, for example, **wffs** and **proofs**; these are specified using rules (**algorithms**) about the manipulation of symbols. Metalogic studies the logical consequences of such rules.

Here is a simple example. Our entry for *propositional logic* defines a wff as a sequence that can be constructed using these rules:

1.   Any capital letter is a wff.
2.   The result of prefixing any wff with "~" is a wff.
3.   The result of joining any two wffs by "·" or "∨" or "⊃" or "≡" and enclosing the result in parentheses is a wff.

It follows from these that there is no longest wff—since, if there were a longest wff, then we could make a longer one by adding another "~." This simple proof is about a logical system, so it is part of metalogic.

Metalogic mostly deals with proof systems. The key questions are whether a proof system is **sound** (will not prove bad things—so every argument provable in the system is valid) and **complete** (can prove every good thing—so every valid argument expressible in the system is provable in the system).

Consider the last inferential proof system for propositional logic in our proofs entry. Could the following happen?

A student named Logicus found a flaw in our proof system. Logicus produced a formal proof of a propositional argument; and he then showed by a **truth table** that this argument is invalid. So some arguments provable on our proof system are invalid.

People have found such flaws in logical systems. How do we know that our system is free from such flaws? We need to prove soundness: that every propositional argument for which we can give a formal proof is valid (on the truth-table test).

To prove soundness, we would first have to check that all our **inference rules** are *truth preserving* (so that, when applied to true wffs, they would yield only further true wffs). Our 12 simplifying and inferring rules are easy to check out using truth tables; let's assume that we do this. To certify **RAA**, we would first show that the *first* use of RAA in a proof is truth-preserving. Suppose all previous not-blocked-off lines in a proof are true, and we use RAA to derive a further line; we have to show that this further line is true:

$$
\begin{array}{lll}
1 & \ldots = 1 & \Leftarrow \\
2 & \ldots = 1 & \Leftarrow
\end{array} \quad \text{Suppose all these are true.}
$$

$$
\begin{array}{ll}
\text{asm: } \sim A & \Leftarrow \text{ We assume that A is false.} \\
\ldots & \\
\therefore B & \Leftarrow \\
\therefore \sim B & \Leftarrow
\end{array} \quad \text{We derive a contradiction.}
$$

$$
\therefore A \qquad \Leftarrow \text{ Does A then have to be true?}
$$

From previous true lines plus assumption "~A," we derive contradictory wffs "B" and "~B" using the 12 simplifying and inferring rules. But these rules are truth-preserving. So if the lines used to derive "B" and "~B" were all true, then both "B" and "~B" would have to be true—which is impossible. So the lines used to derive them cannot all be true. Since the lines before the assumption are presumed to be true, assumption "~A" has to be false. So its opposite ("A") has to be true. So the first use of RAA in a proof is truth-preserving.

We can similarly show that if the first use of RAA is truth-preserving, then the second must be too. And we can show that if the first n uses of RAA are truth-preserving, then the n+1 use must be too. Then we can apply the rule of **mathematical induction**: "Suppose that something holds in the first case, and that, if it holds in the first n cases, then it holds in the n+1 case; then it holds in all cases." Using this rule, it follows that *all* uses of RAA are truth-preserving.

Now suppose an argument is provable in our system. Then there is some proof that derives the **conclusion** from the **premises** using truth-preserving rules. So if the premises are true, then the conclusion, too, must be true—and so the argument is valid. So if an argument is provable in our system, then it is valid. This establishes soundness.

Is not this reasoning **circular**? Are not we assuming principles of propositional inference (like *modus ponens*) as we defend our proof system? Of course we are. Nothing can be proved without assuming logical rules. We are not attempting the impossible task of proving things about a logical system without assuming any logical rules. Instead, we are trying to show, relying on ordinary reasoning, that we did not make errors in setting up our system.

This soundness proof shows that our proof system will not prove invalid arguments. This is fairly reasonable without proof. A more pressing question is whether our system is strong enough to prove all valid propositional arguments. Maybe Logicus will find a further propositional argument that is valid but not provable; then we would have to strengthen our system still further. To calm these doubts, we need to prove completeness: that any propositional argument that is valid (on the truth-table test) can be proved using a formal proof.

The full completeness argument is too complex to give here (chapter 12 of Harry Gensler's *Introduction to Logic* has the full argument). But we can sketch the line of reasoning. First a distinction is made between *simple wffs* (letters or their negations) and *complex wffs* (all others). It is shown that propositional logic has nine basic forms of complex wffs (where, for example, "(A · B)" and "~(A · B)" would be two such forms). It is then shown that the full proof strategy, if it does not result in a proof, will eventually "break down" all complex wffs into simple wffs, that these simple wffs will be consistent (or else RAA would have been applied according to the strategy), that all not-blocked-off steps of the abortive proof would be true under the truth conditions given by these simple wffs, and that these truth conditions would make the original premises true and the conclusion false (since the assumption of the conclusion's denial will be one of the not-blocked-off steps) and thus show the argument to be invalid. Thus we would establish the premise of the following valid argument:

> If we correctly apply our proof strategy to a propositional argument but do not get a proof, then the argument is invalid.
>
> $((C \cdot \sim P) \supset \sim V)$
>
> ∴ If we correctly apply our proof strategy to a propositional argument and the argument is valid, then we will get a proof.
>
> ∴ $((C \cdot V) \supset P)$

Now it is always possible to apply our proof strategy correctly to a propositional argument (since the strategy is consistent and will terminate after a finite number of steps—instead of perhaps going into an **endless loop**). So it is always possible to generate a proof for a valid propositional argument.

So we can establish both soundness and completeness:

> *Soundness:* Every provable propositional argument is valid.
> *Completeness:* Every valid propositional argument is provable.

From both together, we conclude that a propositional argument is provable, if and only if it is valid.

We can now prove two more things about our propositional system. A wff is a *theorem* if it is derivable from zero premises; "$(P \lor \sim P)$" is an example of a theorem. We can prove that a wff is a theorem, if and only if it is a truth-table tautology (here "α" represents any wff):

1. α is a theorem, if and only if "∴ α" is provable.
2. "∴ α" is provable, if and only if "∴ α" is valid.
3. "∴ α" is valid, if and only if α has an all-1 truth table.
4. α has an all-1 truth table, if and only if α is a truth-table tautology.
   ∴ α is a theorem, if and only if α is a truth-table tautology.

Premise 1 is true by the definition of "theorem." Premise 2 follows from our soundness and completeness proofs. Premise 3 is true because an argument without premises is valid, if and only if its conclusion is true in all possible cases. Premise 4 is true by the definition of "truth-table tautology."

We can also prove that our propositional system is *consistent*—in the sense that no pair of contradictory wffs are both theorems. Our proof goes as follows:

1. All theorems are truth-table tautologies.
2. No pair of contradictory wffs are both truth-table tautologies.
∴ No pair of contradictory wffs are both theorems.

We just proved premise 1. Premise 2 is true; if it were false, then some wff α and its contradictory $\sim\alpha$ would both have all-1 truth tables— which is impossible. The conclusion follows. So our propositional system is consistent.

While we have focused on one particular propositional proof system, we could investigate also other proof systems and other branches of logic or **mathematics**. The key issue in metalogic is the harmony, in a specific formal system, between the **syntax** and the **semantics**. What is *provable* is specified by notational (syntactic) rules. What *truths* or *valid arguments* are asserted by the formulas is determined by semantic rules about what these formulas mean. The basic problem is whether the two fit together—for example, whether all the provable arguments are valid and all the valid arguments are provable. *See also* ARITH-METIC; CHURCH; GÖDEL; TRUTH FUNCTIONS.

METAPHYSICS. Connects with logic in various ways. First, logic leads to metaphysical questions about the existence of **abstract entities**. Second, some **logical atomist** logicians, like **Bertrand Russell** and **Ludwig Wittgenstein**, suggest that logical analysis reveals metaphysi-cal structures—and that the proper way to do metaphysics is to ask "What language resources suffice to express every fact about the world?" Third, **quantified modal logic** lends itself easily to a meta-physics that involves **Aristotelian essentialism** and the distinction between necessary and contingent properties; **Saul Kripke** and **Alvin Plantinga** have endorsed such a metaphysics. Some logicians, espe-cially **logical positivist** ones, have opposed metaphysics as intellectual nonsense; but this view is much less popular now than it once was.

MILL, JOHN STUART (1806–1873). The most influential English philosopher of the 19th century. While known today more for his works on utilitarianism and political theory, his earlier *System of Logic* established his reputation as a logician. This book dealt with deductive and inductive reasoning and informal **fallacies**. It made much use of a connotation/denotation distinction, which resembled the later **sense/reference** distinction of **Gottlob Frege**. Mill was a strong **empiricist** and saw principles of logic and **arithmetic** as empirical generalizations rather than as *a priori* truths; such principles seem necessary only because of our weak minds' inability to imagine their falsity.

Mill made his greatest contribution in the area of **inductive** reasoning. *Mill's methods* are inductive ways to arrive at and justify beliefs about causes. The basic idea is that factors that regularly occur together are likely to be causally related. This entry will sketch three of his methods. The first is his *method of agreement*:

> A occurred more than once.
> B is the only additional factor that occurred,
> if and only if A occurred.
> ∴ Probably B caused A, or A caused B.

Suppose several people got food poisoning and we are looking for the cause. We would look for some factor that applies to all and only the sick people. For example, maybe all and only the sick people ate pie. Following the method of agreement, we would conclude that probably either (1) eating pie caused the sickness, or (2) sickness caused the eating of pie. Here (2) is interesting but implausible. So we would conclude that the people *probably* got sick because of eating pie.

The "probably" is important. Eating pie and getting sick might just happen to have occurred together; maybe there is no causal connection. Maybe the sickness had another cause, or maybe different causes in different cases. In applying Mill's methods, we made a simplifying assumption: that the sicknesses had the same cause, which is a single factor and always causes sickness. Our investigation may force us to give up this assumption and consider more complex solutions. But it is good to try simple solutions first and avoid complex ones if we can.

Suppose two factors—eating pie and eating hamburgers—occurred in just those cases where someone got sick. Then the method of agreement would not lead to any definite conclusion about what caused the sickness. To make sure it was the pie, we might do an experiment. We take two people who are as alike as possible in health and diet. We give them all the same things to eat, except that we give pie to one but not the other. (This is unethical, but it makes a good example.) Then we see what happens. Suppose the person eating pie gets sick, but the other person does not. We can conclude that pie probably caused the sickness. This follows Mill's *method of difference*:

> A occurred in the first case but not the second.
> The cases are otherwise identical, except that B also
>   occurred in the first case but not in the second.
> ∴ Probably B is (or is part of) the cause of A, or A is
>   (or is part of) the cause of B.

Here we conclude that probably either (1) eating pie is (or is part of) the cause of the sickness, or (2) the sickness is (or is part of) the cause of eating pie. We reject (2) since we gave the person the pie. So probably eating pie is (or is part of) the cause of the sickness. The cause might simply be the eating of the pie (which was contaminated). Or the cause might be this combined with one's poor physical condition.

Another unethical experiment illustrates Mill's *method of variation*:

> A changes in a certain way, if and only
>   if B also changes in a certain way.
> ∴ Probably B caused A, or A caused B, or
>   some C caused both A and B.

Here we feed several victims varying amounts of pie—and find that they get sicker the more pie they eat. We conclude that probably (1) eating pie caused the sickness, or (2) the sickness caused the eating of pie, or (3) something else caused both the eating and the sickness. Since (2) and (3) are implausible, we conclude that eating pie probably caused the sickness. *See also POST HOC ERGO PROPTER HOC.*

MIND, PHILOSOPHY OF. Three areas of logic raise questions about the nature of the mind.

Identity logic uses a substitute-equals rule that seems to hold universally about matter or mathematics; but this rule can fail with mental phenomena like beliefs. So the mental seems to follow somewhat different logical patterns from the physical. Does this refute the materialist project of reducing the mental to the physical?

Let us assume something that seems to be true of current computing devices, namely that the operation of a computing **machine** can be mirrored in an axiomatic system. Then it follows, using **Gödel**'s theorem, that the design of any specific **computer** prevents it from reaching correct answers about some specific **arithmetic** questions. So there could not be a "Universal Arithmetic Machine" that would always give the right answer on questions of arithmetic. Does this limitation also apply to human beings? Given any human being (with its thought mechanisms and brain structures), is there some arithmetic truth that it could not in principle reach—could not reach because of some built-in limitations on human thought mechanisms, and not because the person would run out of time or paper? We do not know the answer. We know, presumably, that computing machines have this limitation but we do not know whether the human mind has this same limitation.

Computers, which grew out of modern logic, raise further questions: Can all human intellectual abilities be simulated by computer? Can computers be developed whose typed answers to questions cannot be distinguished from those of humans (the **Turing** test)? If so, is there any reason not to ascribe intelligence and consciousness to computers? And do brains and computers use similar mechanisms?

MODAL LOGIC. A branch of logic that studies arguments whose validity depends on "necessary," "possible," and similar notions. (Sometimes "modal logic" is used more broadly, to include areas like **deontic** and **epistemic logics**.) While modal logic in this entry builds on **propositional logic**, it could also be built on the more powerful **quantificational logic** (*see* **quantified modal logic**).

Modal logic here adds to the symbols of propositional logic two new vocabulary items: "◇" and "□" (diamond and box):

$$\diamond A \quad = \quad \text{A is possible (true in some possible world).}$$
$$A \quad = \quad \text{A is true (true in the actual world).}$$
$$\Box A \quad = \quad \text{A is necessary (true in all possible worlds).}$$

Calling something *possible* is a weak claim, weaker than calling it *true*. Calling something *necessary* is a strong claim; it says not just that the thing is true, but that it has to be true—it could not be false.

"Possible" here means *logically possible* (*logically* **consistent**, *not self-contradictory*). "I run a mile in two minutes" may be physically impossible; but there is no self-contradiction in the idea, so it is logically possible. Likewise, "necessary" means *logically necessary* (*self-contradictory to deny*). "2+2=4" and "All bachelors are unmarried" are examples of necessary truths; such truths are based on logic, the meaning of concepts, or necessary connections between properties (*see* **analytic/synthetic**).

We can rephrase "possible" as *true in some possible world*—and "necessary" as *true in all possible worlds*. A **possible world** is a consistent and complete description of how things might have been or might in fact be. Picture a possible world as a *consistent story*. The story is *consistent*, in that its statements do not entail self-contradictions; it describes a set of possible situations that are all possible together. The story is *complete*, in that it is imagined to include every statement or its negation. (This "completeness" of course is an idealization; in practice, we cannot write down an infinite list of sentences.) The story may or may not be true. The *actual world* is the story that is true—the description of how things in fact are.

To propositional logic, we add this new way to form **wffs**:

1.   The result of writing "◇" or "□," and then a wff, is a wff.

This rule lets us construct wffs like these:

$$◇A \quad = \quad \text{A is possible (consistent, could be true).}$$

$$□A \quad = \quad \text{A is necessary (must be true, has to be true).}$$

$$\mathord{\sim}◇A \ = \ □\mathord{\sim}A \quad \begin{aligned} &= \ \text{A could not be true (has to be false).} \\ &= \ \text{A is impossible (self-contradictory).} \end{aligned}$$

An impossible statement (like "2≠2") is one that is false in every possible world. These examples are more complicated:

$$◇(A \cdot B) \quad \begin{aligned} &= \ \text{It is possible that A and B are both true.} \\ &= \ \text{A is consistent (compatible) with B.} \end{aligned}$$

$$□(A \supset B) \quad \begin{aligned} &= \ \text{It is necessary that if A then B.} \\ &= \ \text{A entails B.} \end{aligned}$$

"Entails" makes a stronger claim than plain "if-then." Compare these:

□(R ⊃ P) = "There is rain" entails "There is precipitation."

(S ⊃ ~C) = If it is Saturday, then you have no class.

The first IF-THEN is logically necessary; every conceivable situation with rain also has precipitation. The second IF-THEN may just happen to be true; we can consistently imagine you having a class on Saturday—even if in fact you never do. Some logicians symbolize "A entails B" as "(A ⥽ B)," which is read as "A *strictly implies* B."

These common forms negate the whole wff:

~◇(A · B) = A is inconsistent with B.
= It is not possible that A and B are both true.

~□(A ⊃ B) = A does not entail B.
= It is not necessary that if A then B.

Here is how we translate "contingent":

(◇A · ◇~A) = A is a contingent statement.
= A is possible and not-A is possible.

(A · ◇~A) = A is a contingent truth.
= A is true but could have been false.

Statements are necessary, impossible, or contingent. But truths are only necessary or contingent (since impossible statements are false).

The English sentence "*If A is true, then it is necessary that B*" is ambiguous; it could mean "(A ⊃ □B)" or "□(A ⊃ B)." The first form posits an *inherent necessity* to B, given that A is true. The second posits a *relative necessity*; what is necessary is not either part by itself but only the connection between the parts. This subtle **box-inside/box-outside ambiguity** is important for understanding a number of plausible but unsound philosophical arguments.

By adding four **inference rules**, we can extend to modal logic the last method of **proofs** sketched in our *proofs* entry. First there are two "reverse squiggle" rules; these hold regardless of what pair of contradictory wffs replaces "A"/"~A" (here "→" means we can infer whole lines from left to right):

Reverse Squiggle	$\sim\Box A \quad \rightarrow \quad \Diamond\sim A$
	$\sim\Diamond A \quad \rightarrow \quad \Box\sim A$

These let us go from "not necessary" to "possibly false"—and from "not possible" to "necessarily false." These rules can be based on the interdefinability of the two modal operators:

$$\Box A \quad = \quad \sim\Diamond\sim A$$
It is necessary that A $\quad = \quad$ It is not possible that not-A

$$\Diamond A \quad = \quad \sim\Box\sim A$$
It is possible that A $\quad = \quad$ It is not necessary that not-A

Our final two inference rules use the notion of a *world prefix*, which is a string of zero or more instances of "W." So " " (zero instances), "W," "WW," and so on are world prefixes; these represent possible worlds, with the blank world prefix (" ") representing the actual world. A *derived step* of a proof is now a line consisting of a world prefix and then "∴" and then a wff. And an *assumption* is now a line consisting of a world prefix and then "asm:" and then a wff. Here are examples of derived steps and assumptions:

∴ A　　　　(So A is true in the actual world.)
W ∴ A　　　(So A is true in world W.)
WW ∴ A　　(So A is true in world WW.)

asm: A　　　(Assume A is true in the actual world.)
W asm: A　　(Assume A is true in world W.)
WW asm: A　(Assume A is true in world WW.)

The reverse-squiggle and propositional inference rules can be used only within a given world; so if we have "(A ⊃ B)" and "A" in the same world, then we can infer "B" in this same world. **RAA** needs additional wording (*italicized below*) for world prefixes:

RAA: If we assume a formula A and then later derive a pair of not-blocked-off contradictory formulas (e.g., "B" and "~B") *using the same world prefix*, then we can derive a contradictory of A *using its original world prefix*, provided that A is the last assumption that has not been blocked-off and we now block off all the formulas from (and including) this assumption to the formula just before the newly derived contradictory of A.

To apply RAA, lines with the same world prefix must have contradictory wffs. Having "W ∴ B" and "WW ∴ ~B" is not enough; "B" may well be true in one world but false in another. But "WW ∴ B" and "WW ∴ ~B" give a contradiction. The line derived using RAA must have the same world prefix as the assumption; if "W asm: A" leads to a contradiction in any world, then RAA lets us derive "W ∴ ~A."

Our final two inference rules (which hold regardless of what wff replaces "A") let us drop modal operators and move between worlds. Here is the drop-diamond rule:

Drop Diamond	◇A → W ∴ A, use a *new* string of Ws

Here the line with "◇A" can use any world prefix—and the line with "∴ A" must use a *new* string (one not occurring in earlier lines) of one or more Ws. If "A" is possible, then "A" is thereby true in *some* possible world; we can give this world a name—but a *new* name, since "A" need not be true in any of the worlds used in the proof so far.

Here is the drop-box rule:

Drop Box	□A → W ∴ A, use any world prefix

Here the line with "□A" can use any world prefix—and the line with "∴ A" can use any world prefix too (including the blank one). If "A" is necessary, then "A" is true in *all* possible worlds, and so we can put "A" in any world we like. (This rule assumes **modal system S5**.)

The general proof strategy is to reverse squiggles first, then drop diamonds using new worlds, and then drop boxes using all the old worlds—using propositional rules wherever possible. Here is an example of a modal proof (read the argument as "It is necessary that if there is rain then there is precipitation; it is possible that there is rain; therefore it is possible that there is precipitation"):

```
1      □(R ⊃ P)
2      ◇R
  [∴ ◇P
3   ┌ asm: ~◇P
4   │  ∴ □~P   {from 3}
5   │  W ∴ R   {from 2}
6   │  W ∴ (R ⊃ P)   {from 1}
7   │  W ∴ P   {from 5 and 6}
8   └  W ∴ ~P   {from 4}
9   ∴ ◇P   {from 3; 7 contradicts 8}
```

After assuming the opposite of the conclusion (line 3), we reverse the squiggle to move the modal operator to the outside (line 4). We drop the diamond using a new world (line 5). We drop the box in line 1 using this same world (line 6) and then use a propositional rule to get "P" in world W (line 7). Then we drop the box in line 4 using this same world, to get "~P" in world W (line 8), which gives us a contradiction. RAA gives us the original conclusion (line 9).

Trying to construct a proof for an invalid argument using this strategy will normally lead to a *refutation*—truth conditions making the premises true and conclusion false. Here is an example (read it as "It is necessary that the coin is heads or tails; it is possible that it is not heads; therefore it is necessary that it is tails"):

```
1      □(H ∨ T)                    Invalid
2      ◇~H
  [∴ □T                    W   ┌─────────┐
                               │ H, ~T   │
3      asm: ~□T             WW  ├─────────┤
4      ∴ ◇~T   {from 3}         │ T, ~H   │
5      W ∴ ~T   {from 4}        └─────────┘
6      WW ∴ ~H   {from 2}
7      W ∴ (H ∨ T)   {from 1}
8      WW ∴ (H ∨ T)   {from 1}
9      W ∴ H   {from 5 and 7}
10     WW ∴ T   {from 6 and 8}
```

After assuming the opposite of the conclusion (line 3), we reverse a squiggle to move the modal operator to the outside (line 4). Then we drop the two diamonds, using a new and different world each time (lines 5 and 6). We drop the box twice, using first world W and then world WW (lines 7 and 8). Since we reach no contradiction, we gather the simple pieces to give a refutation. Here our refutation has two possible worlds: one with heads-and-not-tails and another with tails-and-not-heads. In this galaxy of possible worlds, the premises are true, since every world has "(H ∨ T)" true (making "□(H ∨ T)" true) and at least one world has "~H" (making "◇~H" true). But in this galaxy of possible worlds the conclusion "□T" is false, since not every world has "T" true. So our argument is invalid.

MODAL SYSTEMS. When C. I. Lewis in 1932 introduced **proof** procedures for **modal logic**, he regarded the validity of some modal forms as controversial. Accordingly, he presented various modal systems, called S1 to S9, which differ on the validity or invalidity of some arguments. The proof system in our modal logic entry accords with his S5, which is the easiest system to understand and formalize. Here we will contrast S5 with three alternative systems: S4, B, and T.

The disputed argument forms are ones in which one modal operator occurs within the scope of another, such as these three:

$$\begin{array}{ccc} \Box A & \Diamond A & A \\ \therefore \Box\Box A & \therefore \Box\Diamond A & \therefore \Box\Diamond A \end{array}$$

The minimal system T rejects all three. The liberal S5 accepts all three. S4 and B are intermediate, each accepting just one of the three; S4 accepts just the first while B accepts just the second.

In terms of how we set up the proof system in the *modal logic* entry, these disputes reflect differences in how to formulate the box-dropping rule. Our rule assumed the liberal S5, which lets us go from any world to any world when we drop a box:

| Drop Box | □A → W ∴ A, use any world prefix | The line with "□A" can use any world prefix—and so can the line with "∴ A." |

This assumes that whatever is necessary in *any* world is thereby true in *all* worlds without restriction. A further implication is that whatever is necessary in one world is thereby necessary in all worlds.

Our three weaker systems reject these ideas. On these systems, what is necessary is not identified with what is "true in *all* possible worlds."

Instead, what is necessary in a possible world W1 is identified with what is "true in all possible worlds *accessible* from W1"; so necessity on your planet is identified, so to speak, with truth in all the neighboring planets that you can travel to. The three weaker systems are distinguished by what this "accessibility" relation is like. System B sees accessibility as *symmetrical*: if W1 is accessible from W2, then W2 is accessible from W1. System S4 sees accessibility as *transitive*: if W1 is accessible from W2, and W2 is accessible from W3, then W1 is accessible from W3. The liberal S5 sees accessibility as both symmetrical and transitive, while the minimal T sees it as neither. These conditions translate into different restrictions on the drop-box rule. While we will not go into the details here, this proof of "□A ∴ □□A" in S4 may make these matters more intuitive:

```
1     □A
  [∴ □□A
2   ┌ asm: ~□□A
3   │ ∴ ◇~□A   {from 2}
4   │ W ∴ ~□A   {from 3}          actual ⇒ W
5   │ W ∴ ◇~A   {from 4}
6   │ WW ∴ ~A   {from 5}          W ⇒ WW
7   └ WW ∴ A    {from 1}          need S4 or S5
8   ∴ □□A   {from 2; 6 contradicts 7}
```

Steps 4 and 6 drop diamonds and introduce world W (which is accessible from the actual world) and WW (which is accessible from world W). (When we drop a diamond-formula from world W1 into a new world W2, then W2 is accessible from W1.) Step 7, which drops a box from the actual world into world WW, requires that WW be accessible from the actual world. This requires that accessibility be transitive: so if WW is accessible from W, and W from the actual world, then WW is accessible from the actual world. And this requires system S4 or S5.

S5 is the simplest system in several ways:

• We can formulate S5 more simply. The box-dropping rule need not mention accessibility; if we have "□A" in *any* world, then we can put "A" in *any* world (the same or a different one).

• S5 expresses simpler intuitions about necessity and possibility: what is *necessary* is what is true in *all* worlds, what is *possible* is what is true in *some* worlds, and what is necessary or possible does not vary between worlds.

- On S5, any string of boxes and diamonds simplifies to the last element of the string. So "□□" and "◇□" simplify to "□"—and "◇◇" and "□◇" simplify to "◇."

Which is the best system? This depends on what we take the box and diamond to mean. If we take them to be about the logical necessity and possibility of *ideas*, then S5 is arguably the best system. If an idea (for example, the claim that 2=2) is logically necessary, then it could not have been other than logically necessary. So if A is logically necessary, then it is logically necessary that A is logically necessary. Similarly, if an idea is logically possible, then it is logically necessary that it is logically possible. Of the four systems, only S5 accepts both claims.

Alternatively, we could take the box to be about the logical necessity of *sentences*. Now the sentence "2=2" just happens to express a necessary truth; it would not have expressed one if English had used "=" to mean "≠." So the sentence is necessary, but it is not necessary that it is necessary; this makes "(□A ⊃ □□A)" false. But the idea that "2=2" now expresses is both necessary and necessarily necessary—and a change in how we use language would not make this idea false. So whether S5 is the best system can depend on whether we take the box to be about the necessity of ideas or of sentences.

There are still other ways to take "necessary." Calling something "necessary" might mean that it is *physically necessary* (*see* **causal logic**), *obligatory* (*see* **deontic logic**), *proved*, *known*, or whatever. Some logicians like the weak system T because it holds for various senses of "necessary"; such logicians might still use S5 for arguments about the logical necessity of ideas.

The differences among T, B, S4, and S5 are important for a few philosophical arguments. For example, this argument for the existence of a necessary being (often identified with **God**) requires B or S5:

> It is necessary that if there is a necessary
> being at all then "There is a necessary   □(N ⊃ □N)
> being" is inherently necessary.   ◇N
> It is possible that there is a necessary being.   ∴ N
> ∴ There is a necessary being.

And this simplified version of **Alvin Plantinga**'s ontological argument also requires B or S5:

"Someone is unsurpassably great" is logically possible.

"Everyone who is unsurpassably great is, in every possible world, omnipotent, omniscient, and morally perfect" is necessarily true.

∴ Someone is omnipotent, omniscient, and morally perfect.

$$◇(\exists x)Ux$$
$$□(x)(Ux \supset □Ox)$$
$$\therefore (\exists x)Ox$$

MODEL. *See* INTERPRETATION.

***MODUS PONENS* (Latin for "affirming mode").** An inference that goes "If A, then B; A; therefore B." *See also* CONDITIONAL.

***MODUS TOLLENS* (Latin for "denying mode").** An inference that goes "If A, then B; not-B; therefore not-A." *See also* CONDITIONAL.

– N –

NAMES. Raise various questions. In addition to referring to an object, do names (like "Bertrand Russell") also have **meaning** or **sense**? Are names equivalent in meaning to **definite descriptions** (so "Bertrand Russell" might mean "the author of the 1905 article on denoting") or do they refer to objects more directly (as what **Saul Kripke** called "rigid designators")? Are all true identity statements that use just names **necessary truths**? Can there be empty names that do not refer to anything? *See also* FREE LOGIC; GENERAL/SINGULAR TERM.

NECESSARY/SUFFICIENT CONDITION. "A is a sufficient condition for B" means "If A then B." "A is a necessary condition for B" means "If not-A then not-B" or, equivalently, "If B then A." "A is a necessary and sufficient condition for B" means "A is true, if and only if B is true." *See also* IDIOMS.

NECESSITY. *See* ANALYTIC/SYNTHETIC; MODAL LOGIC.

NEGATION. Statement of the form "Not-P." **Propositional logic** symbolizes this as "~P." An English example is "I did not go to Paris." "~P" has this **truth table** (where T = true, F = false):

P	~P
F	T
T	F

A NOT has the opposite truth
value as the corresponding
positive statement.

"~P" has the opposite value of "P." If "P" is false then "~P" is true,
and if "P" is true then "~P" is false.

The main inference rule governing negation is that from "P" we can
conclude "~~P," and from "~~P" we can conclude "P":

$$\frac{P}{\sim\sim P} \qquad \textit{Double} \qquad \frac{\sim\sim P}{P}$$
$$\textit{Negation}$$

While **classical symbolic logic** accepts both forms, **intuitionist logic**
rejects the latter in some mathematical cases involving infinite sets.
While English slang and Spanish use a double negative as an emphatic
single negative (so "I don't say nothing" means "I say nothing"), logicians see this as an idiomatic usage rather than as a **deviant logic**.

In symbolizing English arguments, "it is false that" and "it is not the
case that" translate as "not." "~P" must symbolize the **contradictory**
of "P" and not some other negation. So if "P" represents "Everyone
was there" then "~P" must represent "It is not the case that everyone
was there" (= "Not everyone was there")—and not "Everyone was not-
there" (= "No one was there"). *See also* MANY-VALUED LOGIC;
PARACONSISTENT LOGIC; TRUTH FUNCTIONS.

NOMINALISM. The belief that there are no **abstract entities**. *See also*
MEDIEVAL LOGIC.

NON-MONOTONIC (defeasible). A valid argument is *non-monotonic* if
adding more premises may weaken the argument—as in the second
example below (where adding the premise "Socrates has a Swedish
mother" would weaken the argument):

All Greeks are mortal.	Most Greeks have dark hair.
Socrates is Greek.	Socrates is Greek.
∴ Socrates is mortal.	∴ Socrates has dark hair.

Inductive arguments tend to be non-monotonic.

NON-WESTERN LOGIC. *See* BUDDHIST LOGIC.

NOT. *See* NEGATION.

NOTHING. To grasp the concept of nothingness, meditate on what is inside the box below. Clap with one hand when you understand.

```
┌─────────────────────────────────────┐
│                                     │
│                                     │
│                                     │
│                                     │
└─────────────────────────────────────┘
```

The noun "nothing" is what **Bertrand Russell** called a **logical construct**; while "nothing" refers to no entities, sentences that use it can make definite claims and can be rephrased to show their meaning in a less confusing way. "Nothing is in the box" means "It is not the case that there is something in the box," or "$\sim(\exists x)Bx$" in **quantificational logic**. The noun "nothing," far from referring to a mysterious entity, is simply a *quantifier*; it says that a certain phrase is true in no cases.

Ignoring this point can produce delightful nonsense, as in Peter Heath's classic and hilarious article on "Nothing" in the *Encyclopedia of Philosophy*. Heath was a great fan of **Lewis Carroll**, than whom nobody knew more about nothing. But, of course, *nobody* would be *expected* to know *everything* about *nothing*.

Jean-Paul Sartre used "nothingness" to refer to our personhood. "Thing" is sometimes opposed to "person." In this sense, persons are non-things; from this, we could slide to saying that persons are "nothing." But, of course, you may get equally upset if you are told "You are a thing" or "You are nothing"; so it may be better to avoid the whole subject. *Nobody* knows what to do about this problem; of course, *he should know*.

NUMBERS. *See* ARITHMETIC.

– O –

OCKHAM, WILLIAM OF (c. 1285–1349). A Franciscan logician, philosopher, and theologian of the late **medieval** era. He wrote an influential *Summa Logicae* ("Summary of Logic"). While known for his nominalist analyses of terms like "man" and "humanity," he also developed principles of **modal logic**; for example, he suggested that a **disjunction** is possible if either part is possible—and that whatever follows from something possible is itself possible. Modal ideas were important for Ockham's view of **God**, which emphasized God's freedom and power to bring about whatever is possible; since God could have created the world in any possible way, he argued, the structure of the world has no necessity and thus cannot be known by abstract speculation apart from experience. He also criticized arguments for the existence of God; he based belief in God on faith rather than on reason.

Ockham made much use of a traditional *simplicity criterion* that is now called *Ockham's razor*. This states that, other things being equal, we ought to prefer the simplest theory that adequately explains the data. Another formulation says, "Do not multiply entities (or posit plurality) without necessity." This principle is important in many areas, but especially in **inductive** and **scientific reasoning**.

ONTOLOGY. The study of being, and especially of what sorts of things ultimately exist. *See also* ABSTRACT ENTITIES; LEŚNIEWSKI; LOGICAL CONSTRUCT; QUINE.

OPPOSITION. The **fallacy** of arguing that what an opponent says must thus be false. For example, one might argue that a view must be false because it is supported by "those blasted liberals" or "those blasted conservatives." But our opponents might on occasion be right.

OR. *See* DISJUNCTION.

– P –

PARACONSISTENT LOGIC. Any alternative to classical **propositional logic** that rejects the *explosion principle* "A, ~A ∴ B"—that every statement can be deduced from a self-contradiction.

In classical logic we can, from a single self-contradiction, deduce the truth of every statement and its denial; so admitting the truth of a single self-contradiction would bring intellectual chaos. Here is an intuitive version of how, given contradictory premises "A is true" and "A is not true," we can deduce any arbitrary statement "B":

1   A is true. {premise}
2   A is not true. {premise}
3 ∴ At least one of these two is true: A or B.   {from 1: if A is true then at least one of the two, A or B, is true}
4 ∴ B is true.   {from 2 and 3: if at least one of the two is true and it is not the first, then it is the second}

While it seems difficult to deny any of these steps, paraconsistent logic must do just this. There are three options:

- Deny step 3 and thus the principle called "addition" (if A is true, then at least one of the two, A or B, is true).
- Deny step 4 and thus the principle called "**disjunctive** syllogism" (if at least one of the two, A or B, is true and the first one, A, is not true, then the second one, B, is true).
- Deny that deducibility is *transitive* (if 3 follows from our premises and 4 follows from our premises plus 3, then 4 follows from our premises).

Paraconsistent logicians generally take the second option and deny the validity of disjunctive syllogism. Suppose B is false and A is both-true-and-false (!); then, they say, "(A ∨ B)" is true (since "A" is true), "~A" is true (since "A" is also false), but "B" is false. Some, instead of rejecting disjunctive syllogism, qualify it to apply only when the smaller premise is not both-true-and-false.

There are several ways to work out a paraconsistent logic from a technical standpoint. One way uses three or four **truth values** (*see* **many-valued logic**); we might let "1" stand for "true and not false," "0" for "false and not true," "B" for "both true and false," and perhaps "N" for "neither true nor false"—and then work out **truth tables** and definitions of validity based on this. Or we might keep to two truth values, "1" and "0," but allow "A" and "not-A" to both have the truth value "1"; this rejects the usual understanding of **negation**, by which "not-A" has only one truth value, and the opposite truth value as "A."

Some develop paraconsistent logic to defend **dialethism**, which holds that a statement and its negation are sometimes both true. By classical logic, as we saw above, dialethism entails that every statement and its opposite is true—which brings intellectual chaos. Dialethists respond by rejecting classical logic; instead, they defend a paraconsistent logic that lets us contain an occasional self-contradiction. And they carefully distinguish "being false" from "not being true," since they think that something that is false might also be true.

Others develop paraconsistent logic because people or computers sometimes have to derive conclusions from inconsistent premises. Suppose our best set of data about a crime or some area of science is flawed and inconsistent; we still might want to derive the best conclusions we can from this data. The "anything and its opposite follows from inconsistent data" approach of classical logic is unhelpful here.

PARADIGM-CASE ARGUMENT. One that says that a given term (e.g., "knowledge" or "free choice") must have actual instances, since otherwise it could not be a meaningful part of language. We can construe this reasoning as follows:

> Every meaningful term is explainable by definition or examples.
> The term "knowledge" is not explainable by definition. (There is no agreed upon definition.)
> The term "knowledge" is meaningful.
> If the term "knowledge" is explainable by examples, then there are cases of knowledge.
> ∴ There are cases of knowledge.

Critics tend to dispute one of the premises; they might hold, for example, that "knowledge" (like "unicorn") is empty but definable in words, or that there is some third way to explain the term.

PARADOX. An apparent **self-contradiction**, especially one that seems to be based on correct reasoning. The paradoxes most important to logic are **Russell**'s paradox of **set theory**, the **liar paradox** (which goes back to Epimenides, a sixth century BC Cretan), the **sorites paradox** (which goes back to Eubulides, a fourth century BC Greek), and the *lottery paradox* (*see* **belief logic**).

PEANO, GIUSEPPE (1858–1932). An Italian mathematician and logician. He wanted to reduce areas of **mathematics** to strict axioms and to make explicit the logical rules implicit in mathematical practice.

Peano proposed five axioms about *natural numbers* (0, 1, 2, and so on), from which he hoped that all the truths of **arithmetic** could be derived (but *see* **Gödel**):

1. 0 is a natural number.
2. The successor of any natural number is a natural number.
3. No two natural numbers have the same successor.
4. 0 is not the successor of any natural number.
5. If a set K of natural numbers contains 0, and if K contains the successor of every number that it contains, then K contains all natural numbers. (principle of mathematical induction)

These are called the *Peano axioms* for arithmetic, even though J. W. R. Dedekind (1831–1916) had proposed them earlier.

Peano invented a new symbolism for logic and set theory. **Bertrand Russell**, who called his meeting with Peano in 1900 the turning point of his intellectual life, adapted Peano's symbolism instead of **Gottlob Frege**'s; since then, typographical variations on Peano's style of notation have become standard. Peano used the same symbols for **propositional logic** and for **set theory**:

		propositions	*sets*
$-a$	=	not-a	the complement of set a
$a \cap b$	=	a and b	the intersection of a and b
$a \cup b$	=	a or b	the union of a and b
$a \supset b$	=	if a then b	set a is contained in set b
$a = b$	=	a if and only if b	sets a and b are identical
$\wedge$	=	falsity	the null set
$\vee$	=	truth	the universal set

Peano used "$a \in b$" for "a is a member of b"—and "$(a \supset_{x,y} b)$" for "whatever x and y may be, if a then b" (where x and y are variables that may occur in a and b). He tended to use dots instead of parentheses; we first take together signs not separated by dots, then those separated by one dot, then those separated by two dots, and so on:

$$-:-a \cap b . \cup c \;=\; -((-a \cap b) \cup c)$$

Russell, too, liked dots for grouping, but few people use them today.

PEIRCE, CHARLES SANDERS (1839–1914). American pragmatist philosopher and logician. Peirce added *class inclusion* to **Boolean** algebra. If we let "A" stand for the **set** of animals and "C" for the set of cats, then "C ⊂ A" claims that the set of cats is included in the set of animals. We can express this in Boole's notation as "C = CA," which claims that the set of cats = the set of things that are both cats and animals; but using a separate symbol for class inclusion is more intuitive. Peirce wrote the corresponding statement as "C ⊃ A," which claims in a material **conditional** way that if C then A (e.g., if this is a cat then this is an animal).

In **propositional logic**, Peirce worked out a way to test validity by seeing whether, for every possible way to assign "true" or "false" to the letters, the conclusion is true whenever the premises are; this anticipated the later **truth-table** test. He also invented **logic gates**, which simulate truth-table functions electrically; Claude Shannon independently reinvented the same idea 60 years later.

Peirce also, independently of **Gottlob Frege** but slightly later, invented **quantifiers**, using "$\prod x$" for the universal and "$\sum x$" for the existential; and he worked out a logic of **relations**. Just as Boole interpreted properties as sets of objects (so the property of being Greek is interpreted as the set of Greeks), so also Peirce interpreted relations as ordered sequences of objects (so the relation of x loving y is interpreted as the set of ordered pairs like <Romeo, Juliet> and <Adam, Eve>).

In inductive logic, Peirce was a strong defender of *abduction*, which is today often called "**inference to the best explanation.**" And he analyzed **truth** as that on which all investigators are bound to agree in the long run, if they carry their investigations far enough.

PHILOSOPHICAL LOGIC. This term sometimes is used interchangeably with "**philosophy of logic**," sometimes refers to a philosophical study of the main concepts involved in thinking (like **truth**, **meaning**, **definition**, **reference**, and **necessity**), and sometimes (contrasting with one use of "**mathematical logic**") refers to areas of logic that are of greater interest to philosophers than to mathematicians.

PHILOSOPHY AND LOGIC. There are three classic views on how philosophy and logic relate:

- **Aristotelian**: Logic, while not strictly part of philosophy, is an important tool ("Organon") for doing philosophy.
- Stoic: Logic is one of the three parts of philosophy; the other parts are physics and **ethics**.
- **Bertrand Russell**: Philosophy *is* logic: any genuine philosophical

problem, if understood correctly, resolves into a problem of logic.

The Aristotelian view holds that logic relates to philosophy as statistics relates to psychology. Doing psychology well requires knowing statistics, even though statistics is not part of psychology. Similarly, doing philosophy well requires knowing logic; this is true because philosophy involves *reasoning* about the big questions of life. Logic can help us express reasoning clearly, determine whether a conclusion follows from the premises, and focus on key premises to defend or criticize. Logic, while not itself resolving the big issues, gives us intellectual tools to reason better about such issues. But still, logic is not itself part of philosophy, since the validity of inferences is not one of the big questions of life that philosophy is concerned with.

The Stoic view (*see* **ancient logic since Aristotle**), which prevails today, says that logic *is* an important part of philosophy—indeed, one of the major parts. Contemporary defenders of this view would add that logic is also part of **mathematics** and **computer** science, and is taught in these departments as well, but from a different perspective; and they would want to refine the Stoic logic-physics-ethics divisions of philosophy. But they would agree that questions about how reasoning works, which logic deals with, are among the big questions of life that philosophy is concerned with.

Russell's view, that any genuine philosophical problem resolves into a problem of logic, is somewhat like that of the **logical positivists**, who limited philosophy to the analysis of concepts and logical connections. Such views are implausible, unless we eliminate as illegitimate most of the big questions that philosophers have dealt with, like "Is there a God?" "Do humans have free will?" and "How ought we to live?" While logic and conceptual analysis are important as we try to answer such questions, it is difficult to see how such questions entirely resolve into logic problems. So Russell's view is not very popular today.

Logic is important in the analytic tradition of philosophy, although it is more important to some analytic philosophers than to others. Consider the three founders of the analytic tradition: Bertrand Russell, G. E. Moore, and **Ludwig Wittgenstein**. Logic and the issues that it raised were extremely important to Russell, at least in his earlier years; even his **metaphysics** was much influenced by logic. With Moore, logic was less important; while he reasoned much about philosophical issues (often using *modus tollens* arguments), he only occasionally refers to points about logic. Russell and Moore represent the two ends of the spectrum that we can see in the whole analytic tradition, including today. Wittgenstein himself represents both ends of the spectrum, but at different times; in his early years, he was more like Russell (with logic being very important for doing philosophy)—while in his later years, he was more like Moore (with logic being less important, although reasoning remains important). So, while there is great interest among analytic philosophers in the topics covered in this book, this interest is not spread out equally over everyone.

Logic is less important in the continental tradition of philosophy, especially among those who, like Friedrich Nietzsche, emphasize rhetoric more than reasoning. But we must not overgeneralize here. **Edmund Husserl**, the founder of phenomenology, had a great interest in logic; his *Logical Investigations* criticized the then-popular **psychologism**, which saw logic as a description of how people think. And several of Martin Heidegger's early works dealt with logic; his doctoral dissertation, influenced by Husserl, was entitled "The Doctrine of the Judgment in Psychologism," and other early works include "Duns Scotus's Doctrine of Categories and of Meaning" and "The Metaphysical Foundations of Logic." Still, there is much more interest in logic in the analytic tradition. *See also* PHILOSOPHY OF LOGIC.

PHILOSOPHY OF LANGUAGE. *See* LANGUAGE, PHILOSOPHY OF.

PHILOSOPHY OF LOGIC. The study of philosophical questions raised by **logic**. To some extent, philosophy of logic overlaps with *philosophy of language* (*see* **language, philosophy of**), since both are concerned with symbols (**syntax**), their relationship to reality (**semantics**), and their use by people (*pragmatics*). So both, for example, deal with topics like **abstract entities** and **ontology**, the **analytic/synthetic** distinction, **ambiguity**, **definite descriptions**, **definitions**, **general/singular terms**, **meaning**, **truth** and **paradoxes** about truth, and the **sense/reference** distinction. But philosophy of logic deals also with more specific questions about logic, such as its definition and scope (*see* **logic, scope of**), the analysis of its key concepts, the relationship between logic and ordinary language (*see* the controversy between **Bertrand Russell** and Peter Strawson in the **definite description** entry), the possibility of **deviant logics**, and the status and justification of basic logical principles (*see* **logical principles, status and justification of**). *See also* PHILOSOPHICAL LOGIC; PHILOSOPHY AND LOGIC; QUINE.

PHYSICS. Raises questions like these about logic: What is the logical relationship in **scientific reasoning** between observations and theories? Does physics require a **temporal logic**? Does quantum physics require statements with an indeterminate **truth value**, and thus force us to reject **bivalence** in favor of **many-valued logic**?

PLANTINGA, ALVIN (1932–). While more of a logically minded philosopher of religion than a logician, Plantinga contributed to logic in his *Nature of Necessity*. Here he explains logical **necessity**, argues that it is not based on convention, and defends **quantified modal logic** and **Aristotelian essentialism**; he relies on **Saul Kripke** and argues against **Willard Van Orman Quine**. Plantinga uses these logical ideas in proposing an ontological argument for the existence of **God** (*see* **modal systems**) and trying to show that the evil in the world does not disprove the existence of God.

POLISH NOTATION. A method of writing formulas, invented by **Jan Łukasiewicz**, that avoids parentheses and has shorter formulas. For **propositional logic**, "K," "C," "A," and "E" go in place of the left-hand parentheses for "·," "⊃," "∨," and "≡"; and "N" is used for "~." Here are examples:

(P · Q)	=	Kpq	(P ⊃ Q)	=	Cpq
~P	=	Np	((P · Q) ⊃ R)	=	CKpqr
~(P · Q)	=	NKpq	(P · (Q ⊃ R))	=	KpCqr
(~P · Q)	=	KNpq	(P ⊃ (Q · R))	=	CpKqr

"Πx" and "Σx" are the universal and existential **quantifiers**: "(x)Fx" becomes "ΠxFx" and "(∃x)Fx" becomes "ΣxFx."

Polish notation can be used for arithmetic. Then "+" and "·" would be used in place of the corresponding left-hand parenthesis; so "(2 · (3 + 4))" would be written as "· 2 + 3 4." *Reverse Polish notation* (which is used in some pocket calculators) is similar, except that "+" and "·" go in place of the corresponding right-hand parenthesis; so "(2 · (3 + 4))" would be "2 3 4 + ·."

POSSIBILITY. *See* ANALYTIC/SYNTHETIC; MODAL LOGIC.

POSSIBLE WORLD. A consistent and complete description of how things might have been or might in fact be.

We might picture a possible world as a *consistent story*. The story is *consistent*, in that its statements do not entail self-contradictions; it describes a set of possible situations that are all possible together. The story is *complete*, in that it includes every statement or its negation. (This "completeness" is an idealization; in practice, we cannot write down an infinite list of sentences.) The story may or may not be true. The *actual world* is the story that is true—the description of how things in fact are. Possible worlds are used much in **modal logic** and in contemporary analytic philosophy.

David Lewis is notorious for proposing a *modal realism* that holds that every possible world is equally real. So the possible world where Michigan won every Rose Bowl is just as real as our own world, where Michigan is less successful. We live in the latter world, but other people, who are equally real from their perspective, live in the former world. It is sheer provincialism to hold that our existence is, in some objective sense, more real than theirs. Such a modal realism, while difficult to refute, is even more difficult for most of us to believe.

Paraconsistent logics sometimes use *impossible worlds*, which are inconsistent stories that violate the normal rules of logic. These may be needed to analyze conditionals with impossible antecedents, like "If Jane proved that π equals some fraction n/m of positive integers, she would be promoted to full professor." *See also* KRIPKE; LEIBNIZ.

***POST HOC ERGO PROPTER HOC* (Latin for "after this therefore because of this").** The **fallacy** of arguing that since A happened after B, thus A was caused by B. For example, you have a beer and then get a good grade on your test; you conclude incorrectly that getting a good grade was caused by having a beer. Proving causal connections requires more than just the fact that one factor occurred after another; the two factors might just *happen* to have occurred together. It is not even enough that two factors *always* occur together; day always follows night, and night follows day, but neither causes the other. Proving causal connections is difficult (*see* Mill's methods under **Mill**).

PRAGMATISM (pragmatist theory of meaning). The view, proposed by William James, that the meaning of a synthetic statement (*see* **analytic/synthetic**) is determined by what practical differences the truth or falsity of the statement could make. If these *practical differences* include not just what would be experienced under various circumstances but also how we ought to live, then this view would be broader and more tolerant than **logical positivism**'s approach to language.

PREDICATE LOGIC. *See* QUANTIFICATIONAL LOGIC.

PREMISE. Statement that is supposed to support another statement (the conclusion). *See also* ARGUMENT; INDUCTIVE/DEDUCTIVE.

PRIEST, GRAHAM (1945–). An English logician who teaches in Australia. He is best known for his clear and vigorous defenses of the unorthodox view that some contradictions are true (*see* **law of non-contradiction**). He is a major defender of **dialethism**, **paraconsistent logic**, and various sorts of **deviant logic**—and a major foe of **classical symbolic logic**.

PROBABILITY. The likelihood of a statement being true. Probability is often expressed as a number in the range of 100 percent to 0 percent, or 1 to 0. "The probability of A" is sometimes abbreviated "p(A)."
   Here are some common principles about probability:

- The probability of a necessary truth (e.g., that Michigan will win or not win) = 100 percent.
- The probability of a self-contradiction (e.g., that Michigan will win and not win) = 0 percent.
- If A is one of a number of equally likely alternatives, then the probability of A = the number of cases in which A is true divided by the total number of cases. (E.g., the probability of drawing an ace in one random try from a 52 card deck = 4 aces/52 cards = $4/52 = 1/13 = 7.69$ percent.)
- The probability of not-A = 1 – the probability of A. (E.g., the probability of not drawing an ace in the previous example = $1 - 4/52 = 1 - 1/13 = 12/13 = 92.31$ percent.)
- If A and B are independent, then the probability of (A and B) = the probability of A · the probability of B. (E.g., the probability of drawing an ace from a 52 card deck and then shuffling and doing it again = $4/52 · 4/52 = 1/13 · 1/13 = 1/169 = 0.59$ percent.)
- If A and B are not independent, then the probability of (A and B) = the probability of A · the probability of B after A occurs. (E.g., the probability of drawing an ace out of a 52 card deck and then drawing another ace from the remaining 51 cards = $4/52 · 3/51 = 1/13 · 1/17 = 1/221 = 0.45$ percent.)

- If A and B are mutually exclusive, then the probability of (A or B) = the probability of A + the probability of B. (E.g., the probability of drawing an ace or king out of a 52 card deck = 4/52 + 4/52 = 8/52 = 2/13 = 15.38 percent.)
- If A and B are not mutually exclusive, then the probability of (A or B) = the probability of A + the probability of B − the probability of (A and B). (E.g., the probability of drawing an ace or diamond out of a 52 card deck = 4/52 + 13/52 − 1/52 = 16/52 = 4/13 = 30.77 percent.)
- The probability of A, given B = the probability of (A and B) divided by the probability of B. (E.g., the probability of drawing a king, given that you drew a face card = 4/52 divided by 12/52 = 1/3 = 33.33 percent.) "The probability of A, given B" is sometimes written "p(A|B)."

Gambling odds are convertible to probability. Suppose you bet on a certain outcome. Then:

- If your chance of winning is 50 percent (e.g., you are flipping a coin), then the odds are *even*; your odds of winning are 1 to 1.
- If your chance of winning is less than 50 percent, then the odds are *against* you. The odds against you equal the ratio of your chance of losing to your chance of winning. (E.g., if you bet you will draw an ace out of a 52 card deck, then the odds are 12 to 1 against you.)
- If your chance of winning is greater than 50 percent, then the odds are in *in your favor*. The odds in your favor equal the ratio of your chance of winning to your chance of losing. (E.g., if you bet you will draw a non-ace out of a 52 card deck, then the odds are 12 to 1 in your favor.)

In a *mathematically fair monetary bet*, the amount you get for winning is a function of what you bet and the odds. If you bet $100 that you will draw an ace out of a 52 card deck, the odds against you here being 12 to 1, then you should get $1,200 if you win. At a casino, the house takes a cut; so you would get less than $1,200 if you drew an ace.

The *expected gain* of a bet is the sum of probability-times-gain of the various possible outcomes. Suppose you find a game of dice that pays $3,536 on a $100 bet if you throw a 12. You would work out the expected gain of playing or not playing in this way:

- If you PLAY, winning is 1/36 likely and gains $3,536 (and so is worth 1/36 · $3,536, or $98.22) and losing is 35/36 likely and loses $100 (and so is worth 35/36 · –$100, or –$97.22). So the expected gain of PLAYING is $98.22 – 97.22, or $1.
- If you DO NOT PLAY, you will win or lose nothing; and so the expected gain of NOT PLAYING is 100 percent · $0, or $0.

If you acted to maximize expected gain, then you would play—unless you found another game with a greater expected gain. If you played this dice game only once, you would be 97 percent likely to lose money. But the occasional payoff is great; you would likely gain about a million dollars if you played a million times.

Some see an *ideally rational gambler* as one who acts to maximize expected gain; but this notion is highly artificial. Such an gambler would bet if the odds were favorable, but not otherwise. Since Las Vegas casinos take their cut, their odds are against the individual gambler; so an ideally rational gambler would not gamble at these places. But people have interests other than money; for many, gambling is great fun—and they are willing to pay for the fun. And some whose only concern is money refuse to gamble even when the odds are in their favor. Their concern may be to have enough money. They may better satisfy this goal by being cautious instead of risking what they have for the sake of gaining more. Few people would endanger their total savings for the 1-in-900 chance of gaining a fortune 1,000 times as great.

Some see an *ideally rational believer* as one who believes all and only those things that are more probable than not given the evidence; but this notion too is highly artificial. Suppose that Austria, Brazil, and China each has an equal 33 1/3 percent chance to win the World Cup. Then each of these three statements is 66 2/3 percent probable:

> "Austria will not win it, but Brazil or China will."
> "Brazil will not win it, but Austria or China will."
> "China will not win it, but Austria or Brazil will."

On the view just described, an ideally rational person would believe all three; but only a very confused person could do this. There are further problems if sometimes (or usually?) there is no way to work out numerical probabilities.

"Probable" has various senses. "The *probability* of heads is 50 percent" could be taken in at least four ways:

- *Ratio of observed frequencies:* We have observed that coins land heads about half of the time.
- *Ratio of abstract possibilities:* Heads is one of the two equally likely abstract possibilities.
- *Measure of actual confidence:* We have the same confidence in the toss being heads as we have in it not being heads.
- *Measure of rational confidence:* It is rational to have the same confidence in the toss being heads as in it not being heads.

Sometimes the two ratio approaches can give numerical probabilities. But sometimes they cannot. Neither ratio approach gives a numerical probability to "Michigan will pass" relative to information about ancient Greek philosophy. Only in special cases do the ratio approaches give numerical probabilities.

The measure of actual confidence sometimes yields numerical probabilities. Consider these statements:

> "There is life on other galaxies."
> "Michigan will beat Ohio State this year."
> "There is a God."

If you regard 1-to-1 betting odds on one of these as fair, then your actual confidence in the statement is 50 percent. But maybe you are unwilling to commit yourself on fair betting odds. Maybe you cannot even say if your confidence in the statement is less or greater than your confidence that a coin toss will be heads. Then we likely cannot assign numbers to your actual confidence. The rational confidence view, too, would have trouble assigning numerical probabilities in these cases.

Some doubt whether probability as rational confidence satisfies the standard probability rules mentioned above. These rules say that all necessary statements have a probability of 100 percent relative to any data. But consider a complicated propositional logic formula that is a necessary truth, even though your evidence suggests that it is not; perhaps your normally reliable logic teacher tells you that it is not a necessary truth—or perhaps in error you get a **truth-table** line of false. Relative to your data, it seems rational not to put 100 percent confidence in the formula. So is probability theory wrong?

Probability theory is perhaps idealized rather than wrong. It describes the confidence an ideal reasoner would have, based on an ideal analysis of the data; an ideal reasoner would always recognize necessary truths and put 100 percent confidence in them. So we have to be careful in applying probability theory to the beliefs of non-ideal humans; we must be like physicists who give simple equations for frictionless bodies and then make allowances for the idealization when applying the equations to real cases. *See also* BAYES; GAMBLER FALLACY.

PRO-CON ARGUMENT. One that bases a conclusion on reasons in favor and reasons against. This can be either correct or fallacious; the *correct pro-con form* goes as follows:

> The reasons in favor of act A are . . . .
> The reasons against act A are . . . .
> The former reasons outweigh the latter.
> ∴ Act A ought to be done.

Benjamin Franklin suggested this strategy for deciding whether to do something. Take a sheet of paper, divide it into two columns, and label these PRO and CON. Over the next three or four days, list the reasons in each column. Then try to weigh the relative strength of the reasons. If a reason pro is equal in weight to a reason con, then cross both out; if a reason pro is equal in weight to a combination of two reasons con, then cross out all three. In the end we may see which course of action has the weight of reasons behind it and thus ought to be done.

It is a **fallacy** (sometimes called "stacking the deck") if we look only at reasons pro, or only at reasons con, or do not try to give as complete a list of both as we can, or draw the conclusion without weighting the reasons on both sides against each other.

PROOF. The word "proof" has a cluster of related meanings.

In ordinary usage, a *proof* is a very strong **argument**—roughly a non-circular, non-ambiguous argument in which the **conclusion** clearly follows from the **premises** and the premises are clearly true. But how "clearly true" do the premises have to be? This varies with context. In American law, guilt has to be proved "beyond reasonable doubt." This means that the premises have to be so strong that a *reasonable person* would feel obliged to accept them; while this is vague, the idea has been somewhat workable. In other contexts, some have required premises that are "impossible to doubt"; this could mean that non-belief is either psychologically or logically impossible, or perhaps just totally unreasonable, and this or for most people or for everyone. If we pitch the standards too high, then likely nothing can be proved. Because of ambiguities in the term "proof," it is wise, in discussing whether something (e.g., evolution or the existence of God) can be *proved*, to first specify how "proof" is to be taken in this context and how evident the premises will have to be.

In systems of logic, "proof" (or "formal proof") has a more precise and technical meaning; it is a derivation done according to the notational rules of a given system. The two most common styles of proof are axiomatic and inferential. While these can be developed for any branch of logic, we will here give examples for **propositional logic**.

An *axiomatic proof* is a sequence of formulas in which each formula is (1) an **axiom** of the system, (2) a premise (of a particular argument whose validity is being investigated), or (3) follows from previous formulas of the sequences by an **inference rule** or the substitution of definitional equivalents. Our example here of an axiomatic system, roughly based on the *Principia Mathematica* of **Bertrand Russell** and Alfred North Whitehead, has four axioms:

Axiom 1. $((A \lor A) \supset A)$
Axiom 2. $(A \supset (A \lor B))$
Axiom 3. $((A \lor B) \supset (B \lor A))$
Axiom 4. $((A \supset B) \supset ((C \lor A) \supset (C \lor B)))$

These four formulas are axioms, as is the result of substituting any well-formed formulas uniformly for the letters "A," "B," and "C"; so "$((\sim P \lor \sim P) \supset \sim P)$" also would be an axiom, since it follows from our axiom 1 by substituting "$\sim P$" for "A." There is one rule of inference (*modus ponens*), which also is assumed to hold (as are the other principles of this entry) if we substitute any well-formed formulas uniformly for the capital letters:

Inference Rule: $(A \supset B)$, $A \rightarrow B$

Here "$\rightarrow$" means that given all of the formulas on the left, we can derive any or all of the formulas on the right. So given that we have an IF-THEN as a previous formula, and also have the if-part as a previous formula, we can derive the then-part as a further formula. Following *Principia Mathematica*, our system takes "$\lor$" and "$\sim$" as undefined and defines "$\supset$," "$\cdot$," and "$\equiv$" as follows (*see* **truth functions**):

Definition 1. $(A \supset B) = (\sim A \vee B)$
Definition 2. $(A \cdot B) = \sim(\sim A \vee \sim B)$
Definition 3. $(A \equiv B) = ((A \supset B) \cdot (B \supset A))$

So "$(P \supset Q)$" and "$(\sim P \vee Q)$" are equivalents and can be interchanged anywhere in a formula.

Here is a proof of "$(P \vee \sim P)$" using this system:

1   $\therefore (((P \vee P) \supset P) \supset ((\sim P \vee (P \vee P)) \supset (\sim P \vee P)))$
    {from axiom 4, using "$(P \vee P)$" for "A," "P" for "B," and "$\sim P$" for "C"}

2   $\therefore ((P \vee P) \supset P)$ {from axiom 1, using "P" for "A"}

3   $\therefore ((\sim P \vee (P \vee P)) \supset (\sim P \vee P))$ {from 1 and 2}

4   $\therefore (P \supset (P \vee P))$ {from axiom 2, using "P" for "A" and "P" for "B"}

5   $\therefore (\sim P \vee (P \vee P))$ {from 4, substituting things equivalent by definition 1}

6   $\therefore (\sim P \vee P)$ {from 3 and 5}

7   $\therefore ((\sim P \vee P) \supset (P \vee \sim P))$ {from axiom 3, substituting "$\sim P$" for "A" and "P" for "B"}

8   $\therefore (P \vee \sim P)$ {from 6 and 7}

Creating such proofs requires guesswork and intuition. And we might work for hours trying to prove an argument that is actually invalid. Axiomatic systems tend not to be user-friendly.

Axiomatic systems were the standard approach in the early part of the 20th century. Many such systems were developed for standard propositional logic, using different axioms, inference rules, and definitions—some using only a single complex axiom—but all resulting in the same things being provable.

Inferential systems were developed later, starting with Gerhard Gentzen in 1934; these use strong inference rules and usually no axioms. Irving Copi's logic books popularized a widely used inferential system for propositional logic. This system uses only inference rules and has no axioms or definitions. It uses three groups of inference rules. The first group has nine general rules that let us go from previous steps to new derived steps:

$$(A \cdot B) \rightarrow A$$
$$A, B \rightarrow (A \cdot B)$$
$$(A \vee B), \sim A \rightarrow B$$
$$A \rightarrow (A \vee B)$$

$$(A \supset B), A \rightarrow B$$
$$(A \supset B), \sim B \rightarrow \sim A$$
$$(A \supset B), (B \supset C) \rightarrow (A \supset C)$$
$$(A \supset B) \rightarrow (A \supset (A \cdot B))$$

$$((A \supset B) \cdot (C \supset D)), (A \vee C) \rightarrow (B \vee D)$$

The second group has 16 equivalence rules; these let us replace parts of formulas with equivalent parts (here outer parentheses are dropped to promote readability):

$$A \equiv \sim\sim A$$
$$A \equiv (A \cdot A)$$
$$A \equiv (A \vee A)$$
$$(A \cdot B) \equiv (B \cdot A)$$
$$(A \vee B) \equiv (B \vee A)$$
$$\sim(A \cdot B) \equiv (\sim A \vee \sim B)$$
$$\sim(A \vee B) \equiv (\sim A \cdot \sim B)$$
$$(A \supset B) \equiv (\sim B \supset \sim A)$$

$$(A \supset B) \equiv (\sim A \vee B)$$
$$(A \cdot (B \cdot C)) \equiv ((A \cdot B) \cdot C)$$
$$(A \vee (B \vee C)) \equiv ((A \vee B) \vee C)$$
$$(A \cdot (B \vee C)) \equiv ((A \cdot B) \vee (A \cdot C))$$
$$(A \vee (B \cdot C)) \equiv ((A \vee B) \cdot (A \vee C))$$
$$(A \equiv B) \equiv ((A \supset B) \cdot (B \supset A))$$
$$(A \equiv B) \equiv ((A \cdot B) \vee (\sim A \cdot \sim B))$$
$$((A \cdot B) \supset C) \equiv (A \supset (B \supset C))$$

Finally, we have conditional proof (and then RAA):

Conditional Proof: If we assume a formula A and then later derive a not-blocked-off formula B, then we can derive "$(A \supset B)$," provided that A is the last assumption that has not been blocked-off and we now block off all the formulas from (and including) A to the formula just before the newly derived "$(A \supset B)$."

The "blocking off" of formulas shows that they cannot be used for deriving further steps. So with conditional proof, if we assume A and then derive B, then we can derive "$(A \supset B)$"—but we cannot then use this "A" or "B" or any intermediate steps in deriving further steps:

asm: A
. . .
∴ B

→

⌐ asm: A
⎢ . . .
⌐ ∴ B
∴ $(A \supset B)$

Finally, we have **RAA** (*reductio ad absurdum*—which is Latin for "reduction to absurdity"):

> *RAA:* If we assume a formula A and then later derive a pair of not-blocked-off contradictory formulas (e.g., "B" and "~B"), then we can derive a contradictory of A, provided that A is the last assumption that has not been blocked-off and we now block off all the formulas from (and including) this assumption to the formula just before the newly derived contradictory of A.

Here a pair of formulas are *contradictories* if they are exactly the same except that one starts with an additional "~." This diagram illustrates how RAA works:

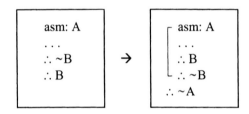

A *proof* on this approach is a sequence of formulas in which every assumption is blocked-off using hypothetical proof or RAA and each formula is (1) a premise (of a particular argument whose validity is being investigated), (2) an assumption (write "asm:" and then the formula), or (3) follows from previous not-blocked-out formulas of the sequences by means of an inference rule.

Here is an inferential proof of "(P ∨ ~P)" using this system:

    1 ⎡ asm: ~~P
    2 ⎣ ∴ P   {from 1, substituting equivalents}
    3 ∴ (~~P ⊃ P)   {from 1 and 2, using conditional proof}
    4 ∴ (~P ∨ P)   {from 3, substituting equivalents}
    5 ∴ (P ∨ ~P)   {from 4, substituting equivalents}

The inferential approach permits proofs that are simpler and more intuitive than on the axiomatic approach.

This inferential system still has some shortcomings. It uses a large and unsystematic set of inference rules, constructing proofs is largely guesswork, and the procedure will not tell us if an argument that we are trying to prove is invalid. There are alternative inferential systems that try to avoid these problems. One such approach, that of Harry Gensler, uses this systematic set of 12 inference rules:

Simplifying Rules	Inferring Rules
$(A \cdot B) \rightarrow A, B$	$\sim(A \cdot B), A \rightarrow \sim B$
$\sim(A \vee B) \rightarrow \sim A, \sim B$	$\sim(A \cdot B), B \rightarrow \sim A$
$\sim(A \supset B) \rightarrow A, \sim B$	$(A \vee B), \sim A \rightarrow B$
$\sim\sim A \rightarrow A$	$(A \vee B), \sim B \rightarrow A$
$(A \equiv B) \rightarrow (A \supset B), (B \supset A)$	$(A \supset B), A \rightarrow B$
$\sim(A \equiv B) \rightarrow (A \vee B), \sim(A \cdot B)$	$(A \supset B), \sim B \rightarrow \sim A$

This approach uses RAA exclusively and drops conditional proof. Here is its proof of "$(P \vee \sim P)$":

```
1   ┌ asm: ~(P ∨ ~P)
2   │ ∴ ~P     {from 1, using a simplifying rule}
3   └ ∴ ~~P    {from 1, using a simplifying rule}
4   ∴ (P ∨ ~P)   {from 1; 2 contradicts 3}
```

The proof strategy is fairly simple: we assume the opposite of what we want to prove, use the simplifying and inferring rules when possible, and apply RAA when we get a contradiction. Multiple assumptions are sometimes needed to break up complex formulas; but this is too complex to explain here. If we try to prove an invalid argument, then we will not get a proof; instead we will get a refutation—truth conditions that make the premises true and conclusion false—thus showing that the argument is invalid. Here is an example:

1   T   *Invalid*

2   (T ⊃ (B ∨ M))

3   (M ⊃ H)   T, M, H, ~B

    [∴ B   {blocked-off conclusion to be proved}

4   asm: ~B

5   ∴ (B ∨ M) {from 1 and 2}

6   ∴ M {from 4 and 5}

7   ∴ H {from 3 and 6}

Here we get no contradiction. But the letters or denials of letters that we end up with make the premises true and conclusion false; to see this, in the premises and conclusion plug in true for "T," "M," and "H"—and plug in false for "B"—and then evaluate the result.

Other proof methods include *natural deduction, semantic tableau,* **truth trees,** and *model sets* (*see* **Jaakko Hintikka**). While this entry focused on propositional logic, similar proof methods can be given for other branches of logic, including **quantificational** and **modal logic.**

PROPOSITIONAL LOGIC (also called sentential or truth-functional logic). A branch of logic that studies arguments whose validity depends on "if-then," "and," "or," "not," and similar notions. Further systems typically build on propositional logic but add other logical notions, like "all" and "some" for **quantificational logic,** or "necessary" and "possible" for **modal logic.** While the ideas behind propositional logic go back to the Stoics (*see* **ancient logic since Aristotle**), our modern approach to it comes from **Gottlob Frege** and **Bertrand Russell.**

It is convenient to present propositional logic as an artificial language set up to express logical relationships precisely; we can test English arguments by translating them into our artificial language and then using its rules to determine validity. Here we use capital letters for true-or-false statements, parentheses for grouping, and five special symbols for logical notions:

Negation:	~P	=	Not-P
Conjunction:	(P · Q)	=	Both P and Q
Disjunction:	(P ∨ Q)	=	Either P or Q
Conditional:	(P ⊃ Q)	=	If P then Q
Biconditional:	(P ≡ Q)	=	P if and only if Q

Some logicians use other symbols (*see* the notation section).

Grammatically correct formulas are called **wffs**, or *well-formed formulas*. Propositional logic wffs are sequences that we can construct using these rules (here, and elsewhere in this dictionary, we will consider letters with primes to be distinct additional letters):

1. Any capital letter is a wff.
2. The result of prefixing any wff with "~" is a wff.
3. The result of joining any two wffs by "·" or "∨" or "⊃" or "≡" and enclosing the result in parentheses is a wff.

These rules let us build wffs like the following:

$$
\begin{array}{rcl}
P & = & \text{I went to Paris.} \\
\sim Q & = & \text{I did not go to Quebec.} \\
(P \cdot \sim Q) & = & \text{I went to Paris and I did not go to Quebec.} \\
(N \supset (P \cdot \sim Q)) & = & \text{If I am Napoleon, then I went to Paris and} \\
& & \text{I did not go to Quebec.}
\end{array}
$$

Our rules for forming wffs force us to use enough parentheses to avoid **scope ambiguity**. Since it lacks parentheses, the malformed "~P · Q" is ambiguous between these two correct wffs:

$$
\begin{array}{rcl}
(\sim P \cdot Q) & = & \text{Both not-P and Q} \\
\sim (P \cdot Q) & = & \text{Not-both P and Q}
\end{array}
$$

The first says definitely that P is false and Q is true; the second just says that not both are true (at least one is false). In English, "both" is often a grouping device, the equivalent of a left-hand parenthesis; "either" and "if" can serve the same function.

We can use wffs to construct propositional arguments:

$$
\begin{array}{ll}
((A \cdot B) \supset C) & \text{If both A and B, then C} \\
\sim C & \text{Not-C} \\
A & \text{A} \\
\therefore \sim B & \therefore \text{Not-B}
\end{array}
$$

We can establish the validity of such argument using **truth tables**, **proofs**, **truth trees**, and other methods. All of these methods can capture the standard and widely accepted views of **classical symbolic logic** about which arguments of propositional logic are valid. Some rebels dispute aspects of this standard approach and instead support some version of **deviant logic**.

PSYCHOLOGISM. The view that the laws of logic describe how people think. This formerly popular view faces the problem that people can reason invalidly; so we cannot define "valid" to mean "in accord with how people think." Also, at least some laws of logic are much more clearly true than any psychological view; so it seems wrong to analyze logic in terms of **psychology**. *See also* HUSSERL.

PSYCHOLOGY. Raises questions like these about logic and reasoning: To what extent can people distinguish between **valid** and invalid reasoning? To what extent do people reason correctly? What influences people to make logical errors? Are humans uncomfortable when they discover inconsistency in their beliefs (as the theory of *cognitive dissonance* says)? Are there **gender** differences in the ability to distinguish between valid and invalid reasoning, and in how much males and females rely on reasoning? How does the ability to reason develop as children grow into adults? How can reasoning ability be stimulated or retarded by various social and educational conditions? Do the laws of logic describe how people think (as **psychologism** holds)?

It would be interesting to do a psychological study of the major figures in the history of logic; some of the greatest geniuses, like **Gottlob Frege**, **Charles Sanders Peirce**, and **Ludwig Wittgenstein**, were very unusual personalities.

– Q –

QED (*quod erat demonstrandum*, **Latin for "what was to be proved"**). A old-fashioned phrase to indicate the end of a proof.

QUANTIFICATIONAL LOGIC (also called predicate, functional, elementary, or first-order logic). A branch of logic that builds on **propositional logic** and studies arguments whose validity depends on "all," "no," "some," and similar notions. While somewhat resembling **syllogistic logic**, quantificational logic is much more powerful. Its strength comes from how it uses relatively simple devices to symbolize a wide range of statements.

Quantificational logic adds small letters and "∃" to the symbols of propositional logic. Capital letters can represent statements, general terms, or relations:

- A capital letter alone (not followed by small letters) represents a **statement**. So "S" represents "It is snowing."
- A capital letter followed by a single small letter represents a **general term**. So "Ir" represents "Romeo is *Italian*."
- A capital letter followed by two or more small letters represents a **relation**. So "Lrj" represents "Romeo *loves* Juliet" and "Gxyz" represents "x *gave* y to z."

Similarly, small letters can be constants or variables:

- A small letter from "a" to "w" is a **constant**—and represents a **singular term** that stands for a specific person or thing. So "Ir" represents "*Romeo* is Italian."
- A small letter from "x" to "z" is a **variable**—and stands for an unspecified member of a class of things. So "Ix" represents "*x* is Italian."

"Ix" ("x is Italian") is incomplete, and thus not true or false, since we have not said whom we are talking about; but we can add a quantifier to complete the claim.

A *quantifier* is a sequence of the form "(x)" or "(∃x)"—where any variable can replace "x":

- "(x)" is a *universal quantifier*. It claims that the formula that follows is true for *all* values of x. So "(x)Ix" represents "All are Italian" or "For all x, x is Italian."
- "(∃x)" is an *existential quantifier*. It claims that the formula that follows is true for *at least one* value of x. So "(∃x)Ix" represents "Some are Italian" or "For some x, x is Italian."

Quantifiers express "all" and "some" by saying in how many cases the following formula is true.

To propositional logic, we add two new ways to form **wffs**:

1. The result of writing a capital letter and then one or more small letters is a wff.
2. The result of writing a quantifier and then a wff is a wff.

These rules let us construct the wffs that we just mentioned: "Ir," "Lrj," "Gxyz," "Ix," "(x)Ix," and "(∃x)Ix."

Here are some further wffs:

∼(x)Ix	=	Not all are Italian.
	=	It is not the case that, for all x, x is Italian.
∼(∃x)Ix	=	No one is Italian.
	=	It is not the case that, for some x, x is Italian.
(Ix ⊃ Lx)	=	If x is Italian then x is a lover.

These wffs are a little more complicated:

(x)(Ix ⊃ Lx)	=	All Italians are lovers.
	=	For all x, *if* x is Italian *then* x is a lover.
(∃x)(Ix · Lx)	=	Some Italians are lovers.
	=	For some x, x is Italian *and* x is a lover.
∼(∃x)(Ix · Lx)	=	No Italians are lovers.
	=	It is not the case that, for some x, x is Italian *and* x is a lover.
(x)((Rx · Ix) ⊃ Lx)	=	All rich Italians are lovers.
	=	For all x, *if* x is rich *and* Italian, *then* x is a lover.
((x)Ix ⊃ (x)Lx)	=	If everyone is Italian, then everyone is a lover.

Here are some simple relational translations without quantifiers:

Ljr	=	Juliet loves Romeo.
Ljj	=	Juliet loves herself.
(Ljr · Lja)	=	Juliet loves Romeo and Antonio.

These use single quantifiers:

(x)Lxj	=	Everyone loves Juliet.
	=	For all x, x loves Juliet.

$(\exists x)Ljx$ = Juliet loves someone.
= For some x, Juliet loves x.

$(x)(Ix \supset Lxj)$ = All Italians love Juliet.
= For all x, if x is Italian then x loves Juliet.

$(\exists x)(Ix \cdot Ljx)$ = Juliet loves some Italians.
= For some x, x is Italian and Juliet loves x.

Here are formulas with two quantifiers:

$(x)(y)Lxy$ = Everyone loves everyone.
= For all x and for all y, x loves y.

$(\exists x)(\exists y)Lxy$ = Someone loves someone.
= For some x and for some y, x loves y.

$(\exists x)Lxx$ = Some love themselves.
= For some x, x loves x.

This next pair, which differs only in the order of quantifiers, is often confused (*see* **quantifier-shift fallacy**):

$(x)(\exists y)Lxy$
= Everyone loves someone.
= Everyone loves at least one person.
= For all x there is some y, such that x loves y.

$(\exists y)(x)Lxy$
= There is someone that everyone loves.
= There is some one person that everyone loves.
= There is some y such that, for all x, x loves y.

In the first case, we might love different people (perhaps we all love our mothers); in the second, we all love the same person (perhaps we all love God). These next examples are more difficult:

$(x)(Ix \supset (\exists y)Lxy)$
= Every Italian loves someone.
= For all x, if x is Italian then there is some y such that x loves y.

$(x)(\exists y)(Iy \cdot Lxy)$
= Everyone loves some Italian.
= For all x there is some y such that y is Italian and x loves y.

$$(x)((\exists y)Lxy \supset (z)Lzx)$$

= Everyone loves a lover.
= For all x, if x loves someone then everyone loves x.

$$(x)(Lrx \equiv {\sim}Lxx)$$

= Romeo loves all and only those who do not love themselves.
= For all x, Romeo loves x if and only if x does not love x.

Most English arguments can be symbolized in quantificational logic. By adding four **inference rules**, we can extend to quantificational logic the last method of **proofs** sketched in our *proofs* entry. First there are two "reverse squiggle" rules; these hold regardless of what variable replaces "x" and what pair of contradictory wffs replaces "Fx"/"~Fx" (here "→" means that we can infer whole lines from left to right):

Reverse Squiggle	${\sim}(x)Fx \quad \rightarrow \quad (\exists x){\sim}Fx$
	${\sim}(\exists x)Fx \quad \rightarrow \quad (x){\sim}Fx$

So "Not everyone is funny" entails "Someone is not funny." Similarly, "It is not the case that someone is funny" entails "Everyone is non-funny." These rules can be based on the interdefinability of the universal and existential quantifiers:

$(x)Fx$ = ${\sim}(\exists x){\sim}Fx$
Everyone is F. = It is not the case that someone is non-F.

$(\exists x)Fx$ = ${\sim}(x){\sim}Fx$
Someone is F. = It is not the case that everyone is non-F.

We drop quantifiers using the next two rules (which hold regardless of what variable replaces "x" and what wffs replace "Fx"/"Fa"— provided that the wffs are identical except that wherever the variable occurs not bound to a quantifier in the former the same constant occurs in the latter). Here is the drop-existential rule:

Drop Existential	$(\exists x)Fx \quad \rightarrow \quad Fa,$
	use a *new* constant

Suppose someone robbed the bank; we can give this robber (or one such robber, in case there is more than one) a name—like "Albert"— but it must be an arbitrary name that we make up. Likewise, when we drop an existential, we will label this "someone" with a *new constant*— one that has not yet occurred in earlier lines of the proof.

Here is the drop-universal rule:

Drop Universal	$(x)Fx \rightarrow Fa$, use any constant

If everyone is funny, then Albert is funny, Bob is funny, and so on. From "(x)Fx" we can derive "Fa," "Fb," and so on—using any constant. With this and the previous rule, the quantifier must begin the wff and we must replace the variable with the same constant throughout.

We can here sketch (omitting some details) a fairly automatic proof strategy that will work most of the time, when used with the propositional proof strategy. First reverse squiggles; go from formulas starting with "~(x)" to corresponding ones starting with "(∃x)~" and from ones starting with "~(∃x)" to ones starting with "(x)~." Next drop any initial existential quantifiers, using a new constant letter each time. Lastly, drop any initial universal quantifiers, using all the old constant letters used in the proof so far. Here is an example of a quantificational proof (read the argument as "All logicians are funny; someone is a logician; therefore someone is funny"):

```
1     (x)(Lx ⊃ Fx)
2     (∃x)Lx
   [ ∴ (∃x)Fx
3   ┌ asm: ~(∃x)Fx
4   │ ∴ (x)~Fx   {from 3}
5   │ ∴ La   {from 2}
6   │ ∴ (La ⊃ Fa)   {from 1}
7   │ ∴ Fa   {from 5 and 6}
8   └ ∴ ~Fa   {from 4}
9   ∴ (∃x)Fx   {from 3; 7 contradicts 8}
```

After assuming the opposite of the conclusion (line 3), we reverse the squiggle to move the quantifier to the outside (line 4). We drop the existential quantifier using a new constant (line 5). We drop the universal quantifier in line 1 using this same constant (line 6) and then use a propositional rule (line 7). Then we drop the universal in line 4 using this same constant (line 8), which gives us a contradiction. **RAA** gives us the original conclusion (line 9).

Trying to construct a proof for an invalid argument using this strategy will normally lead us to a refutation of the argument—truth conditions making the premises true and conclusion false. Here is an example (read the argument as "All logicians are funny; someone is a logician; therefore everyone is funny"):

1	(x)(Lx ⊃ Fx)	Invalid
2	(∃x)Lx	a, b
	[∴ (x)Fx	
3	asm: ~(x)Fx	La, Fa
4	∴ (∃x)~Fx {from 3}	~Lb, ~Fb
5	∴ La {from 2}	
6	∴ ~Fb {from 4}	
7	∴ (La ⊃ Fa) {from 1}	
8	∴ (Lb ⊃ Fb) {from 1}	
9	∴ Fa {from 5 and 7}	
10	∴ ~Lb {from 6 and 8}	

After assuming the opposite of the conclusion (line 3), we reverse a squiggle (line 4). Then we drop the two existential quantifiers, using a new and different constant each time (lines 5 and 6). We drop the universal quantifier twice, first using "a" and then using "b" (lines 7 and 8). Since we get no contradiction, we gather the simple pieces to give a refutation. Our refutation is a little possible world with two people, a and b; if we read "Lx" as "x is a logician" and "Fx" as "x is funny," then this refutation has us imagine this situation:

a is a logician.	a is funny.
b is not a logician.	b is not funny.

Here the premises are true, since all logicians are funny, and someone is a logician. The conclusion is false, since someone is not funny. Since the premises are all true and conclusion false in this possible situation, our argument is invalid. Such refutations must contain at least one entity; usually two entities suffice, but sometimes more are needed.

Relational arguments use the same inference rules and general proof strategy. But such arguments face two problems, neither of which arises in quantificational arguments without relations. The first problem (which is a consequence of **Church**'s theorem) is that relational proof-construction cannot be reduced to an **algorithm**; there is no mechanical strategy that will always flawlessly give us a proof or refutation in a finite number of steps. So working out relational arguments sometimes requires ingenuity and not just mechanical methods. The problem with our proof strategy is that it can lead into **endless loops**.

Instructions lead into an endless loop if they command the same sequence of actions over and over, endlessly. Consider this argument:

(x)(∃y)Lxy   Everyone loves someone.
∴ (∃y)(x)Lxy   ∴ There is someone that everyone loves.

Here our proof strategy leads into an endless loop. In this attempt at a proof, we will ignore step 2, since step 1 is enough to generate the endless loop (and things just get more complicated if we bring in step 2):

```
1     (x)(∃y)Lxy                Everyone loves someone.
 [∴ (∃y)(x)Lxy
2     asm: ~(∃y)(x)Lxy
3     ∴ (∃y)Lay   {from 1}      ∴ a loves someone.
4     ∴ Lab   {from 3}              ∴ a loves b.
5     ∴ (∃y)Lby   {from 1}      ∴ b loves someone.
6     ∴ Lbc   {from 5}              ∴ b loves c.
7     ∴ (∃y)Lcy   {from 1}      ∴ c loves someone.
8     ∴ Lcd   {from 7}              ∴ c loves d.
           · · ·                        · · ·
```

Since we have no old letters (and would not get one even if we brought in step 2), we drop the universal in step 1 using a new letter; this gives us step 3: "a loves someone." We drop the existential here to get step 4: "a loves b." But then we have to drop the universal in step 1 using "b"; this gives us step 5: "b loves someone." We drop the existential here to get step 6: "b loves c." If we followed our proof strategy strictly, we would forever keep deriving formulas that resemble 3–4 and 5–6: "c loves someone and c loves d," "d loves someone and d loves e," "e loves someone and e loves f," and so on. If we see this coming, we can break out of the loop and improvise this **possible world**, with beings a and b, that makes the premise true and conclusion false:

$$
a, b \quad \boxed{\begin{array}{l} \text{Laa, } \sim\text{Lab} \\ \text{Lbb, } \sim\text{Lba} \end{array}} \quad \begin{array}{l} \text{(Egoistic} \\ \text{World)} \end{array}
$$

Here all love themselves, and only themselves. This makes "Everyone loves someone" true but "There is someone that everyone loves" false.

A second problem is that refuting an invalid relational argument sometimes requires a possible world with an infinite number of entities. (Non-relational quantificational arguments require at most 2^n entities, where n is the number of distinct predicates.) Consider this example:

> In all cases, if x is greater than y and y is greater than z, then x is greater than z.
> In all cases, if x is greater than y then y is not greater than x.
> b is greater than a.
> ∴ There is something than which nothing is greater.

> $(x)(y)(z)((Gxy \cdot Gyz) \supset Gxz)$
> $(x)(y)(Gxy \supset \sim Gyx)$
> Gba
> ∴ $(\exists x)\sim(\exists y)Gyx$

We can imagine a possible world with an infinity of beings, in which for every being there is a greater being. In this world, the premises are true but yet the conclusion is false (since there is no greatest being). So the argument is invalid.

We can refute the argument by giving another one of the same form with true premises and a false conclusion. Let us take the natural numbers (0, 1, 2, . . .) as the **universe of discourse**. Let "a" refer to 0 and "b" refer to 1 and "Gxy" mean "x > y." On this **interpretation**, the premises are all true. But the conclusion, which says "There is a number than which no number is greater," is false. So the form is invalid.

So relational arguments raise problems about infinity—endless loops and infinite worlds—that other quantificational arguments do not raise. *See also* ANY/ALL; DEFINITE DESCRIPTION; FREE LOGIC;

IDENTITY LOGIC; SECOND-ORDER LOGIC.

QUANTIFIED MODAL LOGIC. A branch of logic that combines **quantificational logic** with propositional **modal logic**. Here we will consider first a naïve system that just combines the two systems (adding together the symbols, **wff** rules, and **inference rules** of both) and then a revised system that adds refinements to avoid problems.

Many quantified modal translations follow familiar patterns. For example, "everyone" translates into a universal quantifier that follows the English word order—while "anyone" translates into a universal quantifier at the beginning of the wff (*see* **any/all**):

$$\Diamond(x)Ax$$
= It is possible for *everyone* to be above average.
= It is possible that, for all x, x is above average.

$$(x)\Diamond Ax$$
= It is possible for *anyone* to be above average.
= For all x, it is possible that x is above average.

While the first could not be true (in a non-empty domain), the second could be.

Quantified modal logic can express the difference between necessary and contingent properties. The number 8 seemingly has the necessary properties of *being even* and of *being one greater than seven*; 8 could not have lacked these. It has also contingent properties, ones it could have lacked, such as *being less than the number of apostles*. We symbolize "necessary property" and "contingent property" as follows:

$$\Box Fa$$
= F is a necessary (essential) property of a.
= a is necessarily F.
= In all possible worlds, a would be F.

$$(Fa \cdot \Diamond{\sim}Fa)$$
= F is a contingent (accidental) property of a.
= a is F but could have lacked F.
= In the actual world a is F; but in some possible world a is not F.

Humans have mostly contingent properties. Socrates had contingent properties like *having a beard* and *being a philosopher*; these are contingent, because he could without self-contradiction have been clean-shaven and a non-philosopher. But Socrates had also necessary properties, like *being self-identical* and *not being a square circle*; everything has these properties of necessity.

Aristotelian essentialism is the controversial view that there are properties that some beings have of necessity that some other beings totally lack. **Willard Van Orman Quine** opposed quantified modal logic, in part because it seems to support Aristotelian essentialism. But quantified modal logic, while it enables us to symbolize essentialist claims, does not force us to accept these. And **Alvin Plantinga** defended Aristotelian essentialism in his *The Nature of Necessity*. He suggests that Socrates had these properties, which some other beings totally lack, of necessity: *not being a prime number, being snub-nosed in W* (a specific possible world), *being a person* (capable of conscious rational activity), and *being identical with Socrates*. The last property differs from that of *being named "Socrates."*

Plantinga explains "necessary property" as follows. Suppose "a" names a being and "F" names a property. Then the entity named by "a" has the property named by "F" necessarily, if and only if the proposition expressed by "a is non-F" is logically impossible. Then to say that Socrates necessarily has the property of *not being a prime number* is to say that the proposition "Socrates is a prime number" (with the name "Socrates" referring to the person Socrates) is logically impossible. We must use names (like "Socrates") here and not definite descriptions (like "the entity I am thinking about").

Some arguments about essentialism involve a form of the **box-inside/box-outside ambiguity**. This next sentence could have either of the following two meanings:

"All bachelors are necessarily unmarried."

(x)(Bx ⊃ □Ux)	"All bachelors are *inherently unmarriable*—in no possible world would anyone ever marry them." (This is false.)
□(x)(Bx ⊃ Ux)	"It is necessarily true that all bachelors are unmarried." (This is true because "bachelor" means *unmarried man*.)

The box-inside "(x)(Bx ⊃ □Ux)" predicates, falsely, of each bachelor the necessary property of *being unmarried*; the medievals called this *de re* ("of the thing") necessity. The box-outside "□(x)(Bx ⊃ Ux)," in contrast, is true by definition and attributes necessity to the proposition (or saying) "All bachelors are unmarried"; the medievals called this *de dicto* ("of the saying") necessity.

The moral of the story is that we should avoid this way of arguing from definitions of words to necessary properties of individuals:

> All bachelors are necessarily unmarried.
> You are a bachelor.
> ∴ You have the necessary property of being unmarried.

Since the first premise can have two meanings, the argument is ambiguous between these two symbolizations:

| *Box Inside* | $(x)(Bx \supset \Box Ux)$ Bu ∴ $\Box Uu$ | $\Box(x)(Bx \supset Ux)$ Bu ∴ $\Box Uu$ | *Box Outside* |

The box-inside interpretation, as we have seen, has a false premise and thus is unsound. The box-outside interpretation is invalid; to get the "□Uu" conclusion requires a "□Bu" premise—the "Bu" premise gets us only a "Uu" conclusion. To refute the box-outside argument, we can use this galaxy of **possible worlds**—which has "Bu" and "Uu" both true in the actual world but both false in some alternative possible world W:

	Bu, Uu
W	~Bu, ~Uu

In this galaxy, the box-outside argument has true premises and a false conclusion, making it invalid.

Our *naïve approach* to quantified modal logic simply combined quantificational with modal logic. While this may seem plausible so far, it has two problems. First, it mishandles definite descriptions. Second, it assumes that the same individuals exist in all possible worlds, which leads to faulty arguments being provable.

Here is a clearly invalid argument using definite descriptions (terms of the form "the so and so") that is provable in our naïve approach:

8 is the number that I am thinking of.

It is necessary that 8 is 8.

∴ It is necessary that 8 is the number that I am thinking of.

$$e=n$$
$$\Box e=e$$
$$\therefore \Box e=n$$

Given the "e=n" first premise, we can take the second premise and substitute "n" for the second "e" and derive the conclusion (*see* **identity logic**). But the argument is invalid; even if the premises are true, there is no necessity in 8 being the number that I am thinking of. So something is wrong here.

The solution is easy; in doing quantified modal logic, we have to use **Bertrand Russell**'s analysis of **definite descriptions** to translate terms of the form "the so and so." So we cannot use "n" to translate "the number that I am thinking of"; instead we have to use the longer paraphrase "There is just one number that I am thinking of and it . . . ." If we paraphrase the term that way, then the troublesome argument gets transformed into this (here "Tx" means "I am thinking of number x"):

There is just one number that I am thinking of, and it is 8.

It is necessary that 8 is 8.

∴ This is necessary: "There is just one number that I am thinking of, and it is 8."

$$(\exists x)((Tx \cdot \sim(\exists y)(\sim x=y \cdot Ty)) \cdot x=e)$$
$$\Box e=e$$
$$\therefore \Box(\exists x)((Tx \cdot \sim(\exists y)(\sim x=y \cdot Ty)) \cdot x=e)$$

This is invalid and not provable.

Another approach is to qualify the substitute-equals rule so that it does not apply in modal contexts. Or we might use this in addition to using Russell's analysis of definite descriptions.

The second problem is that our naïve approach assumes that the same individuals exist in all possible worlds, which leads to faulty arguments being provable. Here is an example (read "g" as "Gensler"):

$$[\therefore \Box(\exists x)x=g$$

1   ┌ asm: ~□(∃x)x=g

2   │ ∴ ◇~(∃x)x=g   {from 1}

3   │ W ∴ ~(∃x)x=g   {from 2}

4 | W ∴ (x)~x=g   {from 3}
5 | W ∴ ~g=g   {from 4}   ← ??
6 | W ∴ g=g   {self-identity rule}
7 ∴ □(∃x)x=g   {from 1; 5 contradicts 6}

The conclusion follows according to our quantificational and modal rules. But the conclusion, which says "It is necessary that there is a being who is Gensler," is clearly false—since there are impoverished possible worlds without him. Our naïve system makes Gensler and everyone else into necessary beings. So something is wrong here.

There are two ways out of the problem. One way is to change how we take "(∃x)." The provable "□(∃x)x=g" is false if we take "(∃x)" to mean "for some *existing being* x." But we might take "(∃x)" to mean "for some *possible being* x"; then "□(∃x)x=g" would mean the more plausible "In every possible world, there is a possible being who is Gensler." Perhaps there is a possible being Gensler in every world; in some of these worlds Gensler exists, and in others he does not. This view would need an existence predicate "Ex" to distinguish between possible beings that exist and those that do not; we could use the formula "(∃x)~Ex" to say that there are possible beings that do not exist.

This view is paradoxical, since it posits non-existent beings. It is perhaps more plausible to hold that *to be a being* and *to exist* are the same thing: there are no non-existent beings. Of course there could have been beings other than those that now exist. But this does not mean that there *now* are beings that do not exist.

A second approach rejects the derivation of 5 from 4:

4   W ∴ (x)~x=g   {from 3}
5   W ∴ ~g=g   {from 4}   ← ??

4   In world W, every existing being is distinct from Gensler.
5   ∴ In world W, Gensler is distinct from Gensler.

This inference seems wrong, unless we presuppose a further premise "W ∴ (∃x)x=g"—that in world W Gensler is an existing being. Rejecting this inference requires moving to a **free logic**—one free of the assumption that individual constants like "g" always refer to existing beings. In free logic, we cannot move from "(x)Fx" to "Fa" unless we have a further premise "(∃x)x=a"—that a is an existing being.

A *Barcan inference* (named for **Ruth Barcan Marcus**, a pioneer in quantified modal logic) is an inference between a formula that begins with a quantifier and then a modal operator, and then the same formula with these reversed. Barcan inferences often are controversial and raise interesting issues. Our free-logic approach gives better results on such inferences. Our naïve approach makes this Barcan inference provable:

```
1     (x)□Fx
  [ ∴ □(x)Fx
2   ┌ asm: ~□(x)Fx
3   │ ∴ ◇~(x)Fx   {from 2}
4   │ W ∴ ~(x)Fx   {from 3}
5   │ W ∴ (∃x)~Fx   {from 4}
6   │ W ∴ ~Fa   {from 5}
7   │ ∴ □Fa   {from 1}   ←  ??
8   └ W ∴ Fa   {from 7}
9   ∴ □(x)Fx   {from 2; 6 contradicts 8}
```

But the argument seems invalid. Read "Fx" as "x is an abstract object." Suppose only abstract objects existed (numbers, sets, etc.) and all these had the necessary property of being abstract. Then the premise "Every existing being has the necessary property of being abstract" would be true. But the conclusion "In every possible world, every existing being is abstract" could still be false—if other possible worlds had concrete entities. Free logic would block step 7, since this step would now require a "(∃x)x=a" premise, that a exists in the actual world.

Our revised approach allows different worlds to have different existing entities. Gensler might exist in one world but not another. We should not picture existing in different worlds as anything spooky; it is just a way of talking about different possibilities. Gensler might not have existed. We can tell consistent stories where his parents did not meet and he never came into existence. If these stories had been true, then he would not have existed. So he does not exist in these stories (although he might exist in other stories). Existing in a possible world is much like existing in a story; a "possible world" is a technical analogue of a "consistent story." "Gensler exists in world W" just means "If world W had been actual, then Gensler would have existed."

On this revised approach, we would want to move to these definitions of "necessary property" and "contingent property":

F is a necessary (essential) property of a.
= In all possible worlds where a exists, a is F.
= $\Box((\exists x)x=a \supset Fa)$

F is a contingent (accidental) property of a.
= In the actual world a is F; but in some possible world where a exists, a is not F.
= $(Fa \cdot \Diamond((\exists x)x=a \cdot \sim Fa))$

We previously defined "F is a necessary property of Socrates" to mean that Socrates is F in every possible world. Now it seems that Socrates does not exist in some worlds; but then (assuming that he has no properties in those worlds) he can have no necessary properties. On our revised definition, he could still have necessary properties. The claim that Socrates has the necessary property of *being a person* would then mean that Socrates is a person in every possible world where he exists; equivalently, in no possible world does Socrates exist as a non-person. *See also* IDENTITY OVER POSSIBLE WORLDS.

QUANTIFIER. A symbol or phrase that indicates how many. **Quantificational logic** uses quantifiers for "all" and "some." **Identity logic** can express quantifiers like "at least two" and "all but one"—but not "most" or "a few."

QUANTIFIERS: OBJECTUAL/SUBSTITUTIONAL. "$(\exists x)Fx$" in **quantificational logic** can be understood in an *object* way, as meaning "There is at least one entity x such that x is F," or in a *substitution* way, as meaning "There is at least one true sentence of the form 'Fa,' where 'a' is a **singular term.**" One problem with the latter is that there may be an entity that is F but for which we have no singular term; since **Georg Cantor**, most logicians hold that there are more real numbers, for example, than there are potential referring devices.

QUANTIFIER-SHIFT FALLACY. The **quantificational** fallacy of arguing from "Everything has relation-R to something" to "There is something that everything has relation-R to." Here is an example:

Everyone loves someone. $(x)(\exists y)Lxy$
∴ There is someone that everyone loves. ∴ $(\exists y)(x)Lxy$

Suppose everyone loves at least one person (perhaps we all love our mothers), but we all love different people—so there is no one person that everyone loves. Then the premise is true and the conclusion false. This example is even more obviously invalid:

Everyone lives in some house. $(x)(\exists y)Lxy$
$\therefore$ There is some house that everyone lives in. $\therefore (\exists y)(x)Lxy$

Clearly, we might all live in different houses.

Some great minds have seemingly committed this fallacy. **Aristotle** seemed to argue, "Every agent acts for an end, so there must be some (one) end for which every agent acts." St. Thomas Aquinas seemed to argue, "If everything at some time fails to exist, then there must be some (one) time at which everything fails to exist." And John Locke seemed to argue, "Everything is caused by something, so there must be some (one) thing that caused everything." But in these cases some have contended that the thinker only *appears* to have committed the **fallacy**, and that the thinker's argument is better construed in some other way than as arguing from "$(x)(\exists y) \ldots$" to "$(\exists y)(x) \ldots$"

QUESTION BEGGING. *See* CIRCULAR REASONING.

QUESTIONS, LOGIC OF (erotetic or interrogative logic). While it may not make sense to use questions as premises or conclusions, still we can apply some of the standard logic machinery to questions and regard this as a "logic of questions" in an extended sense. Such a study raises issues like these: Should we regard a statement like "I wonder whether Michigan won" as a question, thus taking "question" as a functional category instead of a grammatical one (*see* **language, uses of**)? Can we correlate each question with a set of propositions that represent possible answers (so "How old are you?" would correlate with "I am one year old," "I am two years old," and so forth)? If so, can we then speak of a question as being *true* if at least one of these propositions is true and being *false* if none of these propositions is true? Then do all "false questions" presuppose something false (as the **complex question** "Did you stop beating your wife?" may presuppose the false belief that you have a wife and used to beat her)? Can we speak of "entailment" between questions, so that one question *logically entails* another if every possible answer to the first is a possible answer to the second, and every true answer to the first would necessarily be a true answer to the second? If so, then would "What is your first and last name?" logically entail "What is your first name?"?

While logicians have traditionally focused on statements, some now study **imperatives** and questions. Could a logic of exclamations be far behind? In some sense, "Hurrah for Michigan!" seems inconsistent with "Boo on Michigan!"; and inconsistency is a logical category.

QUINE, WILLARD VAN ORMAN (1908–2000). An American logician and philosopher who is especially known for his provocative ideas about **philosophy of logic**. He was perhaps the most influential American philosopher of the second half of the 20th century.

Quine criticized the **analytic/synthetic** distinction as unclear, because it depends on the notion of **meaning**. Quine was unhappy with *meaning*, because he thought there was no good behavioral test of what a person means by a given expression; his *indeterminacy of translation* thesis claims there are always divergent ways to translate one language into another that are consistent with all the empirical facts, and so no one translation is uniquely correct. Since the notion of meaning is unclear, so is the analytic/synthetic distinction that depends on it.

Quine reinstated **ontology** (the study of what sorts of things ultimately exist) as a legitimate area of investigation. His slogan "To be is to be the value of a bound variable" was intended to clarify ontological disputes. The slogan means that the entities that our theory commits us to are the entities that our quantified variables (like "for all x" and "for some x"—*see* **quantificational logic**) must range over for the statements of our theory to be true. So if we say, "There is some feature that Shakira and Britney have in common," then we must accept *features* (*properties*) as part of our ontology—unless we can show that our reference to them is just an avoidable way of speaking (*see* **logical construct**). Quine accepted **sets** as part of his ontology, because he thought these **abstract entities** were necessary for mathematics and science; his arguments about picking an ontology appeal to pragmatic considerations and **Ockham**'s razor.

Quine rejected *properties* and *propositions*, because he thought these were unclear. His "No entity without identity" slogan means that for there to be entities of a certain sort (like cows) there must be some clear criterion for deciding whether entities of this sort are identical (like whether we have one or two cows). Some distinguish properties and propositions by their logical relationships; the property of *being a human* differs from that of *being a featherless biped*, because it is consistent to say that someone has one property but not the other. Quinc rejected this because it depends on the notion of "consistent," which is part of the "**analytic**"/"synthetic" family of concepts that he rejected.

Quine could not define a **valid** argument in the normal way, as an argument in which it is *logically impossible* for the premises to be true and the conclusion false; this definition uses the notion of "logical impossibility," which is part of the rejected "analytic"/"synthetic" family of concepts. Instead, he defined a *valid* argument as one having a **form** in which it is always the case that if the premises are true then so is the conclusion. He understood the *form* of an argument as how it is structured in terms of logical notions (like "all" and "not") and content phrases (like "Greek," "mortal," and "Socrates"), where the *logical notions* are those used in **classical symbolic logic**.

For Quine, logic stops with classical symbolic logic (*see* **logic, scope of**). Mathematical extensions, like **set theory**, are part of mathematics, not logic. Philosophical extensions, like **modal** and **deontic logic**, use terms like "necessary" and "ought" that are too colorful and topic-specific to be part of logic. Quine further attacked modal logic on the grounds that "necessary" is part of the "analytic"/"synthetic" family of concepts; and he attacked **quantified modal logic** because it supports an outmoded **Aristotelian essentialism** metaphysics of necessary properties and an unintelligible *de re* modal predication.

Quine opposed **deviant logicians** by appealing to the slogan "When they try to deny the doctrine, they only change the subject." This means that, for example, logicians who deny the "Not-not-P ∴ P" law of double **negation** must mean something different by "not"; and so they are not really denying the law. While this may seem close to saying that the law is true by virtue of the meaning of "not," and hence analytic, Quine cannot say this. He means rather that principles of logic like this one are so obvious that someone who denies them must be taking their terms in some other sense.

Quine thought the *a priori/a posteriori* distinction also is unclear, that logic is a bit of both, and that we need to pick logical principles on pragmatic grounds. Despite the fact that logical principles are neither necessary truths nor known *a priori*, they are all obvious, at least potentially; every logical principle is either obvious in itself or else can be traced back to obvious principles by obvious steps.

Not everyone agrees with Quine; many have argued that he was wrong on various points. **Paul Grice** and Peter Strawson defended the analytic/synthetic distinction. **Ruth Barcan Marcus** and **Saul Kripke** defended quantified modal logic. **Jaakko Hintikka** defended *de re* predication. **Alvin Plantinga** defended *logical necessity*, quantified modal logic, and Aristotelian essentialism; "No entity without identity" is criticized on the grounds that it would equally eliminate entities like mountains, electrons, wars, and sentences (when we have two peaks separated by a saddle, there is no clear way to decide whether we have two mountains or one). Quine's special talent, it seems, was to provoke other people to clarify their thinking in order to attack him.

– R –

RAA (*reductio ad absurdum*, Latin for "reduction to absurdity"). A rule that says that we may derive the opposite of an assumption that leads to a contradiction. **Proofs** using RAA are called *indirect proofs*.

RATIONALIST. One who believes that we have synthetic *a priori* knowledge. More broadly, one who emphasizes *a priori* knowledge.

REASONING. "Reasoning" and its cognate "reason" can be taken in a variety of senses. Taken very strictly, *reasoning* is going from premises to a conclusion; **arguments** are the verbal expression of such acts of **inference**. Taken more broadly, *reasoning* is any basing of a belief on further beliefs, whether or not this can be formulated in a strict premise-conclusion format; for example, you might believe P because it is probable on the basis of your evidence. Taken very broadly, *reasoning* is any kind of thinking or intellectual activity.

When someone talks about "reasoning" or "reason," we can sometimes get clearer on what the person means by seeing what these are contrasted with. "Reasoning" can be opposed to *intuition* (where we directly grasp the truth instead of going at it in a premise-conclusion manner), to *sensation* (where we appeal to experience instead of just thinking), to *faith* (where choose to believe even though doubt is still reasonable), or to *feeling* (where we follow what we feel instead of thinking things out). Without some explanation, a question like "What can we know by reason?" is unclear.

REDUCTIO AD ABSURDUM. See RAA.

REDUCTIONISM, LOGICAL. The view that statements about one sphere of reality are logically equivalent to statements about some other sphere of reality. For example, statements about material objects are analyzed as being about sensations, statements about minds are analyzed as being about behavior, or statements about duties are analyzed as being about social approval. Critics object that the analyses (**definitions**) do not work.

REFERENCE. *See* SENSE/REFERENCE.

REFERENTIALLY OPAQUE/TRANSPARENT. Usually truths about an entity do not depend on how we refer to that entity; so "X was born in 1809" is equally true if for "X" we substitute "Abraham Lincoln" or "The first Republican president." But "Jones believes that X's picture is on the penny" can be true if we substitute "Abraham Lincoln" but false if we substitute "The first Republican president." Such contexts are said to be *referentially opaque* (instead of *referentially transparent*). *See also* IDENTITY LOGIC.

REFUTATION. The refutation of a **statement** is the **proof** of its negation. The refutation of an **argument** is a possible situation making the premises all true and conclusion false (*see* **propositional logic**).

RELATION. A connection of any sort between things. **Quantificational logic** symbolizes relations by predicates having two or more places, like "x loves y" or "x is between y and z" ("Lxy" and "Bxyz"). In contrast, a one-place predicate like "x is white" ("Wx") is seen as representing a property of a thing, not a relationship between things. A predicate like "x loves x" ("Lxx") is sometimes considered to express a relation and sometimes considered to express a property.

Many relations have special properties, such as reflexivity or symmetry. Here are examples:

> "Is identical to" is *reflexive*: everything is identical to itself. (Identity is a relation but uses a special symbol.)
> = (x)x=x
>
> "Taller than" is *irreflexive*: nothing is taller than itself.
> = (x)~Txx

"Being a relative of" is *symmetrical*: in all cases, if x is a
relative of y, then y is a relative of x.
=   (x)(y)(Rxy ⊃ Ryx)

"Being a parent of" is *asymmetrical*: in all cases, if x is a
parent of y then y is not a parent of x.
=   (x)(y)(Pxy ⊃ ~Pyx)

"Being taller than" is *transitive*: in all cases, if x is taller
than y and y is taller than z, then x is taller than z.
=   (x)(y)(z)((Txy · Tyz) ⊃ Txz)

"Being a foot taller than" is *intransitive*: in all cases, if x
is a foot taller than y and y is a foot taller than z, then x
is not a foot taller than z.
=   (x)(y)(z)((Txy · Tyz) ⊃ ~Txz)

Love fits none of these six categories. Love is neither reflexive nor
irreflexive: sometimes people love themselves and sometimes they do
not. Love is neither symmetrical nor asymmetrical: if x loves y, then
sometimes y loves x in return and sometimes not. Love is neither transitive nor intransitive: if x loves y and y loves z, then sometimes x
loves z and sometimes not.

Augustus De Morgan in the 19th century complained that the logic
of his time neglected relations and could not validate relational arguments like this one (which modern logic can easily handle—here
"Hxy" is for "x is a head of y"):

All dogs are animals.
∴ All heads of dogs are heads of animals.

(x)(Dx ⊃ Ax)
∴ (x)((∃y)(Dy · Hxy) ⊃ (∃y)(Ay · Hxy))

He would be pleased that relations are so important in modern logic.

RELEVANCE LOGIC (relevant logic). An alternative to standard
propositional logic that tries to give a better analysis of **conditionals**.
Relevance logics became important in the 1970s with the work of Alan
Ross Anderson, Nuel Belnap, and Michael Dunn. Today the analysis of
conditionals is a topic of great controversy.
    Standard logic symbolizes "If A then B" as "(A ⊃ B)"; the latter is

equivalent to "~(A · ~B)," which just denies the A-is-true-and-B-is-false combination. So the *material implication* "(A ⊃ B)" is true just if either "A" is false or "B" is true. This analysis leads to the *paradoxes of material implication*:

- From "Not-A" we can infer "If A then B." So from "Pigs do not fly" we can infer "If pigs fly, then I am rich."
- From "B" we can infer "If A then B." So from "Pigs do not fly" we can infer "If I am rich, then pigs do not fly."

While many logicians see such results as harmless, relevance logicians see them as wrong and want to reconstruct logic to avoid them.

Relevance logicians oppose evaluating the truth of "If A then B" just by the truth values of the parts; they say an IF-THEN can be true only if the parts are *relevant* to each other. While they do not spell out this "relevance" requirement in any complete way, they insist that logic should not be able to prove theorems, like "If P-and-not-P, then Q" and "If P, then Q-or-not-Q," in which the antecedent and consequent do not share any letters. They often symbolize their *relevant implication* as "→," to contrast with the "⊃" of *material implication*.

C. I. Lewis developed **modal logic** in the 1930s, largely to find a better analysis of conditionals. He thought he found it in *logical entailment* (*strict implication*), which he analyzed in terms of the necessity of a material implication; "A *logically entails* B" was taken to mean "□(A ⊃ B)"—and is not made true by the falsity of "A" or the truth of "B." But this conditional, unfortunately, had its own oddities, including the *paradoxes of strict implication*:

- From "A is impossible" we can infer "A logically entails B." So if "2≠2" is impossible, then "2≠2" logically entails "I am rich."
- From "B is necessary" we can infer "A logically entails B." So if "2=2" is necessary, then "I am rich" logically entails "2=2."

While many again see these results as harmless, relevance logicians see them as wrong and want to avoid them. Since **paraconsistent logic** also rejects the idea that a self-contradiction entails every statement, there is a natural affinity between the two approaches; many relevance logics are also paraconsistent. Such logics try to deal with conditionals with a logically impossible antecedent—like "If Jane proved that π equals some fraction n/m of positive integers, she would be promoted to full professor"—without saying that such conditionals are automatically true if the antecedent is impossible.

Relevance logicians have found many conditional arguments that, while valid on the traditional view, seem invalid. Here are examples:

If you marry Suzy, then you will not divorce her.	*Contraposition*
∴ If you do divorce Suzy, then you did not marry her.	$(M \supset \sim D)$ ∴ $(D \supset \sim M)$
If you study, then you will pass the course.	*Strengthening the Antecedent*
∴ If you study and die, then you will pass the course.	$(S \supset P)$ ∴ $((S \cdot D) \supset P)$
If I add rum and I add coke, then I will have a Cuba Libre (a rum-coke).	*AND to OR*
∴ Either if I add rum then I will have a Cuba Libre, or if I add coke then I will have a Cuba Libre.	$((R \cdot C) \supset L)$ ∴ $((R \supset L) \vee (C \supset L))$

Some examples even question the validity of ***modus ponens***:

If you have red spots, then you have measles.	*Modus Ponens*
You have red spots.	$(R \supset M)$ R
∴ You have measles.	∴ M

This is claimed to be invalid because you might have red spots for some other reason. Another objection to *modus ponens*, from Vann McGee, is more complex. Back in 1980, three main candidates ran for U.S. president: two Republicans (Ronald Reagan, who won, and John Anderson, who was thought to have no chance to win) and one Democrat (Jimmy Carter). Consider this argument:

If a Republican will win, then if Reagan
does not win then Anderson will win.
A Republican will win.

∴ If Reagan does not win, then Anderson
will win.

Modus Ponens

$(P \supset (\sim R \supset A))$

P

$\therefore (\sim R \supset A)$

Here it seems right to believe the premises but not the conclusion (since clearly if Reagan does not win then Carter will win, not Anderson).

Defenders of material implication say much in response. They often appeal to *conversational implication* to diffuse objections based on the paradoxes of material implication. **Paul Grice** claimed that what is true may not be sensible to assert in ordinary speech. He suggested, as a rule of communication, that we should not make a weaker claim rather than a stronger one unless we have a special reason. Suppose you tell your five children, "At least three of you will get Christmas presents"— while in fact you know that all five will get presents. Saying this suggests or insinuates that it is false or doubtful that all five will get presents. This is due to speech conventions, not logical entailments. "At least three will get presents" does not logically entail the falsity of "All will get presents." But saying the first in most circumstances suggests that the second is false or doubtful. Similarly, there is not much point to telling people "If P then Q" on the basis of knowing "~P" or knowing "Q"—since it is better to tell them straight off that "~P" or that "Q." There is generally a point to telling people "If P then Q" only if there is some special connection between the two, some way of going from one to the other. But, again, this has to do with speech conventions, not with truth conditions for "If P then Q."

Defenders of material implication think the other objections can be answered too. In the marry-divorce example, we can make a similar objection against conjunctive syllogism:

You will not both marry
Suzy and divorce her.
You will divorce her.

∴ You will not marry her.

Conjunctive Syllogism

$\sim (M \cdot D)$

D

$\therefore \sim M$

Since this seems invalid, should we also reject conjunctive syllogism? Here the premises are inconsistent because of the content. Divorcing entails a prior marriage ("$\Box (D \supset M)$"); so it could not be that both premises are true. So this example is not a conjunctive syllogism with true premises and a false conclusion. We can answer the objection to contraposition in a similar way. Here the "(M ⊃ ~D)" premise plus the idea that divorcing entails a prior marriage ("□(D ⊃ M)") entail "~D"; this makes the "(D ⊃ ~M)" conclusion true. So the example is not a contraposition with a true premise and a false conclusion.

The examples about study, measles, and Republicans all seem to confuse a genuine IF-THEN with a statement of conditional **probability** (*see* **conditional, kinds of**). Compare these three:

1.   UNQUALIFIED IF-THEN: "If you have red spots, then you have measles."
2.   CONDITIONAL PROBABILITY: "The probability is high that you have measles, given that you have red spots."
3.   QUALIFIED IF-THEN: "If you have red spots and other causes can be excluded, then you have measles."

The premise about measles, if a genuine IF-THEN, has to mean 1—not 2 or 3; but then its truth excludes the red spots having other causes in such a way that you do not have measles. The truth of the IF-THEN does not entail that we are *certain* that there are no other causes; but if *in fact* there are other causes (so you have red spots but not measles), then the IF-THEN is false.

The rum-coke example is peculiar since "Either if A then B, or if C then D" using regular IF-THENs is unusual and difficult to understand. But we do assert such things using **modal** IF-THENs (using either logical or **causal** modalities). So we tend to read the rum-coke example in a modal way to make sense of it:

Necessarily, if I add rum and I add coke then I will have a Cuba Libre. $\square((R \cdot C) \supset L)$
∴ Either adding-rum entails having   ∴ $(\square(R \supset L) \vee \square(C \supset L))$
a Cuba Libre, or adding-coke entails having a Cuba Libre.

This modal version is understandable and clearly invalid. But the original version, with non-modal IF-THENs, is so difficult to understand that its validity or invalidity is unclear.

Defenders of material implication say we can defend the so-called paradoxes of material implication by intuitive arguments. For example, we can derive "If not-A then B" from "A":

1   A is true. (Premise)
2   ∴ Either A is true or B is true.   {from 1}
3   ∴ If A is not true, then B is true.   {from 2}

Relevance logic must reject this plausible argument; it must deny that 2 follows from 1, that 3 follows from 2, or that deducibility is *transitive* (if 3 follows from 2, and 2 from 1, then 3 follows from 1). Doing any of these violates our logical intuitions at least as much as the material-implication paradoxes. So relevance logics, although they try to avoid unintuitive results about conditionals, cannot achieve this goal; they all result in oddities at least as bad as the ones they are trying to avoid.

Another problem is that there are many relevance logics, and these differ greatly on which conditional arguments are accepted. Some relevance logics accept contraposition, while others reject it. Some accept hypothetical syllogism, while others reject it. Some accept strengthening-the-antecedent, while others reject it. Some accept *modus ponens*, while others reject it. Some accept "(A ⊃ A)" as a logical truth, while others reject it. Relevance logics lead to a tower of Babel about conditionals. If relevance logics ever become dominant, then philosophers giving arguments about areas like **ethics** could not use even a simple *modus tollens* without a long explanation; they would then perhaps do better to avoid the word "if" and instead express their arguments using material implication or "not both A-being-true and B-being-false"—since the logic of these notions is less controversial.

There is much more to be said on both sides. Conditionals have been debated since ancient times (*see* **ancient logic since Aristotle**) and will continue to be debated in the future.

RENAISSANCE-TO-NINETEENTH-CENTURY LOGIC (up to Gottlob Frege). This topic divides into two currents. There is, from our present perspective, the main current that flows from **medieval logic** into **classical symbolic logic**. But there are also various side currents; we will talk about the side currents first.

Many Renaissance humanists, preferring colorful rhetoric to close analysis, downplayed the importance of logic. But logic was carried on by people like Paul of Venice (1372–1429); Philipp Melanchthon (1497–1560); Peter Ramus (1515–1572); the Port-Royal Jansenists

Antoine Arnauld and Pierre Nicole (1612–1694 and 1625–1695); and the Jesuit Gerolamo Saccheri (1667–1733), who first investigated non-Euclidian geometry. All wrote logic textbooks in the Aristotelian tradition. **Immanuel Kant** (1724–1804) echoed the thinking of his time when he stated that **Aristotle** not only was the first to conceive of logic but also had substantially brought the subject to its completion.

In contrast, Georg Wilhelm Friedrich Hegel (1770–1831) thought that a better logic would admit of contradictions in nature as key to understanding how thought evolves historically; one view provokes its opposite, and then the two tend to come together in a higher synthesis. Karl Marx (1818–1883) followed Hegel in seeing contradictions in the world as real; he applied this idea to political struggles. Critics contend that the *dialectical logic* of Hegel and Marx confuses conflicting properties in the world (like hot/cold or capitalist/proletariat) with logical self-contradictions (like the same object being both white and, in the same sense and time and respect, also non-white).

John Stuart Mill (1806–1873) saw logical and arithmetic principles as empirical generalizations; they seem necessary only because of our weak minds' inability to imagine their falsity. He emphasized inductive logic and developed *Mill's methods* for arriving at causal explanations.

Several logicians contributed to **syllogistic logic**. **Leonhard Euler** (1707–1783) diagrammed "all A is B" by putting an A-circle inside a larger B-circle. William Hamilton (1788–1856) added forms like "all A is all B" and "all A is some B": the first affirms while the second denies that all B is A. **Lewis Carroll** (1832–1898) entertained us with his silly syllogisms and his points about logic in *Alice in Wonderland*. **John Venn** (1834–1923) gave us diagrams for testing syllogisms. And **Christine Ladd-Franklin** (1847–1930) introduced antilogisms.

Now we need to talk about the stream that flowed from medieval logic into classical symbolic logic. Important here is **Gottfried Wilhelm Leibniz** (1646–1716), who proposed the idea of an artificial language that would reduce reasoning to arithmetic calculation; if controversies arose, the parties could take up their pencils and say to each other, "Let us calculate." Leibniz created a logical notation much like that of Boole; but his work on this was published much after Boole, in 1903.

Several others, including Gottfried Ploucquet (1716–1790), Johann Heinrich Lambert (1728–1777), and **Augustus De Morgan** (1806–1871), proposed ways to symbolize logic. But the algebraic notation of **George Boole** (1815–1864) was much more powerful, since it allowed something like mathematical calculation to be used to check the correctness of inferences; Boole is considered the father of **mathematical logic**. Further refinements to *Boolean algebra* were made by William Stanley Jevons (1835–1882), Ernst Schröder (1841–1902), and **Charles Sanders Peirce** (1839–1914). Peirce, in particular, came up with many important innovations, including the quantifier, which he invented independently of Frege but slightly later. Also important were **Georg Cantor** (1845–1918) and David Hilbert (1862–1943).

All of this leads up to the revolutionary work of **Gottlob Frege** (1848–1925), who invented classical symbolic logic, and the synthetic work of **Bertrand Russell** (1872–1970), who gave us the definitive formulation of classical symbolic logic.

RHETORIC. The art of persuasion. Rhetoric teaches us to study our audience and to use powerful tools of language, like metaphor, irony, understatement or overstatement, personal examples, eye contact, and tone of voice. The goal is to convince our audience.

From the very beginning, there has been tension between rhetoric and philosophy. The rhetoric teachers in ancient Greece were called "Sophists"; they taught wealthy young men how to speak and argue effectively—which was an important path to power. But Socrates and Plato thought the Sophists cared more for persuasion than for **truth**. **Aristotle**, the first logician, later quoted with disapproval the statement of the Sophist Protagoras, who bragged that he could make the weaker argument sound like the stronger one; Aristotle's book on *Sophistical Refutations* is about understanding **fallacies** and not being misled by them. And philosophy courses today caution students against accepting something just because someone has made it sound good; instead, we are to check out the reasoning and look for objections. Instead of buying a used car because the salesperson gave a convincing pitch, we should kick the tires and shop around.

Of course, this picture of rhetoric is very one-sided. Teachers of rhetoric (e.g., the speech teachers we had at college) do not seem to be evil monsters who teach people to use language to manipulate others. Many teachers of rhetoric try to combine the tools of rhetoric and the tools of logic, and emphasize the need for ethical and logical responsibility. They recognize that sound, logical **argument** can be very persuasive, if presented rightly—and that it is immoral to resort to fallacies in order to get one's conclusion accepted.

Philosophers in ancient India (*see* **Buddhist logic**) distinguished three kinds of debate: (1) an honest one where both sides seek the truth;

(2) one where the goal is to win by fair or foul means; and (3) one where the goal is to demolish the opponent by any means. Much debate today, especially in politics (even presidential debates), unfortunately is of the latter two kinds.

RUSSELL, BERTRAND (1872–1970). British philosopher and logician. His three-volume *Principia Mathematica* (1910–1913), coauthored with his former teacher Alfred North Whitehead (1861–1947), brought together ideas from **Gottlob Frege, Giuseppe Peano, Georg Cantor, George Boole**, and Ernst Schröder. It gave the definitive formulation of **classical symbolic logic**. Its central claim is the *logistic thesis* that **arithmetic** can be reduced to logic: every truth of arithmetic can be formulated using just notions of logic and proved using just **axioms** and **inference rules** of logic. (The arithmetic entry explains and raises problems about this thesis.)

The young Russell found much to admire in Frege and his groundbreaking work in logic; the two minds worked along similar lines. But Russell found a deep flaw in Frege's system, a contradiction called "Russell's paradox." Frege's plausible *axiom of comprehension* said that every condition on x (e.g., "x is a cat") picks out a set containing just those elements that satisfy that condition. So the condition "x is a cat" picks out the set of cats. But consider that some sets are members of themselves (the set of abstract objects is itself an abstract object) while other sets are not (the set of cats is not itself a cat). By Frege's axiom, "x is not a member of itself" picks out the set containing just those things that are not members of themselves. Call this "set R." So any x is a member of R, if and only if x is not a member of x:

For all x, x ∈ R if and only if x ∉ x.

But, Russell asks, what about set R itself? By the principle just given, R is a member of R, if and only if R is not a member of R:

R ∈ R if and only if R ∉ R.

So is R a member of itself? If it is, then it is not—and if it is not, then it is; either way we get a contradiction. Since this contradiction was provable in Frege's system, that system was flawed.

Principia Mathematica tried to reformulate Frege's theory to avoid the contradiction. This was done by adding a *theory of types* that outlaws certain forms of self-reference, including ones that permit the self-contradiction. Very roughly, there are ordinary objects (type 0), properties of these (type 1), properties of these properties (type 2), and so on. Any meaningful statement can talk only about objects of a lower type; so no speech can talk meaningful about itself. So "R ∈ R" is malformed, and thus not true or false.

[The perceptive reader might notice that this theory of types refutes itself. "Any meaningful statement can talk only about objects of a lower type," to be useful, has to restrict *all statements*, of *every type*; but then it violates its own rule and declares itself meaningless. The theory also had other problems; and other ways were found to patch up **set theory** to avoid the contradiction.]

Another goal of *Principia Mathematica* was to rework Frege's logic using a more understandable symbolism, one adapted from Peano. The *Principia* symbolism, with typographical variations, has become somewhat standard and is basically what is used in this dictionary.

Russell also worked in other areas of philosophy. One common theme is that ordinary language can mislead us, especially since we can too easily take it to mirror the nature of reality. His 1905 article "On Denoting," which presented a theory of **definite descriptions** using the tools of logic, is widely considered to be a classic of the new analytic approach to philosophy that Russell promoted.

Russell's approach to **metaphysics**, which was influenced by his student **Ludwig Wittgenstein**, was called *logical atomism*. Russell asked what language structure would suffice to describe reality completely. The answer is that we need the framework of logic plus terms that refer to the ultimately simple elements of reality. So there is a parallel structure between what ultimately exists (objects and facts) and how a logically ideal language, based on modern logic, would describe things (in terms of names and propositions). We reach the ultimately simple elements of reality by analyzing complexes into their simplest parts: atomic sentences mirror atomic facts. But what are the ultimately simple elements of reality? Are these simple elements ordinary material objects like tables and chairs, or perhaps these plus minds (a dualism)? Or perhaps, as Russell at times seemed to think, these commonsense objects are only **logical constructs** that help us talk about experiences, which then become the ultimately simple elements of reality. *See also* GÖDEL; NAMES; NOTHING; PHILOSOPHY AND LOGIC; PROOF; QUANTIFIED MODAL LOGIC; SECOND-ORDER LOGIC; TWENTIETH-CENTURY LOGIC.

– S –

SAMPLE-PROJECTION SYLLOGISM. *See* INDUCTIVE LOGIC.

SCIENTIFIC REASONING. While logic relates to scientific reasoning in many ways (for example, *see* **Bayes**'s theorem, **biology**, **inductive logic**, **Mill**'s methods, **physics**, and **psychology**), here we will focus on the logical relationships between observations and scientific laws.

Aristotle taught that heavy bodies fall faster than light ones and that the speed of fall is proportional to weight. We can put this into the formula "v = gw," where "v" stands for velocity, "g" is a constant, and "w" is for weight. For the next two thousand years or so, most people accepted Aristotle's view. Then Galileo Galilei (1564–1642) came up with an opposing view, that objects of any weight fall equally fast, at a speed proportional to time: "v = gt." To decide between the two views, Galileo, it is said, dropped objects of different weights at the dinner table and at the leaning tower of Pisa; he found that they fell at about the same speed. This refutes Aristotle's view:

> If Aristotle's view is correct and we drop objects of different weights at the same time, then we will observe that heavier objects hit the ground first.
> We drop objects of different weights at the same time.
> We do not observe that heavier objects hit the ground first.
> ∴ Aristotle's view is not correct.

This is valid in **propositional logic**: "((A · D) ⊃ O), D, ~O ∴ ~A." Given that we can trust the truth of the premises, we can conclude that Aristotle's view is wrong.

Can a similar experiment show that Galileo's view is correct? Unfortunately not. Consider this argument:

> If Galileo's view is correct and we drop objects of different weights at the same time, then we will observe that heavier and lighter objects hit the ground at the same time and at a time predicted by Galileo's formula.
> We drop objects of different weights at the same time.
> We observe that heavier and lighter objects hit the ground at the same time and at a time predicted by Galileo's formula.
> ∴ Galileo's view is correct.

This is invalid: "$((G \cdot D) \supset O), D, O \therefore G$." So the premises do not prove Galileo's view, which still might fail for further cases.

So sometimes we can deductively refute a proposed scientific law through a crucial experiment; experimental results, when combined with other suitable premises, can logically entail the falsity of a proposed law. But we cannot deductively prove a proposed law through experiments—since we cannot eliminate the possibility that the proposed law will fail for further cases.

It is plausible to think that our test results give a strong inductive argument for Galileo's view:

> All examined cases of falling objects follow Galileo's view.
> A large and varied group of such cases has been examined.
> $\therefore$ Probably all cases of falling objects follow Galileo's view.

The second premise is weak if we have tried only a few cases. But we can easily perform more experiments; after we do so, Galileo's view would seem to be securely based.

Unfortunately, we can give a similar argument for a second, incompatible view. Let view X be a view about falling objects that accords with all the test results obtained so far but diverges in some other cases not yet tested—maybe where objects have a different color, shape, size, or constitution from those tested so far; view X might say, for example, that all objects except carrots follow Galileo's law while carrots follow a different law. Then we can argue for view X:

> All examined cases of falling objects follow view X.
> A large and varied group of such cases has been examined.
> $\therefore$ Probably all cases of falling objects follow view X.

This inductive argument for view X seems as strong as the one we gave for Galileo's view. Judging just from these arguments and test results, there seems to be no reason for preferring one over the other.

We can conduct another crucial experiment (e.g., by dropping carrots) to try to refute view X. But there is always another view X behind the bush—so our problems are not over. However many experiments we do, there are always alternative theories that agree with all the test cases so far but disagree on some further predictions. In fact, there is always an *infinity* of theories that do this. No matter how many dots we put on a chart (representing test results), we can draw an unlimited number of lines that go through all these dots but otherwise diverge.

Suppose we conduct 1,000 experiments in which Galileo's view works. Still there are an infinity of alternative theories that fit all the test cases but give conflicting predictions about further cases. And each theory seems to have the same inductive support:

> All examined cases of falling objects follow this view.
> A large and varied group of such cases has been examined.
> ∴ Probably all cases of falling objects follow this view.

In practice, though, we would prefer Galileo's view on the basis of *simplicity* (*see* **Ockham**'s razor). Galileo's view is the simplest one that agrees with all our test results. So we prefer Galileo's view to the alternatives and regard it as firmly based.

What is simplicity and how can we decide which of two theories is simpler? We do not have a neat and tidy answer to these questions. In practice, though, we can tell that Galileo's view is simpler than a view X that is similar except for having falling carrots follow a different pattern. We do not have a clear and satisfying definition of "simplicity"; yet we can apply this notion in a rough way in many cases.

The *simplicity criterion* says that, other things being equal, we ought to prefer the simplest theory that adequately explains the data. The "other things being equal" qualification is important. Experiments may force us to accept very complex theories; but we should not take such theories seriously until we are forced into it.

The physics examples discussed in this entry are relatively easy to test by experiments. At the opposite end of the spectrum is evolution, which has a more subtle connection with observations—one perhaps better illuminated by an **inference-to-the-best-explanation** analysis. *See also* DISCOVERY/JUSTIFICATION.

SCOPE AMBIGUITY. An **ambiguity** about how much of a formula or sentence is governed by an operator. The sentence "Not A and B" has scope ambiguity, since the "not" could cover "A" or "A and B." Here are seven ways to make the difference clear:

1.	$(\sim\!A \cdot B)$	$\sim\!(A \cdot B)$
2.	A is false, and B is true.	A and B are not both true.
3.	Not-A and B.	Not A-and-B.
4.	Both not A and B.	Not both A and B.

5.	KNab	NKab
6.	~A .· B	~.A · B
7.	⌐⌐⌐B	⌐⌐⌐B
	└─┬─A	└──A

Pair 1 uses parentheses, which is the usual way to indicate scope in logic; some logicians mix parentheses with brackets and braces (as in "{[(P · Q) ∨ R] ⊃ S}"). Pairs 2–4 give ways to show the difference in written English; in speaking, we can use "Not A [pause] and [pause] B" for the first and "Not [pause] A and B" for the second. Pair 5 uses **Polish notation**. Pair 6 uses **Giuseppe Peano**'s dots; here we first take together signs not separated by dots, then those separated by one dot, then those separated by two dots, and so on. Pair 7 uses **Gottlob Frege**'s symbolism; here the first means "~(~A ⊃ ~B)" while the second means "(A ⊃ ~B)," where these two are **truth-functional** equivalents of "(~A · B)" and "~(A · B)." *See also* AMPHIBOLY; BOX-INSIDE/BOX-OUTSIDE AMBIGUITY.

SECOND-ORDER LOGIC. An extension of standard **quantificational logic** that allows quantifying over properties. "(x)(∃F)Fx" would mean "For every individual x, there is some property F such that x has property F," and "(∃F)(Fs · Fb)" might symbolize "There is some property F that Shakira and Britney have in common"; **Willard Van Orman Quine** would say that such statements commit us to an **ontology** that includes properties. A logic that allows quantifying over properties of individuals is a *second-order logic*, while one that allows quantifying over properties of properties of individuals is a *third-order logic*; or we might allow quantifying over properties of any level. Some logics allow quantifying over propositions; "(P)(P ∨ ~P)" would mean "For every proposition P, either P or not-P." **Identity** is definable in second-order logic, since we can take "x=y" ("x is identical to y") to mean "(F)(Fx ≡ Fy)" ("for every property F, x is F if and only if y is F"); *see* **Gottfried Wilhelm Leibniz**.

When constructing higher-order logics, we must be careful that our system does not lead to an analogue of **Russell**'s paradox. Suppose we include an axiom that says that every condition on x picks out a property. For example, "x is Colombian and x is a singer" picks out the property of *being a Colombian singer*. This axiom would lead us into trouble. A *homological* property is a property that applies to itself; for example, the property of *being an abstract object* is itself an abstract object. A *heterological* property is one that does not apply to itself; for example, the property of *being a Colombian singer* is not itself a Colombian singer. Now "F is not F" (in symbols, "~FF") seemingly picks out the property H of being heterological; so any property F is heterological if and only if F is not F: (F)(HF ≡ ~FF). But does H apply to itself? From "(F)(HF ≡ ~FF)" we can deduce "(HH ≡ ~HH)," that H applies to itself if and only if it does not apply to itself, which is a self-contradiction. So higher-order logics, unless carefully constructed, can lead to provable contradictions.

SELF-CONTRADICTION (logical falsehood). A statement that is not logically **consistent**.

SELF-REFUTING STATEMENT. One that makes such a sweeping claim that it ends up denying itself. Suppose I tell you, "Everything that I tell you is false." Could this be true? Not if I tell it to you; then it has to be false. (*See* **liar paradox**.) Here are other self-refuting statements:

- I know that there is no human knowledge.
- One ought to accept only what is scientifically proved.
- Any statement whose truth or falsity cannot be decided through scientific experiments is meaningless. (*See* **logical positivism**.)
- No statement is true.
- Every rule has an exception.
- It is impossible to express truth in human concepts.

SEMANTICS/SYNTAX. *Semantics* is about the **meaning** aspects of language (involving things like **truth**, **sense**, and **reference**)—while *syntax* is about purely notational aspects, abstracting from meaning:

- *Semantics:* "Fido is black" is true and ascribes a certain color to my dog.
- *Syntax:* "Fido is black" has three words.

Branches of logic, like **propositional** and **quantificational logic**, are generally presented as **formal systems**. In giving a formal system, we specify which symbols will be used, which sequences of symbols will be considered **wffs** (well-formed formulas), and which sequences of wffs will be considered **proofs**. All this is specified in a purely syntactic or notational way, in terms of rules about the manipulation of symbols, in abstraction from what the symbols might mean. So a formal system is all syntax.

To connect a formal system with claims about reality, there is often added a semantics, which is about what the symbols mean. Thus we might specify that certain letters represent true-or-false statements, or objects, or properties of objects. In *formal semantics* (started by **Alfred Tarski**), we give truth conditions for the formulas of the system relative to various *interpretations*, where an "interpretation" is an abstract model of how language could link to a possible situation. Then an argument of our formal system is *valid*, if and only if every interpretation that makes all the premises true also makes the conclusion true.

Metalogic is the study of formal systems and the attempt to prove things about these systems. The key issue in metalogic is the harmony, in a specific formal system, between the syntax and the semantics. What is provable is specified by syntactic (notational) rules. What arguments are valid is specified by the semantic rules about what these formulas mean. The basic problem is whether the two fit together—for example, whether all the provable arguments are valid and all the valid arguments are provable.

The symbols "⊢" and "⊨" are often used to make claims about syntax and semantics. Let us use "α" for any set of wffs and "β" for any individual wff. Then "α ⊢ β" means "β is provable (syntactically) from α"; and "α ⊨ β" means "β validly follows (semantically) from α." The basic problem of metalogic is then to show that the two go together, that "α ⊢ β" is true if and only if "α ⊨ β" also is true.

A formal semantics for propositional logic is rather simple. Here an *interpretation* is a specification that says, for each propositional letter, whether that letter is true or false. Wffs that are more complex than a single letter are evaluated as true or false on the basis of whether their parts are true or false, in accord with the following rules (which accord with the basic **truth tables**):

- A **negation** like "~P" is true if the negated part is false; otherwise it is false.
- A **conjunction** like "(P · Q)" is true if both conjuncts are true; otherwise it is false.
- A **disjunction** like "(P ∨ Q)" is true if at least one disjunct is true; otherwise it is false.
- A **conditional** like "(P ⊃ Q)" is false if the antecedent is true and

the consequent false; otherwise it is true.

- A **biconditional** like "(P ≡ Q)" is true if both parts are true or both are false; otherwise it is false.

Then a propositional argument is *valid*, if and only if every interpretation that makes all the premises true also makes the conclusion true.

A formal semantics for propositional **modal logic** would be more complicated. Let us define an *interpretation* to be a set of one or more **possible worlds**, where a possible world is a specification that says, for each propositional letter, whether that letter is true or false in that world. Wffs that are more complex than single letters are evaluated as true or false in a possible world on the basis of whether their parts are true or false in that possible world, in accord with the rules like these:

- A *negation* like "~P" is true in world W if the negated part is false in world W; otherwise it is false in world W.
- A *conjunction* like "(P · Q)" is true in world W if both conjuncts are true in world W; otherwise it is false in world W.
- (*Disjunction, conditional,* and *biconditional* go the same way.)

Formulas that start with a box or diamond are evaluated according to these two rules:

- A *necessity* like "□P" is true in world W if the part after "□" is true in every possible world; otherwise it is false in world W.
- A *possibility* like "◇P" is true in world W if the part after "◇" is true in at least one possible world; otherwise it is false in world W.

Then a modal argument is *valid* if and only if every interpretation that makes all the premises true in a possible world also makes the conclusion true in that possible world.

The modal semantics just sketched assumes system S5. Other **modal systems** need an *accessibility relation* between possible worlds. Then "□P"/"◇P" would be true in world W just if the part after the box or diamond is true in all/some possible worlds that are accessible to world W. Different systems result if the accessibility relation is specified to be symmetrical, or transitive, or both, or neither.

A formal semantics for quantificational logic would be similar, but the details get more complex. Here an *interpretation* would specify a non-empty domain of objects, which object from the domain each individual constant denotes, which set of objects from the domain each predicate letter is true of, which ordered sequences of objects from the domain each relational letter is true of, and which statement letters are true and which are false.

SENSE/REFERENCE. The *sense* (or *meaning*) of a term is roughly the set of properties that the term ascribes to an object and would be used to define the term; the *reference* of a term is the set of objects that the term can be truly predicated of.

Suppose that "human" and "featherless biped" picked out the exact same set of beings. Then both terms would have the same reference. But the two terms would still differ in meaning, since "human" means "rational animal" (or perhaps "member of *homo sapiens*") while "featherless biped" means "animal with two feet but no feathers." Other terms for "sense" are "meaning," "connotation," and "intension"; other terms for "reference" are "denotation" and "extension."

Gottlob Frege applied the same distinction to statements and proper **names**. He thought the sense of "Snow is white" is a proposition while its reference is to **truth**. And he thought the sense of "Socrates" is the set of properties that we use to pick out this particular person while its reference is to a person who lived in ancient Greece. *See also* DEFINITIONS; KRIPKE; MEANING; SUPPOSITION.

SENSU COMPOSITO/SENSU DIVISO (Latin for "in composite"/ "divided sense"). We can talk about a group of things as a unit (*sensu composito*) or about all its members as individuals (*sensu diviso*). Suppose you say "All these shirts cost $20"; you might mean that the group of shirts costs $20 (*sensu composito*) or that each individual shirt costs $20 (*sensu diviso*). Or suppose you say the Michigan Wolverines are awesome; you might mean that the team is awesome or that each individual player is awesome. *See also* DIVISION-COMPOSITION.

SET THEORY. A set, roughly, is a well-defined collection of objects. Here are some common symbols and ideas used in set theory:

- "$g \in p$" means "g is a member of set p." This might mean "Gerald Ford is a member of the set of U.S. presidents."
- "$g \notin p$" means "g is not a member of set p" and is a common abbreviation for "$\sim(g \in p)$."

- "V" stands for *the universal set*, the set that contains everything. If we restrict our **universe of discourse** to people, then "V" will stand for the set of people. Sometimes "U" is used for "V."

- "Λ" stands for *the null set*, the set that contains nothing. Since a set is determined by its members, there is only one null set; so the set of square circles = the set of earthworms with PhDs = the null set. Despite this, the property of *being a square circle* is different from the property of *being an earthworm with a PhD*—since the two differ in **meaning**. Sometimes "∅" is used for "Λ."

- "–p" means "the complement of set p" or "the set of beings in the universal set that are not in set p." This might stand for the set of beings who are not presidents. Sometimes "p'" is used for "–p."

- "p ∪ n" means "the union of set p and set n" or "the set of beings that are either in set p or in set n." This might stand for the set of beings who are either presidents or were born in Nebraska.

- "p ∩ n" means "the intersection of set p and set n" or "the set of beings that are both in set p and in set n." This might stand for the set of presidents who were born in Nebraska; Gerald Ford is a member of this set ("g ∈ (p ∩ n)") and no one else is ("~(∃x)(~x=g • x ∈ (p ∩ n))"). This second formula uses symbols from **quantificational** and **identity** logic.

- "{g}" means "the set whose only member is g." "{g} = (p ∩ n)" might mean "The set containing only Gerald Ford is identical to the set of presidents born in Nebraska"; these sets are identical because they have the same members. The null set "Λ" is different from the set containing just the null set "{Λ}," or the set containing just the set containing the null set "{{Λ}}"; this trick can create an endless supply of distinct sets *ex nihilo*.

- "a ⊆ b" means "a is a subset of b" or "any member of set a is also a member of set b." Any set x is a subset of the universal set ("x ⊆ V"); the null set is a subset of any set x ("Λ ⊆ x").

- "a ⊂ b" means "a is a proper subset of b" or "any member of set a is also a member of set b, but some members of set b are not members of set a." The set of presidents is a proper subset of the universal set: "p ⊂ V."

- "a = b" means "a is identical to b." If "a" and "b" stand for sets, this means that every member of a is a member of b, and every member of b is a member of a: "(x)((x ∈ a) ≡ (x ∈ b))."

- "{a, b}" means "the set that contains just a and b." The "{ . . . }" notation is used to specify the set containing just the members listed, however many these may be; so "{}" is the null set and "{a, b, c}" is the set containing just a, b, and c. The order of the objects listed is irrelevant: "{a, b} = {b, a}."

- "<a, b>" means "the ordered pair consisting of a and then b." Here the order is important. Sometimes "<a, b>" is defined as equivalent to "{{a}, {a, b}}," which keeps track of the order. The "< . . . >" notation is used to indicated an ordered sequence of the objects listed, however many these may be.

Set theory is fairly straightforward until we get **Russell**'s paradox and **Georg Cantor**'s *transfinite sets*. An approach to set theory that ignores these complications is sometimes called *naïve set theory*; pioneers in this area were **George Boole** and **Gottlob Frege**.

Russell found a deep flaw in Frege's system, a contradiction called "Russell's paradox." Frege's plausible *axiom of comprehension* said that every condition on x (e.g., "x is a cat") picks out a set containing just those elements that satisfy that condition. So the condition "x is a cat" picks out the set of cats. But consider that some sets are members of themselves (the set of abstract objects is itself an abstract object) while other sets are not (the set of cats is not itself a cat). By Frege's axiom, "x is not a member of itself" picks out the set containing just those things that are not members of themselves. Call this "set R." So any x is a member of R, if and only if x is not a member of x:

$$(x)(x \in R \equiv x \notin x)$$

But, Russell asks, what about set R itself? By the principle just given, R is a member of R, if and only if R is not a member of R:

$$R \in R \equiv R \notin R$$

So is R a member of itself? If it is, then it is not—and if it is not, then it is; either way we get a contradiction. Since this contradiction was provable in Frege's system, that system was flawed.

Principia Mathematica tried to reformulate Frege's theory to avoid the contradiction. This was done by adding a *theory of types* that outlaws certain forms of self-reference, including ones that permit the self-contradiction. Very roughly, there are ordinary objects (type 0), properties of these (type 1), properties of these properties (type 2), and so on. Any meaningful statement can talk only about objects of a lower type.

So "R ∈ R" is malformed, and thus not true or false.

However, Russell's theory seems to refute itself. "Any meaningful statement can talk only about objects of a lower type," to be useful, has to restrict *all statements*, of *every type*; but then it violates its own rule and declares itself meaningless. So the paradox reappears.

Another problem with Russell's approach is that we sometimes want to talk about sets being members of themselves. So **Willard Van Orman Quine** suggested that, instead of adding a theory of types, we just restrict Frege's axiom. This axiom said that every condition on x (e.g., "x is a cat") picks out a set containing just those elements that satisfy the condition. Quine suggests that we require that the condition on x be *stratified*, where this means it must be possible to substitute numbers for variables in such a way that "∈" always goes between a number and the next number in order. The system based on this is called NF, for "new foundations." While NF gets rid of Russell's paradox, it is not clear that it is consistent, and there are further problems as well.

Most mathematicians prefer a set theory based on the work of the German Ernst Zermelo (1871–1953), but with refinements from later thinkers. This approach, called ZF (for Zermelo-Fraenkel), avoids Russell's paradox, seems consistent, and seems adequate for mathematics. ZF gives a set of axioms, mostly about which sets exist:

1. *Elementary Sets:* There is a null set. For every x, there is a set containing just x. For every distinct x and y, there is a set containing just x and y.

2. *Extensionality:* Sets x and y are identical if they have all the same members.

3. *Separation:* If y is a set and "x is . . ." is a condition on x, then there is a set containing just those members of y that satisfy this condition on x. (This is a restricted version of Frege's axiom. We could still generate Russell's paradox if we used the universal set for y; but ZF has no universal set.)

4. *Power Set:* Given a set x, there is a set that contains all and only the subsets of x.

5. *Union:* Given a set x, there is a set that contains all and only the members of the members of x.

6. *Choice:* Given an infinite collection of non-empty sets with no common members, there is a further set with exactly one member from each set. (This *axiom of choice* is controversial.)

7. *Infinity:* There is a set with an infinite number of members. (We

can construct this from the null set, the set that contains just the null set, the set that contains the two previous sets, the set that contains just the three previous sets, and so on.)

8. *Replacement:* Given a set x and a function mapping each member of x to a further set, there is a set containing all and only those further sets. (This was added by Thoralf Albert Skolem and Abraham Adolf Fraenkel, and modified by **John von Neumann**.)

9. *Regularity:* Every non-empty set x has some member y such that nothing is both a member of x and a member of y. (This axiom from von Neumann is somewhat controversial.)

Another system of axioms, called NBG, was proposed by John von Neumann, Paul Bernays, and **Kurt Gödel**. NBG distinguishes between sets (which can contain themselves) and classes (which cannot).

Set theory is in a better condition today than in 1902, when Russell's paradox showed that our best set theory was self-contradictory. But it is difficult to be satisfied with set theory's current state, where we have various conflicting systems, each with its own quirks, and each leading to a somewhat different set of theorems. Owing to Gödel's theorem, the consistency of none of these is provable (since all of them are strong enough to encompass **arithmetic**). Which, if any, is the *correct* system of set theory? *Is* there a correct set theory—a set of axioms that systematizes independently existing truths about sets? Or is set theory just a human construct that can be built up in different ways? The idea of *set* seems so simple and useful; a large number of the entries in this dictionary use this idea in some way. It is truly astounding that this idea of *set* is so difficult to pin down. Perhaps future developments will bring set theory to a happier state. *See also* ABSTRACT ENTITIES; DE MORGAN; FUZZY LOGIC; INTUITIONIST LOGIC; LOGIC, SCOPE OF; MEREOLOGY; PEANO; PEIRCE.

SIMPLICITY CRITERION. *See* INDUCTIVE LOGIC; OCKHAM; SCIENTIFIC REASONING.

SINGULAR TERM. A term, like "Bertrand Russell," that stands for a specific person or thing. *See also* GENERAL/SINGULAR TERM.

SOFTWARE FOR LEARNING LOGIC. In 1955, pioneers in artificial intelligence wrote a **computer** program, LOGIC THEORIST, to generate **proofs** in propositional logic using *Principia* axioms. Over the next few decades, practical instructional software would emerge for logic. Today such programs are important in the teaching of logic, since they provide individualized help for students.

LogiCola is an example of an instructional program to help one learn logic; it runs under Windows and can be downloaded for free from http://www.jcu.edu/philosophy/gensler/exercise.htm#S. LogiCola gives exercise sets on various areas of formal and informal logic:

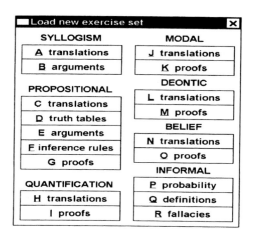

Translation exercises, for example, are provided for **syllogistic, propositional, quantificational, modal, deontic**, and **belief logic**. These can be done in either multiple-choice or type-the-answer format; a typical problem in the latter format looks like this:

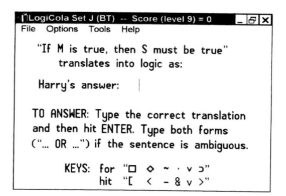

If the student types a wrong answer, the program can often explain why it is wrong and give guidance about the problem. A student gains points for right answers and loses points for wrong answers; when the point total reaches 100 (showing that the student is getting almost all of the problems right), the program records to the disk that the student has successfully completed the exercise. Teachers can use a special scoring program to record scores from the disk.

A typical half-finished proof problem looks like this:

```
┌─────────────────────────────────────────────────────┬──────┐
│ LogiCola Set G (EV) -- Score (level 9) = 54          │_ 🗗 ✕ │
├─────────────────────────────────────────────────────┴──────┤
│ File  Options  Tools  Info  Help                            │
│        1      ~(~(K · J) · ~I)                              │
│        2      (~G ⊃ ~I)                                     │
│      [ .·.  (~G ⊃ J)                                        │
│   *  3      ASM: ~(~G ⊃ J)                                  │
│        4       .·. ~G      {from 3}                         │
│        5       .·. ~J      {from 3}                         │
│                                                             │
│    What's next, Harry?    |                                 │
│                                                             │
│    Type a derived step ("A"), assumption ("ASM: A"),        │
│    or "REFUTE" (to finish an invalid argument).             │
│        KEYS: for  "~ · v ⊃ ≡"  hit  "- & v > 3"             │
└─────────────────────────────────────────────────────────────┘
```

The student keeps typing the next step until the **proof** is completed. Wrong answers will bring a hint on how to proceed, like: "Step 2 is an IF-THEN statement; do you have the first part true or the second part false?" With a second wrong answer on the same step, the student will be told the answer: "TYPE THIS (from 2 and 4): ~I." Since LogiCola randomly generates arguments to prove or disprove, and can generate billions of different ones, different students doing proofs at the same time will get different problems.

Software programs for checking or generating proofs follow similar strategies as would be followed by a human who lacks intuition and needs to have everything spelled out. To cover **arguments** that range from propositional to belief logic, many different strategies have to be put into the program. One such strategy is *modus ponens*; the computer has to examine each step in the proof, identify each step that is an IF-THEN, see if some other line has the if-part, see if the then-part has already been derived, and then, if all goes right, list internally the then-part as something that could be derived. There are complications with modal logic, since here the IF-THEN and the if-part have to be in the same world, and the conclusion has to go into the same world. While the computer goes through many steps to find alternative formulas that could profitably be derived next, it does this very quickly.

One implication of **Church**'s theorem is that constructing proofs and refutations in quantificational logic cannot be reduced to a strict **algorithm**. This has implications for a program like LogiCola, which uses algorithms instead of trial and error. If left to itself, LogiCola would go into an **endless loop** for some invalid quantificational arguments. But LogiCola is told beforehand which arguments would go into an endless loop, so it can stop the proof at a reasonable point, and then give an appropriate message and refutation.

SORITES PARADOX (from the Greek word for "heap"). A **paradox** in which, by using a long string of plausible inferences on a vague predicate, we seem to deduce a falsehood from **truths**.

Let us assume that anyone who is an infant at a given time will be an infant one second later—and that everyone is an infant at birth. We can deduce from these premises the conclusion that everyone is an infant who is 21 years old (662,709,600 seconds old):

> Anyone who is an infant at age n seconds is an
> infant at age n+1 seconds.
> Everyone at age 0 seconds is an infant.
> ∴ Everyone at age 1 second is an infant.
> ∴ Everyone at age 2 seconds is an infant.
> ∴ Everyone at age 3 seconds is an infant.
>   . . .
> ∴ Everyone at age 662,709,600 seconds is an infant.

Since the last conclusion is clearly false, we have to reject some part of the argument. We have various options:

- Our first premise is false; by aging one second an infant can sometimes become a non-infant, even though we cannot say exactly when this occurs.
- The argument assumes that deductive reasoning is transitive— that if A entails B, and B entails C, then A entails C—and that this holds no matter how long the chain of reasoning is. This should be rejected.
- Truth comes in degrees (*see* **fuzzy** and **many-valued logic**). Our

first premise, that anyone who is an infant at a given time will be an infant one second later, is not completely true, but only 99.99+ percent true; by using it repeatedly, we arrive at conclusions that get progressively less true.

- Truth does not come in degrees, but infancy does. Using the "x is an infant" *predicate* is normally a harmless simplification; but here we need the *relation* "x is an infant to degree y." The first premise should be reformulated as "Anyone who is an infant at a given time to a certain degree will be an infant to a slightly lesser degree one second later." Given this, our conclusions will gracefully slide from being an infant at a high degree at 0 seconds to being an infant to no degree at age 21 (or earlier).

The last two proposals face problems of *second-order vagueness*; they seem to assume that precise degrees of truth or infancy can be given without arbitrariness.

SOUND. An **argument** is sound if it is **valid** and has every premise true. A **formal system** is sound if every argument provable in it is valid (*see* **metalogic**).

SQUARE OF OPPOSITION. A chart showing the logical relationships among four statements. The *traditional square of opposition* is about the categorical-statement forms used in **syllogistic logic**:

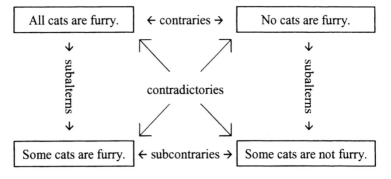

Diagonal pairs (like "All cats are furry" and "Some cats are not furry") are contradictories: it has to be that one is true and the other is false. Contrary pairs ("All cats are furry" and "No cats are furry") cannot both be true, although both can be false. Subcontrary pairs ("Some cats are furry" and "Some cats are not furry") cannot both be false, although both can be true. Subaltern pairs (like "All cats are furry" and "Some cats are furry") have one-way implications; if the first is true then the second must be true, but if the second is true then the first need not be true. This traditional square of opposition, which goes back to **Aristotle**'s *On Interpretation*, assumes that each **general term** ("cat" and "furry" here) refers to at least one entity.

The chart changes if we use an empty term like "unicorn" and then interpret "all," "no," and "some" as modern logic does:

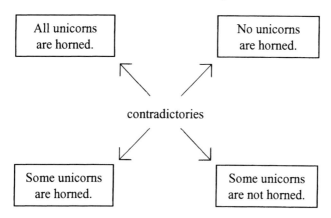

Assume that no unicorns exist. Also assume, with **quantificational logic**, that each "some" statement here asserts the existence of at least one unicorn (and thus is false), that "All unicorns are horned" asserts that anything that is a unicorn is also horned (and thus is vacuously true), and that "No unicorns are horned" just denies that there exists a unicorn that is horned (and thus is true). Then the "all" and "no" statements here are true while the "some" statements are false. Thus the two top statements are no longer contraries, the two bottom ones are no longer subcontraries, and the top/bottom pairs are no longer subcontraries. The diagonal pairs, however, are still contradictories.

STATEMENT. Anything that is **true** or **false**. There is much dispute about what these true-or-false things are. Suppose you point to a green apple and say "This is green." Should we say that what is true is the sentence "This is green," or perhaps the sentence as used on this occasion (where you point to a certain object)? If so, is this sentence concrete physical marks or sounds, or is it a more abstract pattern that has written or auditory instances (*see* **type/token**)? Should we distinguish from these the proposition that this is green, which is what is asserted by various ways to say "This is green" in different languages? If so, then are propositions true-or-false, and not sentences? And are propositions something mental, or are they **abstract entities**, like the meaning of "This is green"?

STATISTICAL SYLLOGISM. *See* INDUCTIVE LOGIC.

STRAW MAN. The **fallacy** of misrepresenting an opponent's views. People who do this are fighting scarecrows of their own invention.

STRICT IMPLICATION (entailment). "A strictly implies B" (or "A entails B") means "It is logically necessary that if A then B." In **modal logic**, this is symbolized as "$\Box(A \supset B)$," or sometimes "$(A \dashv B)$."

SUPPOSITION. A term for "reference" used by **medieval logicians** (*see* **sense/reference**). The medievals would say that in the sentence "Socrates is a man and Socrates has eight letters," the first "Socrates" has *personal supposition* (it refers to the person Socrates) while the second has *material supposition* (it refers to the word "Socrates"). We could more clearly express the sentence as "Socrates is a man and 'Socrates' has eight letters"; quotation marks, which the medievals lacked, make the distinction clearer (*see* **use/mention**). The medievals distinguished many other sorts of supposition, including *simple* (where, depending on the author, we refer either to a concept existing in our mind or to a universal existing outside of our mind) and *distributive* (which is central to the medieval rules for validity in **syllogistic logic**).

SYLLOGISM. The term "syllogism" can be applied broadly, to cover **arguments** of any sort, or narrowly, to cover just categorical syllogisms (*see* **syllogistic logic**).

SYLLOGISTIC LOGIC. A branch of logic that studies arguments using "all," "no," and "some." Syllogistic logic was created by **Aristotle** and was the first branch of logic ever developed. While syllogisms today are often subsumed under **quantificational logic**, they still are sometimes studied by themselves, especially in introductory logic courses.

It is convenient to present syllogistic logic as an artificial language. Let *categorical statements* be sequences of these four forms (where other capital letters can be used in place of "A" and "B"—and capital letters stand for **general terms** like "mortal" or "Greek"):

all A is B	some A is B
no A is B	some A is not B

Let *categorical syllogisms* be sequences of three categorical statements in which each capital letter occurs in exactly two statements. Here are two examples of categorical syllogisms—the first **valid** and the second **invalid**—along with English examples:

no D is F	No determined actions are free.
all H is D	All human actions are determined.
∴ no H is F	∴ No human actions are free.

no D is F	No determined actions are free.
some H is not D	Some human actions are not determined.
∴ some H is F	∴ Some human actions are free.

The premise containing the conclusion's predicate is traditionally put first; but the order of premises does not influence validity.

Aristotle refuted defective forms by giving examples with the same form that clearly had all the premises true and the conclusion false; this example shows that our second form is defective:

no D is F	No dog is a fish.
some H is not D	Some horses are not dogs.
∴ some H is F	∴ Some horses are fish.

And he defended correct forms by arguing that we cannot find examples that refute the form and by deriving the form from ones that are obviously valid (*see* **Barbara, Celarent**).

The **medieval** logician Jean Buridan, building on earlier ideas, presented a convenient set of rules for determining whether a syllogism is valid. This approach uses the notion of *distribution*; a capital letter is *distributed* in a categorical statement if it occurs just after "all" or anywhere after "no" or "not." Here the distributed letters are underlined:

all <u>A</u> is B   some A is B

no <u>A</u> is <u>B</u>   some A is not <u>B</u>

The traditional understanding is that a general term is *distributed* in a statement just if the statement makes some claim about every entity that the general term refers to:

- "All A is B" makes a claim about every A but does not make a claim about every B.
- "No A is B" makes a claim about every A (that it is distinct from this B) and a claim about every B (that it is distinct from this A).
- "Some A is B" makes no claim about every A or every B.
- "Some A is not B" makes no claim about every A, but it does make a claim about every B (that it is distinct from this A).

A syllogism is *valid* just if it satisfies all of these conditions:

1. The middle term must be distributed in at least one premise. (The *middle term* is the term common to both premises; if we violate this rule, we commit the fallacy of the *undistributed middle*.)
2. Every term distributed in the conclusion must be distributed in the premises.
3. If the conclusion is negative, exactly one premise must be negative. (A statement is *negative* if it contains "no" or "not"; otherwise it is positive.)
4. If the conclusion is positive, both premises must be positive.

Our first example above (no D is F, all H is D ∴ no H is F) satisfies all five rules and thus is valid. Our second example (no D is F, some H is not D ∴ some H is F) violates rule 4, and thus is invalid.

When we evaluate the validity of syllogisms, we can take two standpoints. The *Aristotelian view* (which goes back to Aristotle and the medievals) assumes that each general term refers to at least one existing being; the *modern view* allows empty terms like "unicorn" that do not refer to any existing beings. (The **square of opposition** nicely brings out the difference.) Thus some arguments are valid on the Aristotelian view (that is, valid assuming that all the general terms refer to at last one existing being) but invalid on the modern view (that is, invalid if we do not add this assumption). Here is an example:

all U is H	All unicorns are horned animals.
all H is A	All horned animals are animals.
∴ some A is U	∴ Some animals are unicorns.

The premises here are true by definition, and the conclusion (which asserts the existence of an animal that is a unicorn) is false. Yet this argument comes out as *valid* according to the traditional rules just given. And indeed this argument would be valid if we assumed, as an implicit further premise, that there exists at least one unicorn; and that is just what the traditional rules assume.

The rules given above take the Aristotelian view and thus assume that the syllogism does not use empty terms. We get the modern view if we reformulate the first two rules as follows:

1m. The middle term must be distributed in *exactly* one premise.
2m. Each term in the conclusion is distributed in the conclusion if and only if it is distributed in the premises.

The *star test* for syllogisms gives the same result as the traditional medieval rules but is easier to use:

> Star the distributed letters in the premises and undistributed letters in the conclusion. Then the syllogism is VALID if and only if every capital letter is starred exactly once and there is exactly one star on the right-hand side.

Here is how the star test would apply to our two examples:

no D* is F*	VALID: After we star, each capital
all H* is D	letter is starred exactly once and there is
∴ no H is F	exactly one star on the right-hand side.

no D* is F*	INVALID for multiple reasons: "D" and
some H is not D*	"F" are starred twice and there are three
∴ some H* is F*	stars on the right-hand side.

This formulation of the star test assumes the modern interpretation; for the Aristotelian interpretation, substitute "every capital letter is starred *at least once*" for "every capital letter is starred *exactly once*."

Traditional logic tends to see singular statements like "Socrates is human" as having the "all A is B" form ("All Socrates is human"). Another approach is to expand syllogistic logic by using small letters for **singular terms** (like "s" for "Socrates") and these four additional statement forms (where any capital letter can substitute for "A" and any small letters for "x" and "y"):

x is A   x is y

x is not A   x is not y

We could then define "syllogism" in a broader way to allow for these new forms and for any number of premises: a *syllogism* is a sequence of one or more statements (from our eight forms) in which each letter occurs twice and the letters "form a chain" (each statement has at least one letter in common with the statement just below it, if there is one, and the first statement has at least one letter in common with the last statement). The star test would still work for these expanded syllogisms, as in this VALID argument:

all H* is M   All humans are mortal.
all G* is H   All Greeks are humans.
s is G   Socrates is Greek.
∴ s* is M*   ∴ Socrates is mortal.

Venn diagrams give another way to test syllogisms. *See also* BOOLE; IDIOMS; ŁUKASIEWICZ; RENAISSANCE-TO-NINETEENTH-CENTURY LOGIC.

SYMBOLIC LOGIC. A term for contemporary systems of logic, like **propositional** and **quantificational logic**, that represent ideas like "not," "and," and "if-then" by special symbols like "~," "•," and "⊃." **Traditional logic** did not use such symbols.

SYNTAX. *See* SEMANTICS/SYNTAX.

SYNTHETIC STATEMENT. *See* ANALYTIC/SYNTHETIC.

– T –

TARSKI, ALFRED (1901–1983). An influential Polish logician. Tarski was of Jewish descent, although a convert to Catholicism; he moved to the United States to avoid Nazi persecution and he continued his career at the University of California at Berkeley.

Tarski was a prolific writer and contributed much to logic, **mathematics**, and **semantics**. He is most remembered for his clear-headed ideas about **truth**, and especially for *Tarski's convention T*, which he proposed as an adequacy condition that any definition of truth must satisfy; here is an example of convention T:

> The sentence "Snow is white" is *true*,
> if and only if snow *is* white.

This equivalence raises problems for many definitions of "true" that try to water down the notion's objectivity. Suppose someone proposes that "true" just means "accepted in our culture." This definition leads to an absurdity, since we can easily imagine a situation in which snow is white and yet "Snow is white" is not accepted in our culture (perhaps because we live far from snow and have been deceived about its color); in this situation, on the proposed definition, snow would *be* white but "Snow is white" would not be *true*—which is absurd.

TAUTOLOGY. An **analytic truth**, especially one that repeats some of the same information in different words, like "All bachelors (single men) are single." A *truth-table tautology* is a statement, like "If A and B, then A," whose **truth table** is true in every case.

TEMPORAL LOGIC (tense or chronological logic). The logic of time. Logicians tend to slur over tense problems; in "All men are mortal; Socrates is a man; therefore Socrates is mortal," for example, should we not say "Socrates *was* a man" and rephrase "All men *are* mortal" to cover past men too? Contemporary work in temporal logic started with the New Zealand logician Arthur Prior in the 1950s.

One approach uses operators much like **modal logic**'s "◇" and "□":

FA = A will be true at some future time.
GA = A is going to be true at all future times. = ~F~A

PA   =   A was true at some past time.
HA   =   A has been true at all past times.   =   ~P~A

Another approach uses **quantificational logic** and time variables (here "At" is for "A is true at time t" and "n" is for "now"):

(∃t)(t>n · At)	=	For some time t, t is greater
= FA		than now and A is true at t.
(t)(t>n ⊃ At)	=	For every time t, if t is greater
= GA		than now then A is true at t.
(∃t)(t<n · At)	=	For some time t, t is less than
= PA		now and A is true at t.
(t)(t<n ⊃ At)	=	For every time t, if t is less
= HA		than now then A is true at t.

The modal approach resembles what John Ellis McTaggart (1866–1925) called the *A-series* view of time, which is based on the past-present-future distinction. The quantificational approach is closer to his *B-series* view of time, which is based on the before-after relationship.

Temporal logic was discussed in ancient times (*see* **ancient logic since Aristotle**). Many thinkers analyzed "possible" as "true at some time," and "necessary" as "true at all times." But Diodorus Cronus saw "possible" as "true now or at some future time" and "necessary" as "true now and at all future times." His *master argument* claimed that fatalism must be true, because all **truths** about the past are unalterable, and thus necessary. Prior analyzed his argument roughly as follows:

If A is not true and never will be true, then it was true in the past that A never will be true.

If it was true in the past that A never will be true, then it is necessary in itself that it was true in the past that A never will be true.

Necessarily, if it was true in the past that A never will be true, then A is false.

∴ If A is not true and never will be true, then A is impossible.

((~A · ~FA) ⊃ P~FA)
(P~FA ⊃ □P~FA)
□(P~FA ⊃ ~A)
∴ ((~A · ~FA) ⊃ ~◇A)

(This is valid.)

Prior, who wanted to preserve free choice, rejected the first premise; he thought future contingent events can have an indeterminate truth value, being neither true nor false (*see* **many-valued logic**). Prior saw this solution as going back to **Aristotle** and his sea-battle example. But it is equally possible to reject the second premise, which is based on that idea that all truths about the past are unalterable, and thus necessary (presumably in some sense weaker than "logically necessary"); perhaps some truths that are verbally about the past, like "It was true yesterday that this would happen next week," *are* alterable, in the sense that their **truth value** can depend on what we freely choose to do.

THEOREM. A formula **provable** in a **formal system** without assuming additional premises. For example, "(P ∨ ~P)" and "((x)(Fx ⊃ Gx) ⊃ (x)(~Gx ⊃ ~Fx))" are theorems of **classical symbolic logic**.

TRADITIONAL LOGIC. The **Aristotelian** approach to logic that dominated from ancient times to the early 20th century. Traditional logic focused on **syllogistic logic** but included also some **propositional logic** (like *modus ponens*, *modus tollens*, and **disjunctive** syllogism), **informal fallacies**, and **definitions**. *See also* ANCIENT LOGIC SINCE ARISTOTLE; MEDIEVAL LOGIC; RENAISSANCE-TO-NINE-TEENTH-CENTURY LOGIC; SYMBOLIC LOGIC.

TRANSCENDENTAL ARGUMENT. One that says that something must be the case, because it is a necessary condition for the possibility of something else being the case—it being assumed that the "something else" really is the case. For example, **Immanuel Kant** argued that our phenomenal world must include objects and properties—since otherwise conceptualized experience would be impossible. We can construe this reasoning as a *modus tollens* argument:

> If our world does not contain objects and properties, then we have no experience that can be put into concepts.
> We do have experience that we can put into concepts.
> ∴ Our world contains objects and properties.

TRUTH. There are at least six major philosophical questions about truth.

1. What does it mean to call something *true*? To be "true" is perhaps to correspond to the facts, or to cohere with other statements, or to be useful to believe, or to be verified, or to be what we would agree to

under cognitively ideal conditions; or perhaps "It is true that A" is just a verbose way to assert A. The "useful to believe" and "verified" analyses demand giving up the **law of excluded middle**, since it can happen that neither a statement nor its negation is useful to belief, or is verified.

2. What kinds of entity can be true? If we call these entities "**statements**," we can further ask whether these are physical (sounds or marks), or mental, or **abstract entities**.

3. How many **truth values** are there? Are there truth values besides true and false? (*See* **bivalence** and **many-valued logic**.)

4. Can statements be neither true nor false (truth-value *gaps*)? (*See* **law of excluded middle**.)

5. Can statements be both true and false (truth-value *gluts*)? (*See* **dialethism** and **law of non-contradiction**.)

6. How can we deal with the **liar paradox** about truth?

See also PEIRCE; SEMANTICS; TARSKI; TRUTH TABLES.

TRUTH FUNCTIONS. A way to compound statements is *truth functional* if the truth or falsity of the whole depends on the truth or falsity of the parts, and not on their meaning. "A and B" is truth functional; it is true if and only if both parts are true, regardless of what the parts mean. In contrast, "A because B" is not truth functional, since its truth depends on what the parts mean and not just on their truth value.

We can show that every truth function is expressible by a **propositional logic** formula. Consider that each truth function is expressible by a **truth table**, where the left side has all possible truth combinations of the letters and the right side shows whether the statement is true or false on each. Here is one possible truth function with three statements (since each of the 8 letters on the right column could be either T or F, there are 2^8, or 256, truth functions with three statements):

A B C	??
F F F	T
F F T	T
F T F	F
F T T	T
T F F	F
T F T	T
T T F	F
T T T	T

Here "??" stands for some formula that would give the truth-table column on the right. To find such a formula, write a **disjunction** of **conjunctions**; use a disjunct for every "T" in the right column, and make it a conjunction that corresponds to the combination of truth values on the left. In the case above, we would write disjuncts for lines 1, 2, 4, 6, and 8 (here parentheses are omitted to promote readability):

$$(\sim A \cdot \sim B \cdot \sim C) \vee (\sim A \cdot \sim B \cdot C) \vee (\sim A \cdot B \cdot C)$$
$$\vee (A \cdot \sim B \cdot C) \vee (A \cdot B \cdot C)$$

This formula would give the truth-table column on the right, as would the simpler formula "$((A \vee B) \supset C)$." So, given a truth function, it is easy to come up with a formula using "$\sim$," "$\cdot$," and "$\vee$" that expresses that truth function. (If the truth table has "F" in every place in the right column, use a formula like "$(A \cdot \sim A)$," which is always false.)

So we can express all truth functions using just the three connectives "$\sim$," "$\cdot$," and "$\vee$." Actually, we could do this using just "$\sim$" and "$\cdot$"— since "$\vee$" can be defined using "$\sim$" and "$\cdot$":

$$(P \vee Q) \quad = \quad \sim(\sim P \cdot \sim Q)$$
At least one is true. = Not both are false.

We could symbolize the same arguments if we had just "$\sim$" and "$\cdot$"; instead of writing "$(P \vee Q)$," we could then write "$\sim(\sim P \cdot \sim Q)$." So having "$\vee$" is just a matter of convenience. Similarly, we could define "$\supset$" and "$\equiv$" using "$\sim$" and "$\cdot$":

$$(P \supset Q) \quad = \quad \sim(P \cdot \sim Q)$$
If P then Q = We do not have P true and Q false.

$$(P \equiv Q) \quad = \quad (\sim(P \cdot \sim Q) \cdot \sim(Q \cdot \sim P))$$
P if and only if Q = We do not have P true and Q false, and
we do not have Q true and P false.

Or we might define the other connectives using "$\sim$" and "$\vee$":

$$(P \cdot Q) \quad = \quad \sim(\sim P \vee \sim Q)$$
$$(P \supset Q) \quad = \quad (\sim P \vee Q)$$
$$(P \equiv Q) \quad = \quad (\sim(P \vee Q) \vee \sim(\sim P \vee \sim Q))$$

Or we might use just "~" and "⊃," in light of these equivalences:

$$(P \cdot Q) \quad = \quad \sim(P \supset \sim Q)$$
$$(P \vee Q) \quad = \quad (\sim P \supset Q)$$
$$(P \equiv Q) \quad = \quad \sim((P \supset Q) \supset \sim(Q \supset P))$$

Systems with only two symbols are more elegantly simple (*see* **aesthetics**) but harder to use. But logicians are sometimes more interested in proving results about a system than in using it to test arguments; and it may be easier to prove these results if we use fewer symbols.

Another approach uses all five symbols but divides them into undefined (primitive) symbols and defined ones. We could take "~" and either "·" or "∨" or "⊃" as undefined, and define the others using these. We could then view the defined symbols as abbreviations; whenever we liked, we could eliminate them and use only undefined symbols.

It is possible to get along with just one connective. Henry Sheffer in 1910 proposed using the *Sheffer stroke* "(P | Q)" for that purpose. "(P | Q)" means "not both A and B" (sometimes put as "A nand B") and is true if and only if A or B or both are false. We can define "~A" as "(A | A)" and "(A · B)" as "((A | B) | (A | B))"; while these definitions are not very intuitive, the defined equivalents have the same truth table as the original formulas. We could also get along with just "(P ↓ Q)," which means "not either A or B" (sometimes put as "(P nor Q)") and is true if and only if A and B are both false.

TRUTH TABLE. Chart showing how the truth or falsity of a compound statement depends on the truth or falsity of the parts. Normally the parts are assumed to be true or false, but not both, and the larger statement is compounded from these by **truth-functional** connectives of **propositional logic** ("not," "and," "or," "if-then," and "if and only if").

The truth tables for **conjunction** "(P · Q)" ("P and Q") and **disjunction** "(P ∨ Q)" ("P or Q") look like this (where T = true, F = false):

P Q	(P · Q)	P Q	(P ∨ Q)
F F	F	F F	F
F T	F	F T	T
T F	F	T F	T
T T	T	T T	T

The left side of each table gives all possible truth combinations for "P" and "Q": maybe both are false, or just the second is true, or just the first is true, or both are true. "(P · Q)" is false in the first three cases and true in the third; AND claims that both parts are true. Similarly, the OR "(P ∨ Q)" is true if at least one part is true.

Here are the truth tables for **conditional** "(P ⊃ Q)" ("if P then Q") and **biconditional** "(P ≡ Q)" ("P if and only if Q"):

P Q	(P ⊃ Q)		P Q	(P ≡ Q)
F F	T		F F	T
F T	T		F T	F
T F	F		T F	F
T T	T		T T	T

"(P ⊃ Q)" is true unless we have it that A-is-true-and-B-is-false; so an IF-THEN just claims that we do not have the first part true and second part false. The IF-AND-ONLY-IF "(P ≡ Q)" claims that both parts have the same **truth value**: both are true or both are false.

Finally, here is the truth table for **negation** "~P" ("not-P"):

P	~P
F	T
T	F

"~P" has the opposite value of "P." If "P" is false then "~P" is true, and if "P" is true then "~P" is false.

Using these tables, we can calculate the truth or falsity of a complex formula if we know the truth or falsity of its parts. Here is an example:

Suppose P=T, Q=F, and R=F.
Then is "((P ⊃ Q) ≡ ~R)" true or false?

To figure this out, we write "T" for "P," "F" for "Q," and "F" for "R"; then we simplify from the inside out until we get "T" or "F":

((P ⊃ Q) ≡ ~R)	←	original formula
((T ⊃ F) ≡ ~F)	←	substitute "T" and "F" for the letters
(F ≡ T)	←	put "F" for "(T ⊃ F)," and "T" for "~F"
F	←	put "F" for "(F ≡ T)"

So the formula is false. Simplify parts inside parentheses first; so if you have a formula of the form "~(. . .)," first work out the part inside parentheses to get T or F; then apply "~" to the result.

Here is a truth table for a longer formula:

P Q R	$((P \lor Q) \supset R)$
F F F	T
F F T	T
F T F	F
F T T	T
T F F	F
T F T	T
T T F	F
T T T	T

On the left we have all possible truth combinations for the letters. With n letters we get 2^n combinations; so with 3 letters we get 2^3 (8). On the right, we work out the truth or falsity of the formula for each case. In the first case, for example, all three letters are false—so the formula would be $((F \lor F) \supset F)$, which simplifies to $(F \supset F)$ and then T.

Propositional formulas divide up into **tautologies, self-contradictions**, and **contingencies**—depending on whether the truth table for the formula comes out always true, always false, or sometimes true and sometimes false. "$((P \lor Q) \supset R)$" is a *contingency*, since it comes out sometimes true and sometimes false. By contrast, "$(P \lor {\sim}P)$" (that P is either true or not true) is a *tautology*, since it comes out as always true, and "$(P \cdot {\sim}P)$" (that P is both true and not true) is a *self-contradiction*, since it comes out as always false:

P	$(P \lor {\sim}P)$		P	$(P \cdot {\sim}P)$
F	T		F	F
T	T		T	F

"$(P \lor {\sim}P)$" is the **law of excluded middle**; the denial of "$(P \cdot {\sim}P)$" is the **law of non-contradiction**.

We can use truth tables to test **arguments**, such as this one:

If you are a dog, then you are an animal.	(D ⊃ A)
You are not a dog.	~D
∴ You are not an animal.	∴ ~A

To test this, we can construct a truth table showing the truth or falsity of the premises and conclusion for all possible cases; the argument is valid if and only if no possible case has the premises all true and conclusion false. If we do the table, we find that the argument is invalid:

D A	(D ⊃ A),	~D	∴	~A	
F F	T	T		T	
F T	T	T		F	← Invalid
T F	F	F		T	
T T	T	F		F	

The premises are all true and conclusion false in some possible case—namely where you are not a dog but are an animal (perhaps a cat).

A quicker way to test this argument would be to start by setting each premise to T and the conclusion to F. The argument is valid if and only if no consistent way of assigning T and F to the letters will make this work—so we cannot make the premises all true and conclusion false. Let us apply this to the same argument:

$$(D \supset A) = T \qquad (D^F \supset A^T) = T \qquad \text{Invalid}$$
$$\sim D = T \qquad \sim D^F = T$$
$$\therefore \sim A = F \qquad \therefore \sim A^T = F$$

On the left, we set the premises equal to true and conclusion equal to false, to see if this is possible. Since "~D" is T (second premise), "D" has to be F; so we put a "F" superscript above each "D." Likewise, since "~A" is F (conclusion), "A" has to be T; so we put a "T" superscript above each "A." The first premise comes out as "T," giving us true premises and a false conclusion. Since it is possible to have true premises and a false conclusion, the argument is invalid.

Let us change our example to get a valid argument:

If you are a dog, then you are an animal.	(D ⊃ A)
You are not an animal.	~A
∴ You are not a dog.	∴ ~D

This is valid, since we never get true premises and a false conclusion:

D A	(D ⊃ A),	~A	∴ ~D	
F F	T	T	T	Valid
F T	T	F	T	
T F	F	T	F	
T T	T	F	F	

It tests out as valid also on our quicker method:

$$(D \supset A) = T \qquad (D^T \supset A^F) \neq T \qquad \text{Valid}$$
$$\sim A = T \qquad \sim A^F = T$$
$$\therefore \sim D = F \qquad \therefore \sim D^T = F$$

If we try to make the premises all true and conclusion false, then "A" has to be F and "D" has to be T—which makes the first premise false. So we cannot make the premises all true and conclusion false.

Truth tables were invented in 1922 by **Ludwig Wittgenstein** and, independently, by Emil Post. They are used in other areas besides logic, including electronics and **computers** (*see* **logic gates**). Some truth tables use more than two truth values (*see* **many-valued logic**).

TRUTH TREES. A graphical way to test arguments that decomposes formulas into branching sub-cases. Truth trees give another way, besides **truth tables** and **proofs**, to show the validity of **propositional** arguments. The technique can be extended to other areas of logic.

Truth trees use simplifying rules and branching rules. These four simplifying rules let us simplify a formula into smaller parts:

$$\sim\sim P \rightarrow P$$
$$(P \cdot Q) \rightarrow P, Q$$
$$\sim(P \vee Q) \rightarrow \sim P, \sim Q$$
$$\sim(P \supset Q) \rightarrow P, \sim Q$$

Each form that cannot be simplified is branched into the two sub-cases that would make it true; for example, since "~(P · Q)" is true just if "~P" is true or "~Q" is true, it branches into these two formulas. There are five branching rules:

$\sim(P \cdot Q)$	$(P \vee Q)$	$(P \supset Q)$	$(P \equiv Q)$	$\sim(P \equiv Q)$
$\sim P$ $\sim Q$	P Q	$\sim P$ Q	P $\sim P$ Q $\sim Q$	P Q $\sim Q$ $\sim P$

To test an argument, we write the premises, block off the original conclusion (showing that it is to be ignored in constructing the tree), and add the denial of the conclusion. Then we apply the simplifying and branching rules to each formula, and to each further formula that we get, until every branch either dies (contains a pair of contradictory formulas) or contains only letters or the negation of letters. The argument is valid if and only if every branch dies. Here is an example:

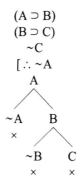

$$(A \supset B)$$
$$(B \supset C)$$
$$\sim C$$
$$[\therefore \sim A$$
$$A$$

Here we write the premises, block off the original "∴ ~A" conclusion (and henceforth ignore it), and add its contradictory "A." Then we branch "(A ⊃ B)" into its two sub-cases: "~A" and "B." The left branch dies, since it contains "A" and "~A"; we indicate this by putting "×" at its bottom. Then we branch "(B ⊃ C)" into its two sub-cases: "~B" and "C." Each branch dies; the left branch has "B" and "~B," while the right has "C" and "~C." Since every branch of the tree dies, no possible truth conditions would make the premises all true and conclusion false, and so the argument is valid.

An argument is invalid if some branch does not die. Then the letters and denial of letters on each live branch give a *refutation* of the argument—truth conditions making the premises true and conclusion false.

TRUTH VALUE. Truth or **falsity**, according to the standard **bivalence** view. **Many-valued logic** adds further alternatives, like maybe half-true, or maybe a continuous scale of truth values from 1.00 (completely true) to 0.00 (completely false). *See also* TRUTH TABLES.

TURING, ALAN (1912–1954). A British logician important for his work with **computers** and **cognitive science.** He extended **Church**'s thesis by proposing that the intuitive idea of an **algorithm** (a finite mechanical procedure) matches what could be done by a simple kind of computer called a *Turing machine.* His work during World War II helped break the code used in German radio transmissions; this helped the Allies gain better intelligence about Nazi activities. Later he worked in computer design; he did much of the work for the first commercial electronic computer, the Ferranti Mark I, which went on sale in 1951 and helped to design the St. Lawrence Seaway. His *Turing test* rephrases "Can computers think?" into "Can computers be developed whose typed answers to questions cannot be distinguished from those of humans?" He predicted that by 2000 some computers would be able to convince most people that they were communicating with a human; but as yet computers cannot do this.

TWENTIETH-CENTURY LOGIC. This divides roughly into two periods. The first period, going up to the 1930s, focused on **classical symbolic logic** and the foundations of **arithmetic.** The second period, since about the 1930s, is more diverse.

    Bertrand Russell in 1902 wrote a letter to **Gottlob Frege** praising his work in logic and the foundations of arithmetic, and showing a contradiction in his axioms. The works of Frege and Russell would create and define classical symbolic logic. Frege's 1879 *Begriffsschrift* ("Concept Writing") had presented the first complete formalization of **propositional** and **quantificational logic**; but Frege's work was largely unknown to the larger world, in part because of its intimidating symbolism. Russell's *Principia Mathematica* (1910–1913), coauthored with Alfred North Whitehead (1861–1947), continued Frege's contention that arithmetic can be reduced to logic; but it avoided the contradictions and used a more intuitive symbolism. *Principia* had a huge influence and became the standard formulation of the new logic.

    The next few decades brought a flurry of activity about classical logic and the foundations of arithmetic. **Ludwig Wittgenstein** and Emil Post independently invented **truth tables.** New forms of **set theory** appeared that avoided the contradictions. Various **metalogical** results were proved. **Formal systems** for propositional and quantificational logic were shown to be consistent, sound, and complete. **Kurt Gödel** showed, in perhaps the most striking result of 20th-century logic, that arithmetic could not be reduced to any formal system. **Alonzo Church** showed that the problem of determining validity in quantificational logic could not be reduced to an **algorithm**. **Alfred Tarski** introduced the **semantic** approach to logic, and Gerhard Gentzen introduced inferential **proof** systems; both were to become increasingly popular.

German logicians were very important in the first part of the 20th century. There was also, between the two world wars, an important Polish school of logicians; while sometimes compared to the Vienna Circle (*see* **logical positivism**), the Polish group was broader in its philosophical commitments (avoiding the "science is everything" mentality), focused more on logic, and carried on a fruitful dialogic with the earlier history of logic. Nazi oppression, and later Marxism, hurt the state of logic in Germany and Poland, as key figures like Rudolph Carnap, Kurt Gödel, **John von Neumann**, and Alfred Tarski fled to the United States. Henceforth, the United States and other English speaking countries would be especially important in the area of logic, even though logic today is pursued on a wide international basis, with people from many nations making important contributions.

Since the 1930s, logicians have diversified their interests from the earlier emphasis on classical logic and the foundations of arithmetic. There has been growing interest in non-classical logics; these include areas like **modal logic**, which builds on but does not contradict classical logic, and **deviant logics**, which conflict with the classical approach.

C. I. Lewis in 1932 started the contemporary interest in modal logic. **Ruth Barcan Marcus** later developed a system of **quantified modal logic**. **Willard Van Orman Quine** argued that modal logic was based on a confusion; he thought that the idea of logical necessity was unclear and that quantified modal logic led to an objectionable **Aristotelian essentialism**, with its **metaphysics** of necessary properties. There was lively debate on modal logic for many years. Then in 1959 **Saul Kripke** presented a **semantics** for modal logic based on **possible worlds**; this seemed to make more sense of modal logic and gave it new respect among logicians. His possible-worlds techniques have proved useful in many other areas and are now a common tool among logicians; **Alvin Plantinga** has used them to give a plausible defense of Aristotelian essentialism. Many other extensions of classical logic have been developed, including **axiological, belief, deontic, epistemic, imperative, infinitary, mereological, question, second-order**, and **temporal logic**. Most logicians today would agree that classical logic needs to be supplemented in order to apply to certain areas or kinds of argument.

There is also much interest in deviant logics, which conflict with classical logic about which arguments are valid. These suggested replacements for parts of classical logic include **free, fuzzy, intuitionist, many-valued, paraconsistent, relative identity**, and **relevance logic**. Some deviant logicians are very radical and question such things as *modus ponens, modus tollens*, and the **law of non-contradiction**; today there is more questioning of basic logical principles than ever before. Some see this as a breath of fresh air; they welcome the fact that logic is becoming, in some circles, as controversial as other areas of philosophy. Others defend orthodox classical logic and see the new trends as dangerous; they wonder what would happen to other areas of philosophy if we could not take for granted that *modus ponens* and *modus tollens* are valid and that contradictions are to be avoided. It will be interesting to see what the future brings on this issue.

There has been much interest in a whole array of other areas, including **philosophy of logic** (where Quine has been especially important), **informal logic** and **fallacies, inductive logic, truth** and **paradoxes**, the history of logic (*see* **logic: history of**), and **ontology**. The bibliography at the end of this book gives a more detailed picture of this range of interests, as does the chronology. As with other areas of study, specialization has increased; so it is not unusual to see whole books now on rather narrow topics.

Logic was important in the development of modern **computers**. The basic insight behind the computer is that **truth functions** can be simulated electrically by **logic gates**; if we hook together a large number of these in the right way and add memory and input-output devices, we get a computer. Logicians like **Alan Turing** and **John von Neumann** were important in the design of the first computers. So, thanks in part to logic, we have moved into the computer age.

Logic today is studied in three main departments: philosophy, mathematics, and computer science. Each of these has a different focus, and the three are becoming increasingly more separate.

TYPE/TOKEN. A *type* is a pattern; a *token* is an instance of a pattern. The word "book" has three letter types but four letter tokens. If we say that English has 26 letters, we mean 26 letter types.

– U –

UNIVERSALIZABILITY PRINCIPLE. *Ethical universalizability* is the principle that whatever is right (wrong, good, bad, etc.) in one case would also be right (wrong, good, bad, etc.) in any exactly or relevantly similar case, regardless of the individuals involved; *see* **formal ethics**. Analogous principles cover other areas; for example, if A logically entails (or caused, or is probable on evidence) B then this is because of certain features and relationships that A and B have, and if any A′ and B′ had these same features and relationships then A′ would logically entail (or would cause, or would be probable on evidence) B′, regardless of the individuals involved.

UNIVERSALS. *See* ABSTRACT ENTITIES; MEDIEVAL LOGIC.

UNIVERSE OF DISCOURSE (domain). Set of entities that variables like "x" or words like "all," "no," and "some" range over in a given context. In English, when we ask "Did anyone see this movie?" we sometimes mean something like "Did anyone from this group see this movie?"—thus momentarily restricting the range of "anyone."

As we symbolize **quantificational** arguments about some one kind of entity, such as persons or statements, we can simplify our formulas by restricting the universe of discourse to that one kind of entity. For example, we can translate "All are Italian" as "(x)Ix"—instead of as "(x)(Px ⊃ Ix)" ("All persons are Italians"); here our "(x)" really means "For all persons x." Consider this argument:

> All in the electoral college who do their jobs are useless.
> All in the electoral college who do not do their jobs are dangerous.
> ∴ All in the electoral college are useless or dangerous.

Here we could use a universe of discourse of persons (as on the left) or of electoral college members (as on the right):

(x)((Ex · Jx) ⊃ Ux)	(x)(Jx ⊃ Ux)
(x)((Ex · ~Jx) ⊃ Dx)	(x)(~Jx ⊃ Ux)
∴ (x)(Ex ⊃ (Ux ∨ Dx))	∴ (x) (Ux ∨ Dx)

In the simpler version on the right, the first premise says, "All who do their jobs are useless"; but it is assumed that "all" here refers only to electoral college members.

UNSOUND ARGUMENT. An **argument** that is not **sound**; equivalently, an argument in which either some premise is false or else the conclusion does not follow from the premises.

USE/MENTION. You *use* a term when you put it in a sentence to talk about objects in the normal way; you *use* "man" when you say "Socrates is a man." You *mention* a term when you put it, usually in quotation marks, in a sentence to talk about the term itself; you *mention* "man" when you say "'Man' has three letters." *See also* SUPPOSITION.

– V –

VALID. Applies primarily to **arguments**, but sometimes also to **statements**. A *valid argument* is one in which it would be self-contradictory to have the premises all true and conclusion false. A *valid statement* is a one whose denial is self-contradictory. Logicians generally avoid using "valid statement" to mean "true statement."

VARIABLE/CONSTANT. A *variable* is a term that can assume different values, while a *constant* is a term with a fixed value. In the mathematical formula "$x = 2$," "x" is a variable while "2" is a constant. In the logical formula "($P \supset Q$)," "P" and "Q" are variables (and can represent different statements), while "$\supset$" is a logical constant.

In **quantificational logic**, a *quantified variable* is a letter that is used in a quantifier or refers back to one, while an *individual constant* is a letter that abbreviates a **singular term** like "Albert." In the formula "($(x)Fx \supset Fa$)" (which means "If everything is F, then a is F"), "x" is a quantified variable, "a" is an individual constant, and "F" abbreviates a predicate (like "French"). In **second-order logic**, we can quantify over properties too; so then "$(x)(\exists F)Fx$" would mean "For every individual x, there is some property F such that x has property F"; both occurrences of "F" in this formula are *quantified variables*.

Some formal systems use *meta-variables*, which range over formulas or other linguistic entities and are often written using Greek letters or a different font. Such meta-variables are often used in presenting an **inference rule** like "($\alpha \supset \beta$), $\alpha \rightarrow \beta$," which is thereby claimed to hold regardless of what **wffs** substitute for "α" and "β."

VENN, JOHN (1834–1923). An English logician, mathematician, and Anglican priest. He introduced *Venn diagrams*, which give an intuitive, graphical way to test categorical **syllogisms**.

Suppose we have a syllogism (like "all B is C; all A is B; therefore all A is C") and want to test whether it is valid. First we draw three overlapping circles, labeling each with one of the syllogism's letters:

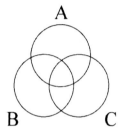

Circle A contains all A things, circle B contains all B things, and circle C contains all C things. Within the circles are seven distinct areas:

- The center is where all three circles overlap; this contains things that have all three features (A, B, and C).
- Further out are areas where two circles overlap; these contain things that have only two features (for example, A and B).
- Furthest out are areas inside only one circle; these contain things that have only one feature (for example, A).

Each of the seven areas can be either empty or non-empty. We shade areas we know to be empty. We put "×" in areas we know to contain at least one entity. An area without either shading or "×" may be either empty or non-empty. After we draw our circles, we diagram the premises following the directions below. The syllogism is valid if and only if drawing the premises automatically draws the conclusion.

We first draw "all" and "no" premises by shading empty areas:

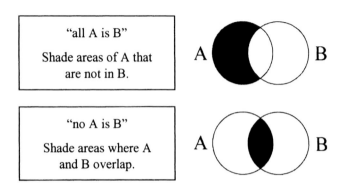

"all A is B"

Shade areas of A that are not in B.

"no A is B"

Shade areas where A and B overlap.

Then we diagram "some" premises by putting "×" (symbolizing an existing thing) in some area that is not already shaded:

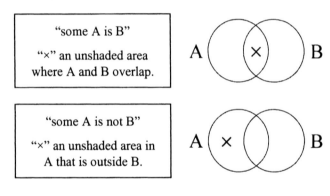

"some A is B"

"×" an unshaded area where A and B overlap.

"some A is not B"

"×" an unshaded area in A that is outside B.

In some cases, consistent with the above directions, we could put "×" in either of two areas. When this happens, the argument will be invalid; to show this, put "×" in an area that *does not* draw the conclusion.

Consider the syllogism on the left and how we diagram it:

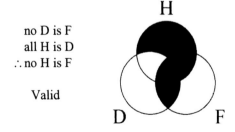

no D is F
all H is D
∴ no H is F

Valid

First we draw three overlapping circles and label each with one of the letters. Then we draw premise "no D is F" by shading where D and F overlap—and premise "all H is D" by shading areas of H that are outside D. Then we have automatically drawn the conclusion "no H is F," since we shaded where H and F overlap. So the argument is valid.

Here is a second example:

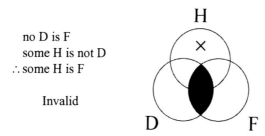

no D is F
some H is not D
∴ some H is F

Invalid

Again, we first draw the circles and label them with letters. Then, following the previous directions, we draw the "no" premise before the "some" premise. So we first draw "no D is F" by shading where D and F overlap. Then, to draw "some H is not D," we need to put "×" in some unshaded area in H that is outside D. Now "×" could go in either of two places: in the part of H that is outside the other circles (as above) or in the part of H that overlaps with just F. When we could put "×" in either of two places, our directions say to put it in some place where it *does not* draw the conclusion; so we do this. The argument is invalid, since we drew the premises without drawing the conclusion.

The test as formulated here permits empty terms like "unicorn" (*see* **square of opposition**). A corresponding test that assumes that there are no empty terms would add the rule that if you end up with a circle with only one non-shaded area, then you must put "×" in this area; this is equivalent to assuming that the circle in question is not entirely empty. *See also* EULER.

VON NEUMANN, JOHN (1903–1957). A versatile Hungarian mathematician and logician who moved to the United States and taught at Princeton. He created game theory, contributed to **set theory**, explored the mathematics and logic of quantum physics, helped in the creation of the atomic bomb, was important in economics, and developed the idea of self-reproducing automata.

Perhaps von Neumann's most important contribution was his work in designing and creating the modern **computer**. He and Arthur Burks were the logicians on the team that built the first large-scale electronic computer, the ENIAC, which was completed in 1946. Unfortunately, the ENIAC had to be rewired physically to change programs. Von Neumann suggested putting programs and data in the same memory space; this *von Neumann architecture* proved successful and is still used in most computers today. While von Neumann is sometimes called the "father of the computer" (a term applied also to several other people), he regarded this as an exaggeration; he saw himself as developing into practical devices basic ideas that came from **Alan Turing**.

– W –

WFF (well-formed formula). Grammatical sentence of a **formal system**. The exact definition of how to form a wff varies with the system in question (e.g., **propositional** or **quantificational logic**).

WITTGENSTEIN, LUDWIG (1889–1951). An eccentric Austrian and English thinker who was arguably the most influential philosopher of the 20th century. His early work had a role in the birth of **logical positivism**, which he scorned; his later work led to *ordinary language philosophy* (*linguistic analysis*), which ruled from about 1940 to 1965.

Wittgenstein's first great work was his 1922 *Tractatus Logico-Philosophicus*, a slim book of 75 pages with numbered theses instead of paragraphs. It was influenced by his teacher and friend at Cambridge, **Bertrand Russell**. The book starts as follows:

1   The world is all that is the case.
1.1   The world is the totality of facts, not of things.
1.11 The world is determined by the facts, and by
      their being all the facts.

Each atomic fact pictures some aspect of the world. If you state all the atomic facts, you have completely described reality. **Language**, when completely analyzed, breaks down into atomic statements that mirror possible atomic facts. Complex statements are formed from atomic ones using **truth-functional** connectives (like "not," "and," and "or"); Wittgenstein invented **truth tables** to show how this works. Anything outside this structure is beyond meaningful speech, and this includes values, the meaning of life, and even his own theory. These cannot be talked about, but they exist and are the most important things. They are the mystical. So Wittgenstein is both a logician and a mystic. He ends the *Tractatus* by saying, "What we cannot speak about we must pass over in silence." Having solved all the problems of philosophy, he left Cambridge and became a village schoolteacher in Austria.

Later he returned to Cambridge, convinced that his earlier work was a big mistake. It came from imposing a framework on reality instead of investigating what reality was like. His new slogan became "Do not think, but look!" Do not say that reality has to be such and such, because that is what your preconceptions say it must be; instead, look and see how it is. He now claimed that many, or perhaps most, concepts did not admit of strict analytic **definitions**. His favorite example was "game"; while games tend to share *family resemblances*, any attempt at a strict definition of "game" is easily refuted by giving examples of games that violate the definition. To understand **meaning**, we have to see how language is used in real-life language games; that is the way to resolve philosophical problems, which come from linguistic confusions. The masterpiece of his later period was the *Philosophical Investigations*, published in 1953 after his death. *See also* LOGICAL ATOMISM; METAPHYSICS; PHILOSOPHY AND LOGIC.

– Z –

ZENO. *See* ANCIENT LOGIC BEFORE ARISTOTLE; ANCIENT LOGIC SINCE ARISTOTLE.

ZERMELO, ERNST. *See* SET THEORY.

Bibliography

Creating a substantial logic bibliography is a humbling experience. There is so much out there—from so many different perspectives—that it would take several lifetimes to read and understand it all. Just listing and organizing the materials is a big task.

I faced two key questions in constructing the bibliography: (1) What is to be included? (2) How is it to be organized?

My general policy was to include materials that reflect general approach of this dictionary, as explained in the preface. So the scope of the bibliography is broad; it includes items on topics like logic's history, its various branches, its controversies, and its relationships to other areas. It emphasizes items that are readable and understandable to beginners over those that are highly technical; but I have included some very technical works that are classics. Except for a couple of important works that have not yet been translated, I included only works available in English.

Logic today is studied in three main departments: philosophy, mathematics, and computer science; if you do a library or Web search for items on "logic," you will find many items from all three areas. Here I have emphasized the *philosophical* branch of logic and the kinds of areas that philosophers tend to be interested in and teach; the *mathematical* and *computer-science* branches tend to be highly specialized and technical.

Even within the philosophical realm, it is not always clear what to include under "logic." This is especially true in the area called "philosophy of logic," which deals with areas like truth, reasoning, meaning, and reference; since most works in philosophy somehow deal with topics like these, philosophy of logic tends to shade off into philosophy-in-general. So what to include at times becomes more arbitrary than I would like. But that is one of the problems in constructing a bibliography.

Another problem is that there is far too much to include everything; so I had to be selective. While some choices were obvious, others were not. Subjective factors come into play here, especially about what readings I happen to have found and what topics I happen to have found interesting. If your writings are not included here, please do not take it as an affront or as a negative evaluation of your work. And bear in mind that many of the works that I include have further bibliographies; most people find readings by combining suggestions from various sources.

A second problem is how to organize the material; the goal here is to provide categories that will help people find things easily. I found it useful to divide the materials into seven main groups (1–7), with subgroups indicated by a letter after the digit (1A, 1B, etc.).

Contents

Earlier sections generally take precedence; so a medieval logic textbook would go under 2E (Medieval Logic) rather than 3 (Textbooks). Websites are included with corresponding written materials—except for interactive websites (e.g., proof checkers), which go under 4C; all websites mentioned in this dictionary were accessed on 14 July 2005, so I do not keep repeating this. Anything that does not fit elsewhere goes under 7 (Other); this big group includes works on other topics or on such a mixture of topics that it would be misleading to put them into one of the previous categories. With a few exceptions, works go into only one section.

It may be helpful to give reading suggestions for some of the sections; I will not try to cover every section here. Group 1 is General Materials. Under 1A (Philosophy Reference Works), I especially recommend the Edwards *Encyclopedia of Philosophy* and the Pappas *Encyclopædia Britannica*; for longer articles, I suggest the Craig *Routledge Encyclopedia of Philosophy* and the Internet *Stanford Encyclopedia of Philosophy*. Under 1C (Logic Journals), I suggest *History and Philosophy of Logic*, since it tends to be broader in scope and less technical than most of the other journals.

Group 2 is History of Logic, which includes anything by or mainly about pre-20th-century logicians (including figures, like Frege, whose work overlaps the 19th and 20th centuries). Under 2A (General History of Logic), the books by the Kneales and by Bocheński are excellent but long and over 40 years old; the Nidditch book is short but equally old. Someone needs to write an up-to-date history of logic of about 200 or 300 pages. The anthologies by Copi, Jager, Manicas, and Runes on the history of logic give a good primary-source introduction to the area. Under 2B (Aristotle), the most important works are Aristotle's *Prior Analytics* and book 4 of his *Metaphysics*. Under 2C (Other Ancient Logic), I suggest looking at one of the works about Stoic logic; unfortunately, most original sources here have been lost. Under 2E (Medieval Logic), I suggest the Kretzmann-Stump anthology and Spade's "Late Medieval Logic" and website. Under 2G (Nineteenth-Century Logic), I suggest Boole's *The Laws of Thought* and Peano's "Set of Axioms for Integers." Under 2H (Frege), I suggest Frege's *Conceptual Notation* (*Begriffsschrift*).

Group 3 is Textbooks. Here Copi is the classic and Hurley is one of the most popular. I suggest also Gensler, Barker (easier), and Boolos-Jeffrey (more technical).

Group 4 is Beyond Text. Under 4B (Logic Software), I suggest Gensler's *LogiCola*, which is free, easy to use, and fits the notation and proof systems used in this dictionary.

Group 5 is Interdisciplinary. The works by Piaget under 5D (Psychology and Other Social Sciences) are especially influential, as are those by Perelman in 5F (Rhetoric, Communications, and Debate). Under 5G (Pre-College Teaching), the fifth-grade logic textbook by Lipman is a real gem.

Group 6 is Individual Figures and Topics. Under 6A (Russell), I suggest browsing the classic *Principia Mathematica* (coauthored with Alfred North Whitehead) and the Russell anthology *Logic and Knowledge* (ed. Marsh). Under 6B (Gödel), I suggest his "On Formally Undecidable Propositions of *Principia Mathematica* and Related Systems I" (skimming over technical parts); for secondary sources, I suggest the Gensler and Nagel-Newman books and the brief article by Boolos. Under 6C (Quine), I suggest his *Philosophy of Logic* and *From a Logical Point of View*. Under 6D (Modal Logic), the standard introduction is the Hughes-Cresswell book; but the Konyndyk book is better for those needing an easier introduction. I suggest also reading the earlier chapters of Plantinga's book and one of the works by Kripke. Under 6E (Ethics and Deontic/Imperative Logic), I suggest von Wright's "Deontic Logic," any of Castañeda's essays, and Gensler's *Formal Ethics*. Under 6F (Logic of Belief and

Knowledge), the classic is Hintikka's *Knowledge and Belief.* Under 6G (Temporal Logic), I suggest Prior's *Time and Modality.* Under 6H (Mathematically Oriented Logic), I suggest the Calinger and Davis anthologies and the Boolos-Jeffrey textbook; the Zuckerman book is a good introduction to set theory. Under 6I (Traditional Logic and Syllogisms), I suggest the Sullivan text, which I used when I was a freshman in college. Under 6J (Deviant Logics and Conditionals), I suggest Priest's *Introduction to Non-Classical Logic*, which also has good suggestions for further reading; I recommend also the Smiley-Priest debate and the Suber Web page. Under 6K (Inductive Logic, Probability, and Science), I recommend Hacking's *Introduction to Probability and Inductive Logic* and Goodman's *Fact, Fiction, and Forecast.* Under 6L (Fallacies, Informal Logic, and Critical Thinking), I recommend Walton's book on *Informal Logic.* Under 6M (Truth and Paradoxes), I recommend the works by Martin, Sainsbury, and Tarski. Under 6N (Ontology and Abstract Entities), I recommend McGinn's book. Under 6R (African Thought and Intercultural Issues), I suggest Wiredu's book.

Group 7 is Other; this includes books that do not fit neatly elsewhere, because they deal with other topics or a wide mixture of topics. Here I suggest the Copi anthology, the Pollock book, Strawson's *Introduction to Logical Theory*, and Wittgenstein's *Tractatus.*

For beginners in logic, a smaller bibliography may be more useful. For these people, I suggest these eight works (the information in brackets tells where they are listed in the bibliography):

- Gensler's *Introduction to Logic* [3]. This is my logic textbook, and it has longer explanations for many areas than what could be included here; it has also many philosophical arguments (which are helpful in showing how logic relates to the rest of philosophy) and exercises (which are essential in mastering techniques like proof construction).
- Gensler's *LogiCola* [4B]. This instructional software is a free download from http://www.jcu.edu/philosophy/gensler/logicola.htm. It covers both formal logic (from syllogisms and classical symbolic logic through modal, deontic, and belief logic—including proofs and translations) and informal logic (informal fallacies, definitions, inductive logic, and probability); it runs under Windows and my students say that it is useful, entertaining, and addictive.
- Nidditch's *Development of Mathematical Logic* [2A]. This gives a brief sketch of the history of logic and its most important results, from Aristotle through the 20th century.
- Quine's *Philosophy of Logic* [6C]. This is a good introduction to this

area from a very influential logician and philosopher.

- McGinn's *Logical Properties* [6N]. This gives an opposite view on philosophy of logic to that of Quine's book; reading them both should give an idea of alternative perspectives in this area.
- Priest's *Introduction to Non-Classical Logic* [6J]. This is a defense of deviant logic by its most eloquent defender, who thinks the standard systems of logic are flawed and need to be revamped. The book is written so that technical parts may be skipped.
- Hacking's *Introduction to Probability and Inductive Logic* [6K]. This a lively introduction to inductive logic and its applications.
- Boolos and Jeffrey's *Computability and Logic* [6H]. This is a technical treatment of areas like Turing machines, uncomputable functions, the Skolem-Löwenheim theorem, and Gödel's theorem. While you have to read this book more slowly than the others, it is clear and interesting and does not presume much mathematics.

1. General Materials

1A. Philosophy Reference Works

Alexander, Dey, ed. *Philosophy in Cyberspace*. 2nd ed. Bowling Green, Ohio: Philosophy Documentation Center, 1998. [See especially "Logic & Philosophy of Science," pp. 85–92.]

Angeles, Peter Adam, ed. *The HarperCollins Dictionary of Philosophy*. 2nd ed. New York: HarperPerennial, 1992.

Audi, Robert, ed. *The Cambridge Dictionary of Philosophy*. 2nd ed. Cambridge: Cambridge Univ. Pr., 1999.

Bertman, Martin A. *Research Guide in Philosophy*. Morristown, N.J.: General Learning Pr., 1974.

Blackburn, Simon, ed. *The Oxford Dictionary of Philosophy*. Oxford: Oxford Univ. Pr., 1994.

Craig, Edward, ed. *Routledge Encyclopedia of Philosophy*. 10 vols. London: Routledge, 1998.

Edwards, Paul, ed. *Encyclopedia of Philosophy*. 8 vols. New York: Macmillan and the Free Press, 1967. Also *Encyclopedia of Philosophy Supplement*, ed. Donald M. Borchert. New York: Macmillan and Simon & Schuster, 1996.

EpistemeLinks.com (public domain books). http://www.epistemelinks.com/Main/MainText.aspx

EServer (public domain philosophy writings). http://eserver.org/philosophy

Flew, Antony, ed. *A Dictionary of Philosophy*. 2nd ed. New York: St. Martin's,

1984.

Free Online Dictionary of Philosophy. http://www.swif.it/foldop

Guerry, Herbert. *A Bibliography of Philosophical Bibliographies.* Westport, Conn.: Greenwood, 1977. [Pp. 258–63 list specialized logic bibliographies.]

Gutenberg Project (public domain books). http://www.gutenberg.org/catalog

Honderich, Ted, ed. *Oxford Companion to Philosophy.* Oxford: Oxford Univ. Pr., 1995.

Iannone, A. Pablo. *Dictionary of World Philosophy.* London: Routledge, 2001.

Internet Encyclopedia of Philosophy. http://www.utm.edu/research/iep

Lacey, Alan Robert. *A Dictionary of Philosophy.* 3rd ed. London: Routledge, 1996.

Lineback, R. H., ed. *The Philosopher's Index.* Bowling Green, Ohio: Philosophy Documentation Center, 1967– .

Mautner, Thomas. *A Dictionary of Philosophy.* Oxford: Blackwell Reference, 1996.

Online Books Page (public domain books). http://digital.library.upenn.edu/books/authors.html

Pappas, Theodore, ed. *Encyclopædia Britannica.* Chicago: Encyclopædia Britannica, 2005.

Reese, William L. *Dictionary of Philosophy and Religion: Eastern and Western Thought.* 2nd ed. Atlantic Highlands, N.J.: Humanities, 1996.

Runes, Dagobert D., ed. *Dictionary of Philosophy.* Rev. ed. New York: Philosophical Library, 1983.

Stanford Encyclopedia of Philosophy. http://plato.stanford.edu/contents.html

Urmson, J. O., and Jonathan Rée. *The Concise Encyclopedia of Western Philosophy and Philosophers.* Rev. ed. London: Unwin Hyman, 1989.

Wikipedia. http://en.wikipedia.org

1B. Logic Reference Works

Barwise, Jon, ed. *Handbook of Mathematical Logic.* Amsterdam: North-Holland, 1977.

Birthdays of Famous Logicians. http://www.volny.cz/logici/vyroci/english.html

Church, Alonzo. *A Bibliography of Symbolic Logic: 1666–1935,* with additional listings. Providence, R.I.: Association for Symbolic Logic, 1984. [This is an updated version of what was to be a complete symbolic logic bibliography for those years, from *The Journal of Symbolic Logic* 1, no. 4 (December 1936): 121–216, with additions in vol. 3, no. 4 (January 1939): 178–212.]

Detlefsen, Michael, David Charles McCarty, and John B. Bacon. *Logic from A to Z.* London: Routledge, 1999. [very brief entries]

EpistemeLinks.com. *Logic and Philosophy of Logic* (general logic links). http://www.epistemelinks.com/Main/Topics.aspx?TopiCode=Logi

Feys, Robert, and Frederic B. Fitch, eds. *Dictionary of Symbols of Mathematical Logic.* Amsterdam: North-Holland, 1969.

Gabbay, D., and Guenthner, F., eds. *Handbook of Philosophical Logic.* 1st ed. 4 vols. Boston and Dordrecht: D. Reidel: 1983. Also 2nd ed., 11 vols., Boston and Dordrecht: Kluwer, 2001. [a multi-volume encyclopedia of logic, covering a wide range of topics, with long articles, some over 100 pages]

Greenstein, Carol Horn. *Dictionary of Logical Terms and Symbols.* New York: Van Nostrand, 1978.

Iridis. *List of Some Famous Logicians.* http://www.iridis.com/List_of_logicians

Janz, Bruce. *The Reasoning Page* (general logic links). http://pegasus.cc.ucf.edu/~janzb/reasoning

Johnson, Peter. *Logic Links.* http://www.dmat.ufpe.br/~peterj/Logica.html

King, Peter. *Philosophy Sites by Topic: Logic.* http://users.ox.ac.uk/~worc0337/phil_topics.html#logic

Lamarque, Peter V., ed. *Concise Encyclopedia of Philosophy of Language.* Oxford: Elsevier Science, 1997.

Marciszewski, Witold, ed. *Dictionary of Logic as Applied in the Study of Language.* Boston: M. Nijhoff, 1981. [technical]

Quine, Willard Van Orman. *Quiddities: An Intermittently Philosophical Dictionary.* Cambridge, Mass.: Harvard Univ. Pr., 1987. [mostly logic]

Risse, Wilhelm, ed. *Bibliographia Logica.* 4 vols. Hildesheim, Germany: Georg Olms, 1965–1979. [huge]

Suber, Peter. *Logical Systems* (general logic links). http://www.earlham.edu/~peters/courses/logsys/lslinks.htm

Warburton, Nigel. *Thinking from A to Z.* 2nd ed. London: Routledge, 1998. [brief entries on critical thinking vocabulary]

Windelband, Wilhelm, Arnold Ruge, and Henry Jones, eds. *Logic: Volume I of the Encyclopædia of Philosophical Sciences.* Trans. B. Ethel Meyer. London: Macmillan, 1913. [This has seven large articles; while a series of volumes were planned, only the logic one was actually published.]

Wong, Paul. *Logic Links.* http://users.rsise.anu.edu.au/~wongas

1C. Logic Journals

Argumentation. Dordrecht: Kluwer Academic.

Australasian Journal of Logic. Melbourne, Australia: Univ. of Melbourne.

Bulletin of Symbolic Logic. Champaign, Ill.: Association for Symbolic Logic.

History and Philosophy of Logic. London: Taylor & Francis.

Informal Logic. Windsor, Ont.: P. F. Wilkinson.

Journal of Applied Logic. Amsterdam: Elsevier Science.

Journal of Logic, Language and Information. Dordrecht: Kluwer Academic.

Journal of Philosophical Logic. Dordrecht: Kluwer Academic.

Journal of Symbolic Logic. Pasadena, Calif.: Association for Symbolic Logic. [more philosophical earlier, now mostly mathematical]

Logic Journal of the IGPL (Interest Group in Pure and Applied Logics). Oxford:

Oxford Univ. Pr. http://jigpal.oupjournals.org

Logique et Analyse. Brussels, Belgium: Centre National de Recherches de Logique. [some articles in English]

Nordic Journal of Philosophical Logic. London: Taylor and Francis. Early articles at http://www.hf.uio.no/filosofi/njpl/read.html

Notre Dame Journal of Formal Logic. Notre Dame, Ind.: Univ. of Notre Dame. [mix of mathematical and philosophical]

Studia Logica. Warsaw: D. Reidel. [some philosophical]

1D.   Logic Organizations

Association for Informal Logic & Critical Thinking (McMaster Univ., Hamilton, Ont.). http://ailact.mcmaster.ca

Association for Symbolic Logic. http://www.aslonline.org

British Logic Colloquium. http://www.cs.bham.ac.uk/~exr/blc

Center for Critical Thinking. http://www.criticalthinking.org

Centre for Logic, Epistemology and the History of Science (Brazil). http://www.cle.unicamp.br

Center for Philosophy of Logic, Language, Mathematics and Mind (Univ. of St. Andrews, Scotland). http://www.st-andrews.ac.uk/~arche

European Association for Computer Science Logic. http://www.dimi.uniud.it/~eacsl

European Foundation for Logic, Language, and Information. http://www.folli.org

Helsinki Logic Group (Univ. of Helsinki, Finland). http://www.logic.math.helsinki.fi

Institut für Mathematische Logik und Grundlagen der Mathematik. http://logik.mathematik.uni-freiburg.de

Institute for Logic, Language and Computation (Univ. of Amsterdam). http://www.illc.uva.nl

Interest Group in Pure and Applied Logics. http://theory.doc.ic.ac.uk/tfm/igpl.html

International Federation for Computational Logic. http://www.ifcolog.org

International Society for the Study of Argumentation. http://cf.hum.uva.nl/issa

Kurt Gödel Society (Technical Univ. of Vienna). http://kgs.logic.at

Mathematical Logic Group (Univ. of Bonn). http://www.math.uni-bonn.de/people/logic

Ontario Society for the Study of Argumentation. http://venus.uwindsor.ca/ossa

PHILOG (Danish Network for Philosophical Logic and Its Applications). http://www.philog.ruc.dk/phiindex.html

Research Groups in Logic (Univ. of Wales). http://www.cs.swan.ac.uk/~csetzer/logic-server

Swiss Society for Logic and Philosophy of Sciences. http://www.sslps.unibe.ch

Technical Committee on Multiple-Valued Logic. http://wwwj3.comp.eng.himeji-tech.ac.jp/mvl

2. History of Logic

2A. General History of Logic

Bocheński, Joseph M. *A History of Formal Logic.* Ed. and trans. Ivo Thomas. Notre Dame, Ind.: Univ. of Notre Dame Pr., 1961.

Copi, Irving M., and James A. Gould, eds. *Readings on Logic.* New York: Macmillan, 1964.

Dumitriu, Anton. *History of Logic.* 4 vols. Tunbridge Wells, England: Abacus, 1977.

Haaparanta, Leila, ed. *Development of Modern Logic: A Philosophical and Historical Handbook.* Oxford: Oxford Univ. Pr., 2005.

Jager, Ronald, ed. *Essays in Logic from Aristotle to Russell.* Englewood Cliffs, N.J.: Prentice-Hall, 1963.

Kneale, William Calvert, and Martha Kneale. *The Development of Logic.* Oxford: Clarendon, 1962.

Lewis, Clarence Irving. *A Survey of Symbolic Logic.* New York: Dover, 1960.

Manicas, Peter T., ed. *Logic as Philosophy.* New York: Van Nostrand, 1971.

Nidditch, P. H. *The Development of Mathematical Logic.* London: Routledge & Kegan Paul, 1962.

Runes, Dagobert D., ed. *Classics in Logic.* New York: Philosophical Library, 1962.

2B. Aristotle

Aristotle. *The Online Books Page.* http://onlinebooks.library.upenn.edu/webbin/book/search?author=Aristotle [The six books that are mostly on logic form the "Organon": *Categories, On Interpretation, Prior Analytics, Posterior Analytics, Topics,* and *Sophistical Refutations.* Also, book 4 of the *Metaphysics* talks about the law of non-contradiction.]

———. *The Works of Aristotle.* 12 vols. Ed. W. D. Ross. Oxford: Clarendon, 1908–1952.

Barnes, Jonathan. "Aristotle and Stoic Logic." In *Topics in Stoic Philosophy,* ed. Katerina Ierodiakonou, 23–53. Oxford: Clarendon, 1999.

Hass, Marjorie. "Feminist Readings of Aristotelian Logic." In *Feminist Interpretations of Aristotle,* ed. Cynthia A. Freeland, 19–40. Univ. Park: Pennsylvania State Univ. Pr., 1998.

Lear, Jonathan. *Aristotle and Logical Theory.* Cambridge: Cambridge Univ. Pr., 1980.

Łukasiewicz, Jan. *Aristotle's Syllogistic from the Standpoint of Modern Formal Logic.* 2nd ed. Oxford: Clarendon, 1957.

McCall, Storrs. *Aristotle's Modal Syllogisms.* Amsterdam: North-Holland, 1963.

Patterson, Richard. *Aristotle's Modal Logic: Essence and Entailment in the Organon.* Cambridge: Cambridge Univ. Pr., 1995.

Patzig, Günther. *Aristotle's Theory of the Syllogism: A Logicophilological Study of Book A of the Prior Analytics.* Trans. Jonathan Barnes. Dordrecht: D. Reidel, 1969.

2C. Other Ancient Logic

Barnes, Jonathan. "Aristotle and Stoic Logic." In *Topics in Stoic Philosophy*, ed. Katerina Ierodiakonou, 23–53. Oxford: Clarendon, 1999.

Bocheński, Joseph M. *Ancient Formal Logic.* Amsterdam: North-Holland, 1963.

Euclid. *Euclid's Elements: All Thirteen Books Complete in One Volume.* Ed. Dana Densmore. Trans. Thomas L. Heath. Santa Fe, N.M.: Green Lion, 2002. [While not a logician, Euclid systematized geometry and provided geometric proofs that stimulated interested in logic.]

Flannery, Kevin L. *Ways into the Logic of Alexander of Aphrodisias.* Leiden, Netherlands: Brill, 1995. [Alexander was a second century AD Aristotelian.]

Freeman, Kathleen, ed. and trans. *Ancilla to the Pre-Socratic Philosophers.* Cambridge, Mass.: Harvard Univ. Pr., 1956. [Several early figures expressed ideas relating to logic: Heraclitus (especially fragment 49a on p. 28), Parmenides (especially fragments 7–8 on p. 43), Zeno of Elea (p. 47), Melissus of Samos (pp. 48–51), and Gorgias (especially fragment 3 on pp. 128–29).]

Galen. *Galen's Institutio Logica: English Translation, Introduction, and Commentary.* Ed. John Spangler Kieffer. Baltimore, Md.: Johns Hopkins, 1964. [Dating from the second century AD, this is the oldest surviving logic textbook.]

Johnson, David Martel. "The Greek Origins of Belief." *American Philosophical Quarterly* 24, no. 4 (October 1987): 319–27. [Johnson says the ancient Greeks "introduced a new ideal of thought—viz., that of complete mental consistency, necessitating a forced choice among incompatible alternatives" (p. 323).]

Londey, David, and Carmen Johanson, eds. *The Logic of Apuleius: Including a Complete Latin Text and English Translation of the Peri Hermeneias of Apuleius of Madaura.* New York: E. J. Brill, 1987. [This is the oldest surviving logic textbook in Latin.]

Mates, Benson. *Stoic Logic.* Berkeley: Univ. of California, 1953.

Mueller, Ian. "An Introduction to Stoic Logic." In *The Stoics*, ed. John M. Rist, 1–26. Berkeley: Univ. of California, 1978.

Plato. *The Collected Dialogues.* Ed. Edith Hamilton and Huntington Cairns. New York: Pantheon Books, 1961. [Plato's emphasis on reasoning prepared for the first systematization of logic, by his student Aristotle.]

Saunders, Jason L., ed. "Early Stoic Logic." In *Greek and Roman Philosophy after Aristotle*, 60–79. New York: The Free Press, 1966. [This has paragraphs from Diogenes Laertius, Sextus Empiricus, Cicero, and others.]

Sharples, R. W., ed. "[Theophrastus's] Logic." In *Theophrastus of Eresus: Sources for His Life, Writings, Thought, and Influence.* 5 vols., 1:114–275. Leiden, Netherlands: E. J. Brill, 1995.

Speca, Anthony. *Hypothetical Syllogistic and Stoic Logic.* Leiden, Netherlands: Brill, 2001.

Sullivan, Mark W. *Apuleian Logic.* Amsterdam: North-Holland, 1967.

2D.   Buddhist/Indian/Chinese Logic

Bhatt, Siddheswar Rameshwar, and Anu Mehrotra. "The Buddhist Theory of Inference." In *Buddhist Epistemology*, foreword by the Dalai Lama, 49–99. Westport, Conn.: Greenwood, 2000.

Bocheński, Joseph M. "The Indian Variety of Logic." In *A History of Formal Logic*, ed. and trans. Ivo Thomas, 416–47. Notre Dame, Ind.: Univ. of Notre Dame Pr., 1961.

Dreyfus, Georges B. J. *Recognizing Reality: Dharmakīrti's Philosophy and Its Tibetan Interpretation.* Albany, N.Y.: SUNY, 1997. [This covers logical and epistemological problems like universals and realism in the seventh century Buddhist philosopher and logician Dharmakīrti.]

Jackson, Roger R., ed. "The Turn to Epistemology." In *Is Enlightenment Possible? Dharmakīrti and rGyal tshab rje on Knowledge, Rebirth, No-Self and Liberation*, 99–107. Ithaca, N.Y.: Snow Lion, 1993.

Matilal, Bimal Krishna. *The Character of Logic in India.* Ed. Jonardon Ganeri and Heeraman Tiwari. Albany, N.Y.: SUNY, 1998.

———. *Logic, Language and Reality: An Introduction to Indian Philosophical Studies.* Delhi, India: Motilal Banarsidass, 1985.

Matilal, Bimal Krishna, and Robert D. Evans, eds. *Buddhist Logic and Epistemology.* Dordrecht: D. Reidel, 1986.

Shcherbatskoi, Fedor I. *Buddhist Logic.* 2 vols. New York: Dover, 1962.

Shi, Hu. *The Development of the Logical Method in Ancient China.* New York: Paragon Book Reprint Corp., 1963.

Tillemans, Tom J. F. *Scripture, Logic, Language: Essays on Dharmakīrti and His Tibetan Successors.* Boston, Mass.: Wisdom, 1999.

Vidyabhusan, Satischandra Mahāmahopādhyāya. *A History of Indian Logic: Ancient, Medieval and Modern Schools.* Calcutta, India: Calcutta Univ. Pr., 1921.

2E.   Medieval Logic

Al-Tayyib, Ibn. *Arabic Logic: Ibn al-Tayyib's Commentary on Porphyry's Eisagoge.* Trans. Kwame Gyekye. Albany, N.Y.: SUNY, 1979.

Anselm. *The De Grammatico of St. Anselm: The Theory of Paronymy.* Trans. Desmond P. Henry. Notre Dame, Ind.: Univ. of Notre Dame Pr., 1964.

Avicenna (Ibn Sīnā). *Remarks and Admonitions, Part One: Logic.* Trans. Shams Constantine Inati. Toronto, Ont.: Pontifical Institute of Mediaeval Studies,

1984.

Boehner, Philotheus. *Medieval Logic: An Outline of Its Development from 1250 to c. 1400*. Chicago: Univ. of Chicago Pr., 1952.

Boethius, Anicius Manlius Severinus. *Boethius's De Topicis Differentiis*. Trans. Eleonore Stump. Ithaca, N.Y.: Cornell, 1978.

———. *The Consolation of Philosophy*. Trans. Richard Green. Indianapolis, Ind.: Bobbs-Merrill, 1962. [The end of book 5 has important modal distinctions.]

Boh, Ivan. *Epistemic Logic in the Later Middle Ages*. London: Routledge, 1993.

Broadie, Alexander. *Introduction to Medieval Logic*. Oxford: Clarendon, 1993.

Buridan, Jean. *Jean Buridan's Logic: The Treatise on Supposition, the Treatise on Consequences*. Trans. Peter King. Dordrecht: D. Reidel, 1985.

Byrne, Edmund F. *Probability and Opinion: A Study in the Medieval Presuppositions of Post-Medieval Theories of Probability*. The Hague, Netherlands: Martinus Nijhoff, 1968.

Chadwick, Henry. *Boethius: The Consolations of Music, Logic, Theology, and Philosophy*. Oxford: Clarendon, 1981. [Chapter 3, pp. 108–73, is on logic.]

Dürr, Karl. *The Propositional Logic of Boethius*. Amsterdam: North-Holland, 1951.

Grant, Edward. *God and Reason in the Middle Ages*. Cambridge: Cambridge Univ. Pr., 2001. [Chapters 2–4, pp. 31–147, are on logic.]

Henninger, Mark Gerald. *Relations: Medieval Theories, 1250–1325*. Oxford: Clarendon, 1989.

Knuuttila, Simo. *Modalities in Medieval Philosophy*. London: Routledge, 1993.

Kretzmann, Norman, and Eleonore Stump, eds. *Logic and the Philosophy of Language*. Vol. 1 of *The Cambridge Translations of Medieval Philosophical Texts*. Cambridge: Cambridge Univ. Pr., 1988.

Lagerlund, Henrik. *Modal Syllogistics in the Middle Ages*. Boston: Brill, 2000.

Leff, Gordon. *William of Ockham*. Manchester, England: Manchester Univ. Pr., 1979. [Chapters 2–4, pp. 78–317, are on logic.]

Moody, Ernest Addison. *The Logic of William of Ockham*. London: Sheed & Ward, 1935.

———. *Studies in Medieval Philosophy, Science, and Logic*. Berkeley: Univ. of California, 1975.

———. *Truth and Consequence in Medieval Logic*. Amsterdam: North-Holland, 1953.

Ockham, William, of. *Ockham's Theory of Propositions: Part II of the Summa Logicae*. Trans. Alfred J. Freddoso and Henry Schuurman. Notre Dame, Ind.: Univ. of Notre Dame Pr., 1980.

———. *Ockham's Theory of Terms: Part I of the Summa Logicae*. Trans. Michael J. Loux. Notre Dame, Ind.: Univ. of Notre Dame Pr., 1974.

———. *Philosophical Writings*. Trans. Philotheus Boehner. Indianapolis, Ind.: Bobbs-Merrill, 1964. [Pp. 49–97 are on logic.]

Peter of Spain (Pope John XXI). *The Summulae Logicales of Peter of Spain*. Trans.

Joseph P. Mullally. Notre Dame, Ind.: Univ. of Notre Dame Pr., 1945. [This was a very popular logic textbook for several centuries.]

———. *Tractatus Syncategorematum and Selected Anonymous Treatises*. Trans. Joseph P. Mullally. Milwaukee, Wis.: Marquette Univ. Pr., 1964.

Rescher, Nicholas. *The Development of Arabic Logic*. Pittsburgh, Pa.: Univ. of Pittsburgh Pr., 1964.

———. *Studies in the History of Arabic Logic*. Pittsburgh, Pa.: Univ. of Pittsburgh Pr., 1963.

Sherwood, William of. *William of Sherwood's Introduction to Logic*. Trans. Norman Kretzmann. Minneapolis: Univ. of Minnesota Pr., 1966.

———. *William of Sherwood's Treatise on Syncategorematic Words*. Trans. Norman Kretzmann. Minneapolis: Univ. of Minnesota Pr., 1966.

Spade, Paul Vincent. "Late Medieval Logic." In *Medieval Philosophy*, ed. John Marenbon, 402–25. London: Routledge, 1998.

———. *Medieval Logic and Philosophy*. http://pvspade.com/Logic

———, ed. and trans. *Five Texts on the Mediaeval Problem of Universals: Porphyry, Boethius, Abelard, Duns Scotus, Ockham*. Indianapolis, Ind.: Hackett, 1994.

Stump, Eleonore. *Dialectic and Its Place in the Development of Medieval Logic*. Ithaca, N.Y.: Cornell, 1989.

Taymiyya, Ibn. *Ibn Taymiyya Against the Greek Logicians*. Trans. Wael B. Hallaq. Oxford: Clarendon, 1993.

Tournay, Stephen Chad. *Ockham: Studies and Selections*. La Salle, Ill.: Open Court, 1938. [Pp. 1–28 and 91–118 are on logic.]

Trundle, Robert C. *Medieval Modal Logic and Science*. Lanham, Md.: Univ. Pr. of America, 1999.

2F. Renaissance and Enlightenment Logic

Arnauld, Antoine. *The Art of Thinking: Port-Royal Logic*. Trans. James Dickoff and Patricia James. Indianapolis, Ind.: Bobbs-Merrill, 1964.

Ashworth, E. J. *Language and Logic in the Post-Medieval Period*. Dordrecht: D. Reidel, 1974.

Bacon, Francis. *Essays, Advancement of Learning, New Atlantis and Other Pieces*. New York: Odyssey, 1937. [Pp. 266–348 have Bacon's "New Organon."]

Broadie, Alexander. *George Lokert, Late-Scholastic Logician*, Edinburgh, Scotland: Edinburgh Univ. Pr., 1983.

Gassendi, Pierre. *Pierre Gassendi's Institutio Logica (1658): A Critical Edition with Translation and Introduction*. Trans. Howard Jones. Assen, Netherlands: Van Gorcum, 1981.

Gensler, Harry. "Logic and the First Critique." *Kant-Studien* 76, no. 3 (1985): 276–87.

Holzhey, Helmut, and Vilem Mudroch. *Historical Dictionary of Kant and*

Kantianism. Lanham, Md.: Scarecrow Pr., 2005.

Hume, David. *Treatise of Human Nature.* Ed. L. A. Selby-Bigge. Oxford: Clarendon, 1888. [Hume raised problems about inductive reasoning and about deriving "ought" from "is." He talked also about the origin of concepts in experience and the two types of knowledge (matters of fact and relations of ideas).]

John of St. Thomas. *Outlines of Formal Logic.* Trans. F. C. Wade. Milwaukee, Wis.: Marquette Univ. Pr., 1955.

Kant, Immanuel. *Critique of Pure Reason.* Trans. Norman Kemp Smith. New York: St. Martin's, 1929. [Kant saw logic as a closed system, created and perfected by Aristotle, and the key to the categories of the understanding.]

————. *Lectures on Logic.* Ed. and trans. J. Michael Young. Cambridge: Cambridge Univ. Pr., 1992.

————. *Logic.* Trans. Robert S. Hartman and Wolfgang Schwarz. Indianapolis, Ind.: Bobbs-Merrill, 1974.

Leibniz, Gottfried Wilhelm. *Logical Papers.* Ed. and trans. G. H. R. Parkinson. Oxford: Clarendon, 1966.

Martin, Gottfried. *Leibniz: Logic and Metaphysics.* Trans. K. J. Northcott and P. G. Lucas. Manchester: Manchester Univ. Pr., 1964.

Ong, Walter J. *Ramus: Method, and the Decay of Dialogue: From the Art of Discourse to the Art of Reason.* Cambridge, Mass.: Harvard Univ. Pr., 1958.

2G.   Nineteenth-Century Logic

This includes figures whose writings overlap the nineteenth and twentieth centuries, like Dewey, Hilbert, Peano, Peirce, and Royce. Frege gets his own section (2H).

Bolzano, Bernard. *Paradoxes of the Infinite.* Trans. Fr. Prihonský. London: Routledge & Kegan Paul, 1950.

————. *Wissenschaftslehre: Theory of Science, Attempt at a Detailed and in the Main Novel Exposition of Logic with Constant Attention to Earlier Authors.* Ed. and trans. Rolf George. Berkeley: Univ. of California, 1972.

Boole, George. *The Laws of Thought (1854).* London: Open Court, 1916.

————. *The Mathematical Analysis of Logic: Being an Essay towards a Calculus of Deductive Reasoning.* New York: Philosophical Library, 1948.

————. *Selected Manuscripts on Logic and its Philosophy.* Ed. Ivor Grattan-Guinness and Gérard Barnet. Basil, Switzerland: Birkhäuser Verlag, 1997.

————. *Studies in Logic and Probability.* London: Watts, 1952.

Burbidge, John W. "Hegel's Conception of Logic." In *The Cambridge Companion to Hegel,* ed. Frederick C. Beiser, 86–101. Cambridge: Cambridge Univ. Pr., 1993.

————. *On Hegel's Logic: Fragments of a Commentary.* Atlantic Highlands, N.J.: Humanities, 1981.

————. *Real Process: How Logic and Chemistry Combine in Hegel's Philosophy of Nature*. Toronto: Univ. of Toronto Pr., 1996.

Burke, Tom. *Dewey's New Logic: A Reply to Russell*. Chicago: Univ. of Chicago Pr., 1994.

Cantor, Georg, *Contributions to the Founding of the Theory of Transfinite Numbers*. Trans. Philip E. B. Jourdain. La Salle, Ill.: Open Court, 1915.

Carroll, Lewis. *The Philosopher's Alice: Alice's Adventures in Wonderland & Through the Looking-Glass*, illustrations by John Tenniel and notes by Peter Heath. London: Academy Editions, 1974.

———. *Symbolic Logic*. Ed. William Warren Bartley III. New York: C. N. Potter, 1977.

———. "What the Tortoise Said to Achilles." *Mind* 4, no. 14 (April 1895): 278–80.

De Morgan, Augustus. *Formal Logic; or, The Calculus of Inference, Necessary and Probable*. London: Taylor and Walton, 1847.

———. *On the Syllogism, and Other Logical Writings*. Ed. Peter Heath. London: Routledge & Kegan Paul, 1966.

Detlefsen, Michael. *Hilbert's Program: An Essay on Mathematical Instrumentalism*. Dordrecht: D. Reidel, 1986.

Dewey, John. *Essays in Experimental Logic*. New York: Dover, 1953.

———. *Logic: The Theory of Inquiry*. New York: Holt, Rinehart, and Winston, 1960.

Fisher, John, ed. *The Magic of Lewis Carroll*. London: Nelson, 1973.

Harris, Errol E. *An Interpretation of the Logic of Hegel*. Lanham, Md.: Univ. Pr. of America, 1983.

Hegel, Georg Wilhelm Friedrich. *Hegel's Logic: Being Part One of the Encyclopaedia of the Philosophical Sciences (1830)*. Trans. William Wallace. Oxford: Clarendon, 1975.

———. *Hegel's Science of Logic*. Trans. A. V. Miller. London: George Allen & Unwin, 1969.

———. *The Jena System, 1804–1805: Logic and Metaphysics*. Ed. and trans. John W. Burbidge and George di Giovanni. Kingston, Ont.: McGill-Queen's Univ. Pr., 1986.

Hilbert, David. *David Hilbert's Lectures on the Foundations of Mathematics and Physics, 1891–1933*. Ed. William Ewald. New York: Springer, 2004.

Hilbert, David, and W. Ackermann. *Principles of Mathematical Logic*. Ed. Robert E. Luce. Trans. Lewis M. Hammond, George G. Leckie, and F. Steinhardt. New York: Chelsea, 1950.

Husserl, Edmund. *Formal and Transcendental Logic*. Trans. Dorion Cairns. The Hague, Netherlands: Martinus Nijhoff, 1969.

———. *Introduction to the Logical Investigations: A Draft of a Preface to the Logical Investigations (1913)*. Ed. Eugen Fink. Trans. Philip J. Bossert and Curtis H. Peters. The Hague, Netherlands: Martinus Nijhoff, 1975. [Husserl here criticizes his earlier psychologism about logic.]

———. *Logical Investigations*. Trans. J. N. Findlay. Ed. Dermot Moran, preface by Michael Dummett. London: Routledge, 2001. [This covers both volumes of the second German edition.]

———. *Philosophy of Arithmetic: Psychological and Logical Investigations: with Supplementary Texts from 1887–1901*. Trans. Dallas Willard. Dordrecht: Kluwer Academic, 2003.

Jevons, William Stanley. *Elementary Lessons in Logic: Deductive and Inductive*. New ed. London: Macmillan, 1905.

———. *The Principles of Science; A Treatise on Logic and Scientific Method*. 2nd ed. London: Macmillan, 1924.

Kent, Beverley. *Charles S. Peirce: Logic and the Classification of the Sciences*. Kingston, Ont.: McGill-Queen's Univ. Pr., 1987.

Lawrence, Wilde. "Logic: Dialectic and Contradiction." In *The Cambridge Companion to Marx*, ed. Terrell Carver, 275–95. Cambridge: Cambridge Univ. Pr., 1991.

Mill, John Stuart. *A System of Logic*. London: Longmans Green, 1884.

Mohanty, J. N. *Readings on Edmund Husserl's Logical Investigations*. The Hague, Netherlands: Nijhoff, 1977.

Peano, Giuseppe. "Set of Axioms for Integers." In Jean van Heijenoort, *From Frege to Gödel: A Source Book in Mathematical Logic 1879–1931*, 85–95. Cambridge, Mass.: Harvard Univ. Pr., 1967. [also in *Classics of Mathematics*, ed. Ronald Calinger, 663–67. Englewood Cliffs, N.J.: Prentice-Hall, 1995]

Peirce, Charles Sanders. *Collected Papers of Charles Sanders Peirce*. 8 vols. Ed. Charles Hartshorne, Paul Weiss, and Arthur Burks. Cambridge, Mass.: Harvard Univ. Pr., 1931–1958. [See vol. 2: *Elements of Logic*, vol. 3: *Exact Logic*, and vol. 4: *The Simplest Mathematics*.]

———. *Reasoning and the Logic of Things*. Ed. Kenneth Laine Ketner. Cambridge, Mass.: Harvard Univ. Pr., 1992.

Prantl, Carl. *Geschichte der Logik im Abendlande* [History of Western Logic]. 4 vols. Leipzig, Germany: S. Hirzel, 1855–1870. [This was very influential in its time but is much criticized today; there seems to be no English translation.]

Pulkkinen, Jarmo. *The Threat of Logical Mathematism: A Study on the Critique of Mathematical Logic in Germany at the Turn of the 20th Century*. New York: Peter Lang, 1994. [This tries to explain why the new logic was so slow to catch on in Germany as the 19th century ended and the 20th began.]

Royce, Josiah. *Royce's Logical Essays*. Ed. Daniel S. Robinson. Dubuque, Iowa: W. C. Brown, 1951.

Royce, Josiah. *The Principles of Logic*. New York: Wisdom Library, 1961.

Venn, John. *The Logic of Chance*. London: Macmillan, 1888.

———. *The Principles of Empirical or Inductive Logic*. New York: B. Franklin, 1972.

———. *Symbolic Logic*. 2nd ed. London: Macmillan, 1894.

Whewell, William. *The Philosophy of the Inductive Sciences, Founded upon Their*

History. 2 vols. New York: Johnson Reprint Corp., 1966.

2H.   Gottlob Frege

Dummett, Michael A. E. *Frege: Philosophy of Language.* London: Duckworth, 1973.
———. *Frege: Philosophy of Mathematics.* Cambridge, Mass.: Harvard Univ. Pr., 1991.
———. *The Interpretation of Frege's Philosophy.* Cambridge, Mass.: Harvard Univ. Pr., 1981.
Frege, Gottlob. *The Basic Laws of Arithmetic.* Trans. Montgomery Furth. Berkeley: Univ. of California, 1964. [parts of *Grundgesetze*]
———. *Collected Papers on Mathematics, Logic, and Philosophy.* Ed. Brian McGuinness. Oxford: Blackwell, 1984.
———. *Conceptual Notation, and Related Articles.* Translated with a biography and introduction by Terrell Ward Bynum. Oxford: Clarendon, 1972. [*Conceptual Notation* was published in 1879 as *Begriffsschrift.*]
———. *The Foundations of Arithmetic: A Logico-Mathematical Enquiry into the Concept of Number.* 2nd ed. Trans. J. L. Austin. Evanston, Ill.: Northwestern, 1980. [This was published in 1884 as *Die Grundlagen der Arithmetik.*]
———. *The Frege Reader.* Ed. Michael Beaney. Oxford: Blackwell, 1997.
———. *Grundgesetze der Arithmetik* [Basic Laws of Arithmetic]. 2 vols. Hildesheim, Germany: G. Olms, 1962. [This was first published 1893 and 1903; there seems to be no English translation of the whole work.]
———. *Logical Investigations.* Ed. P. T. Geach. Trans. P. T. Geach and R. H. Stoothoff. New Haven, Conn.: Yale, 1977. [This has less important essays published by Frege in 1918 and 1923.]
———. *On the Foundations of Geometry and Formal Theories of Arithmetic.* Trans. Eike-Henner W. Kluge. New Haven, Conn.: Yale, 1971.
———. *Posthumous Writings.* Ed. Hans Hermes, Friedrich Kambartel, and Friedrich Kaulbach. Trans. Peter Long and Roger White. Chicago: Univ. of Chicago Pr., 1979.
———. *Translations from the Philosophical Writings of Gottlob Frege.* Ed. Peter Geach and Max Black. Oxford: Blackwell, 1952.
Klemke, E. D., ed. *Essays on Frege.* Urbana: Univ. of Illinois Pr., 1968.
Wright, Crispin. *Frege's Conception of Numbers as Objects.* Aberdeen, Scotland: Aberdeen Univ. Pr., 1983.

3. Textbooks

Section 61 (Traditional Logic and Syllogisms) has further textbooks.

Ambrose, Alice, and Morris Lazerowitz. *Fundamentals of Symbolic Logic.* New York: Holt, Rinehart and Winston, 1962.

Barker, Stephen Francis. *The Elements of Logic.* 5th ed. New York: McGraw-Hill, 1989.

Barwise, Jon, and John Etchemendy. *Language, Proof and Logic.* New York: Seven Bridges, 1999.

Beth, Evert Willem. *Formal Methods: An Introduction to Symbolic Logic and to the Study of Effective Operations in Arithmetic and Logic.* Dordrecht: D. Reidel, 1962.

Bocheński, Joseph M. *A Precis of Mathematical Logic.* Trans. Otto Bird. Dordrecht: D. Reidel, 1959.

Boolos, George S., and Richard C. Jeffrey. *Computability and Logic.* 3rd ed. Cambridge: Cambridge Univ. Pr., 1989.

Carnap, Rudolf. *Introduction to Symbolic Logic and Its Applications.* Trans. William H. Meyer and John Wilkinson. New York: Dover, 1958.

Cooley, John C. *A Primer of Formal Logic.* New York: Macmillan, 1942.

Copi, Irving M. *Introduction to Logic.* 1st ed. New York: Macmillan, 1953. [This was very influential, the first of many editions and the model for further logic textbooks by other authors; the 2004 edition is coauthored with Carl Cohen.]

———. *Symbolic Logic.* 1st ed. New York: Macmillan, 1954.

Dauer, Francis Watanabe. *Critical Thinking: An Introduction to Reasoning.* New York: Barnes and Noble, 1989.

Fisk, Milton. *A Modern Formal Logic.* Englewood Cliffs, N.J.: Prentice-Hall, 1964.

Fitch, Frederic B. *Symbolic Logic: An Introduction.* New York: Ronald, 1952.

Gensler, Harry J. *Introduction to Logic.* London: Routledge, 2002.

Gustason, William, and Dolph E. Ulrich. *Elementary Symbolic Logic.* 2nd ed. Prospect Heights, Ill.: Waveland, 1989.

Havas, Katalin G. *It Is Logical!* Amsterdam: Rodopi, 1999.

Hughes, G. Bernard, and D. G. Londey. *The Elements of Formal Logic.* London: Methuen, 1965.

Hurley, Patrick J. *A Concise Introduction to Logic.* 8th ed. Belmont, Calif.: Wadsworth, 2003. [a popular textbook]

Jeffrey, Richard C. *Formal Logic: Its Scope and Limits.* New York: McGraw-Hill, 1981.

Kahane, Howard, and Paul Tidman. *Logic and Philosophy: A Modern Introduction.* 9th ed. Belmont, Calif.: Wadsworth, 2002.

Keene, Geoffrey Bourton. *The Foundations of Rational Argument.* Lewiston, N.Y.:

Edwin Mellen, 1992.

Klenk, Virginia. *Understanding Symbolic Logic*. 4th ed. Englewood Cliffs, N.J.: Prentice-Hall, 2001.

Langer, Susanne Katherina Knauth. *An Introduction to Symbolic Logic*. New York: Dover, 1953.

Layman, Charles Steven. *The Power of Logic*. 3rd ed. Mountain View, Calif.: Mayfield, 2004.

Leblanc, Hugues, and William A. Wisdom. *Deductive Logic*. Boston: Allyn & Bacon, 1972.

Lemmon, Edward John. *Beginning Logic*. London: Nelson, 1965.

Lewin, Morton H. *Logic Design and Computer Organization*. Reading, Mass.: Addison-Wesley, 1983.

Lewis, Clarence Irving, and Cooper Harold Langford. *Symbolic Logic*. New York: Dover, 1932. [This introduced various systems of modal logic.]

Łukasiewicz, Jan. *Elements of Mathematical Logic*. Trans. Olgierd Wojtasiewicz. New York: Macmillan, 1963.

Mates, Benson. *Elementary Logic*. 2nd ed. New York: Oxford Univ. Pr., 1972.

Nidditch, P. H. *Propositional Calculus*. London: Routledge & Kegan Paul, 1962.

Prior, Arthur N. *Formal Logic*. 2nd ed. Oxford: Clarendon, 1962.

Purtill, Richard L. *A Logical Introduction to Philosophy*. Englewood Cliffs, N.J.: Prentice-Hall, 1989.

Purtill, Richard L. *Logic for Philosophers*. New York: Harper & Row, 1971.

Quine, Willard Van Orman. *Elementary Logic*. Boston, Mass.: Ginn, 1941.

———. *Mathematical Logic*. Cambridge, Mass.: Harvard Univ. Pr., 1951.

———. *Methods of Logic*. 4th ed. Cambridge, Mass.: Harvard Univ. Pr., 1982.

Reichenbach, Hans. *Elements of Symbolic Logic*. New York: Macmillan, 1947.

Rescher, Nicholas. *Introduction to Logic*. New York: St. Martin's, 1964.

Sainsbury, Richard Mark. *Logical Forms*. 2nd ed. Oxford: Blackwell, 2001.

Salmon, Wesley C. *Logic*. 3rd ed. Englewood Cliffs, N.J.: Prentice-Hall, 1984.

Schagrin, Morton L., William J. Rapaport, and Randall R. Dipert. *Logic: A Computer Approach*. New York: McGraw-Hill, 1985.

Stebbing, L. Susan. *A Modern Elementary Logic*. London: Methuen, 1943.

Suppes, Patrick. *Introduction to Logic*. Princeton, N.J.: Van Nostrand, 1957.

Suppes, Patrick, and Shirley Hill. *First Course in Mathematical Logic*. New York: Blaisdell, 1964.

Tarski, Alfred. *Introduction to Logic*. 2nd ed. Oxford: Oxford Univ. Pr., 1946.

Teays, Wanda. *Second Thoughts: Critical Thinking from a Multicultural Perspective*. Mountain View, Calif.: Mayfield, 1996.

4. Beyond Text

4A. Logic Videos

These are VCR tapes suitable for classroom use.

Magee, Bryan. *Bryan Magee Talks to A. J. Ayer about Frege, Russell and Modern Logic.* London: BBC, 1987.

McGinn, Colin, Hilary Putnam, and Kit Fine. *Great Ideas of Philosophy: Logic, The Structure of Reason.* Ed. Canila O'Donnell. Princeton, N.J.: Films for the Humanities and Sciences, 2004. [from Aristotle to Gödel]

McGinn, Colin, Hilary Putnam, and Scott Soames. *Great Ideas of Philosophy: Analytic Philosophy.* Ed. Chris Scherer. Princeton, N.J.: Films for the Humanities and Sciences, 2004. [much on logic]

4B. Logic Software

These run on personal computers that need not be connected to the Web.

Bailhache, Patrice. *Program to Learn Natural Deduction in Gentzen-Kleene's Style* (DOS, free). http://193.51.78.161/dnfn/deductioneng.html

Barwise, Jon, and John Etchemendy. *Hyperproof, The Language of First-order Logic, Tarski's World,* and *Turing's World* (Windows and Macintosh, you buy the individual programs). http://www-csli.stanford.edu/hp

Barwise, Jon, John Etchemendy, et al. *Language, Proof, and Logic: Boole, Fitch, Tarski's World, and Submit* (Windows and Macintosh, you get the CD when you buy the text-software package). http://www-csli.stanford.edu/LPL

Brady, Rob B. *LogicWorks* (Windows and Macintosh, buy CD). http://www.logicworks4.com

Clark, Austen. *Bertie* (DOS, free). http://vm.uconn.edu/~wwwphil/software.html

———. *Plato* (Windows and Macintosh, free). http://www.utexas.edu/courses/plato

Gelder, Tim van. *Reason!Able* (Windows, free trial). http://www.goreason.com

Gensler, Harry J. *LogiCola* (Windows and classic-Macintosh, free). http://www.jcu.edu/philosophy/gensler/logicola.htm

Herzberg, Larry A. *Bertrand* (Macintosh, free). http://www.uwosh.edu/faculty_staff/herzberg/Bertrand.html

Pole, Nelson. *Logic Coach* (Windows and Macintosh, free). http://academic.csuohio.edu/polen

Rolf, Bertil. *Athena* (Windows and Macintosh, free). http://www.athenasoft.org

4C.   Interactive Logic Websites

These run on a Web browser.

Allen, Colin. *Power of Logic.* http://www.poweroflogic.com
Allen, Colin, and Chris Menzel. *The Logic Machine.* http://logic.tamu.edu
Burris, Stanley N. *Interactive Logic Programs.* http://www.math.uwaterloo.ca/
   ~snburris/htdocs/LOGIC/st_ilp.html
Elder, Frank. *A Tutorial in Critical Reasoning.* http://commhum.mccneb.edu/
   argument/summary.htm
Franconi, Enrico. *The Propositional Logic Calculator.* http://www.inf.unibz.it/
   ~franconi/teaching/propcalc
Gensler, Harry J. *Logic Tutorial.* http://www.jcu.edu/philosophy/gensler/logic.htm
———. *Web Exercises on Quine, Russell, Wittgenstein, and Others.* http://www
   .jcu.edu/philosophy/gensler/exercise.htm#AP
Gottschall, Christian. *Gateway to Logic.* http://logik.phl.univie.ac.at/~chris/
   formular-uk.html
Green, Michael K. *LogicTutor.* http://www.wwnorton.com/college/phil/logic3
Halpin, John F. *The Logic Café* (an online logic textbook). http://www.oakland
   .edu/phil/cafe
Howitt, Corin. *BlobLogic.* http://users.ox.ac.uk/~univ0675/blob
Jaeger, Gerhard. *Logic Workbench.* http://www.lwb.unibe.ch
Mesher, D. *Mission: Critical* (a tutorial in critical thinking). http://www2.sjsu.edu/
   depts/itl/graphics/main.html
Oruç, A. Yavuz. *Logic Calculator.* http://www.ee.umd.edu/~yavuz/logiccalc.html
Pezzullo, John C. *Bayes' Theorem Calculator.* http://members.aol.com/johnp71/
   bayes.html
Stangroom, Jeremy. *So You Think You're Logical?* (a logic game) http://www
   .philosophersnet.com/games/logic_task.htm
Sydow, Björn von. *Alfie.* http://www.cs.chalmers.se/~sydow/alfie
Tempest Media. *The Logic Course.* http://www.thelogiccourse.com
Tobin, Richard. *Euler's Circles.* http://www.cogsci.ed.ac.uk/~richard/Java/Euler
Truth Table Constructor. http://www.libertyk12.org/highschool/academics/math/
   truth/truth.html
Truth Table Practice. http://www.math.csusb.edu/notes/quizzes/tablequiz/
   tablepractice.html
Velleman, David. *Blogic.* http://webapps.itcs.umich.edu/velleman/Logic
Warthman Associates. *Turing Machine Java Animation.* http://www.warthman
   .com/ex-turing.htm
Weber, Helmut. *Turing Machine Simulator.* http://wap03.informatik.fh-wiesbaden
   .de/weber1/turing/tm.html
Yule, Pete. *Syllogism Experiment.* http://www.psyc.bbk.ac.uk/people/associates/
   pgy/experiments/syllog/intro.html

5. Interdisciplinary

5A. Computers and Artificial Intelligence

Boden, Margaret A. *Artificial Intelligence and Natural Man.* New York: Basic Books, 1977.

Boden, Margaret A., ed. *The Philosophy of Artificial Intelligence.* Oxford: Oxford Univ. Pr., 1990.

Burks, Arthur W., Hermann H. Goldstine, and John von Neumann. "Preliminary Discussion of the Logical Design of an Electronic Computing Instrument." This 1946 report to the U.S. Army Ordinance Department outlined the von Neumann architecture that is still used in most computers; it puts programs and data in the same memory space. The paper is on the Web at http://www.cs.unc.edu/~adyilie/comp265/vonNeumann.html and http://research.microsoft.com/~gbell/Computer_Structures__Readings_and_Examples/00000112.htm.

Cleary, J. F. *General Electric Transistor Manual.* 6th ed. Syracuse, N.Y.: General Electric, 1962. [See chapter 12 on logic, pp. 175–89.]

Crocco, G., L. Fariñas del Cerro, and A. Herzig, eds. *Conditionals: From Philosophy to Computer Science.* Oxford: Clarendon, 1995.

Cummins, Robert, and John Pollock, eds. *Philosophy and AI.* Cambridge, Mass.: MIT, 1991.

Fitting, Melvin. *First-Order Logic and Automated Theorem Proving.* New York: Springer-Verlag, 1990.

Gardner, Martin. *Logic Machines and Diagrams.* New York: McGraw-Hill, 1958.

Gazdar, Gerald, and Chris Mellish. *Natural Language Processing in Prolog: An Introduction to Computational Linguistics.* Wokingham, England: Addison-Wesley, 1989.

Haugeland, John. *Artificial Intelligence: The Very Idea.* Cambridge, Mass.: MIT, 1985.

Hilton, Alice Mary. *Logic, Computing Machines, and Automation.* Cleveland, Ohio: World, 1963.

Houghton, Janaye M., and Robert S. Houghton. *Circuit Sense for Elementary Teachers and Students: Understanding and Building Simple Logic Circuits.* Englewood, Colo.: Libraries Unlimited, 1994.

———. *Decision Points: Boolean Logic for Computer Users and Beginning Online Searchers.* Englewood, Colo.: Libraries Unlimited, 1999.

Jacquet, Jean-Marie, ed. *Constructing Logic Programs.* Chichester, West Sussex, England: Wiley, 1993.

Jean H., Gallier. *Logic for Computer Science: Foundations of Automatic Theorem Proving.* New York: N.Y.: Harper & Row, 1986.

Kees, Doets. *From Logic to Logic Programming.* Cambridge, Mass.: MIT, 1994.

Keller, Bill. *Feature Logics, Infinitary Descriptions, and Grammar.* Stanford,

Calif.: Center for the Study of Language and Information, 1993.

Korta, Kepa, Ernest Sosa, and Xabier Arrazola, eds. *Cognition, Agency, and Rationality: Proceedings of the Fifth International Colloquium on Cognitive Science*. Dordrecht: Kluwer Academic, 1999.

Kosko, Bart. *Neural Networks and Fuzzy Systems: A Dynamical Systems Approach to Machine Intelligence*. Englewood Cliffs, N.J.: Prentice-Hall, 1992.

Lavrač, Nada, and Sašo Džeroski. *Inductive Logic Programming*. New York: Ellis Horwood, 1994.

Lewin, Morton H. *Logic Design and Computer Organization*. Reading, Mass.: Addison-Wesley, 1983.

Moore, Robert C. *Logic and Representation*. Stanford, Calif.: Center for the Study of Language and Information, 1995.

Øhrstrøm, Peter, and Per F. V. Hasle. *Temporal Logic: From Ancient Ideas to Artificial Intelligence*. Dordrecht: Kluwer Academic, 1995.

Pollock, John L. *Cognitive Carpentry: A Blueprint for How to Build a Person*. Cambridge, Mass.: MIT, 1995.

Reeves, Steve, and Michael Clarke. *Logic for Computer Science*. Wokingham, England: Addison-Wesley, 1990.

Rine, David C., ed. *Computer Science and Multiple-Valued Logic: Theory and Applications*. Amsterdam: North-Holland, 1984.

Rong, Yang. *P-Prolog: A Parallel Logic Programming Language*. Singapore: World Scientific, 1987.

Roth, Charles H., Jr. *Fundamentals of Logic Design*. St. Paul, Minn.: West, 1992.

Schagrin, Morton L., William J. Rapaport, and Randall R. Dipert. *Logic: A Computer Approach*. New York: McGraw-Hill, 1985.

Sharma, Ashok K. *Programmable Logic Handbook*. New York: McGraw-Hill, 1998.

Thayse, André, ed. *From Standard Logic to Logic Programming: Introducing a Logic Based Approach to Artificial Intelligence*. Chichester, West Sussex, England: Wiley, 1988.

Thomason, Richmond H., ed. *Philosophical Logic and Artificial Intelligence*. Dordrecht: Kluwer Academic, 1989.

Wang, Hao. *Computation, Logic, Philosophy: A Collection of Essays*. Beijing, China: Science, 1990.

5B. Physics

Conen, Robert S., and Marx W. Wartofsky, eds. *Logical and Epistemological Studies in Contemporary Physics*. Dordrecht: D. Reidel, 1970.

Craig, William. "On Axiomatizability within a System." *Journal of Symbolic Logic* 18, no. 1 (May 1953): 30–32. [*Craig's theorem* shows the dispensability in scientific theories of terms that refer to unobservables.]

Fraassen, Bas C. van, and Enrico G. Beltrametti. *Current Issues in Quantum Logic*.

New York: Plenum, 1981.

Mittelstaedt, Peter. *Quantum Logic.* Dordrecht: D. Reidel, 1978.

———, ed. *Philosophical Problems of Modern Physics.* Dordrecht: D. Reidel, 1963.

von Neumann, John, and Garrett Birkhoff. "The Logic of Quantum Mechanics." *Annals of Mathematics* 37, no. 4 (October 1936): 823–43.

5C. Biology

Cooper, William S. *The Evolution of Reason: Logic as a Branch of Biology.* Cambridge: Cambridge Univ. Pr., 2001. [The conventional wisdom sees logic as absolute and independent and explains how animals to survive must evolve into somewhat logical beings. Cooper argues that this puts things backwards; he tries to derive the principles of logic from evolutionary biology.]

5D. Psychology and Other Social Sciences

Arrow, Kenneth Joseph. *Social Choice and Individual Values.* 2nd ed. New Haven, Conn.: Yale, 1963. [a political analogue of Gödel's theorem]

Braine, Martin D. S., and David P. O'Brien, ed. *Mental Logic.* Mahwah, N.J.: Lawrence Erlbaum, 1998.

Festinger, Leon. *A Theory of Cognitive Dissonance.* Stanford, Calif.: Stanford Univ. Pr., 1957.

Langer, Jonas. *The Origins of Logic: Six to Twelve Months.* New York: Academic, 1980.

Macnamara, John. *A Border Dispute: The Place of Logic in Psychology.* Cambridge, Mass.: MIT, 1986.

Macnamara, John, and Gonzalo E. Reyes, eds. *The Logical Foundations of Cognition.* New York: Oxford Univ. Pr., 1994.

Manktelow, K. I., and D. E. Over, eds. *Rationality: Psychological and Philosophical Perspectives.* London: Routledge, 1993.

Overton, Willis F., ed. *Reasoning, Necessity, and Logic: Developmental Perspectives.* Hillsdale, N.J.: Lawrence Erlbaum, 1990.

Piaget, Jean. "Genetic Logic and Sociology." In *Sociological Studies,* ed. Leslie Smith, trans. Terrance Brown et al., 184–210. London: Routledge, 1995.

———. *Logic and Psychology.* New York: Basic Books, 1957.

———. "Logical Operations and Social Life." In *Sociological Studies,* ed. Leslie Smith, trans. Terrance Brown et al., 134–57. London: Routledge, 1995.

Piaget, Jean, and Bärbel Inhelder. *The Early Growth of Logic in the Child.* Trans. E. A. Lunzer and D. Papert. London: Routledge & Kegan Paul, 1964.

———. *The Growth of Logical Thinking from Childhood to Adolescence: An Essay on the Construction of Formal Operational Structures.* Trans. Anne

Parsons and Stanley Milgram. London: Routledge, 1999.

Piaget, Jean, and Rolando Garcia. *Toward a Logic of Meanings*. Ed. Philip M. Davidson and Jack Easley. Hillsdale, N.J.: Lawrence Erlbaum, 1991.

Rips, Lance J. *The Psychology of Proof: Deductive Reasoning in Human Thinking*. Cambridge, Mass.: MIT, 1994.

Sutherland, Stuart. *Irrationality: Why We Do Not Think Straight*. New Brunswick, N.J.: Rutgers Univ. Pr., 1994.

5E. Linguistics

Benthem, Johan van. *Essays in Logical Semantics*. Dordrecht: D. Reidel, 1986.

Dixon, Robert M. W. *Linguistic Science and Logic*. The Hague, Netherlands: Mouton, 1963.

Hornstein, Norbert. *Logic as Grammar*. Cambridge, Mass.: MIT, 1984.

Martin, John N. *Elements of Formal Semantics: An Introduction to Logic for Students of Language*. Orlando, Fla.: Academic, 1987.

May, Robert. *Logical Form: Its Structure and Derivation*. Cambridge, Mass.: MIT, 1985.

McCawley, James D. *Everything that Linguists Have Always Wanted to Know about Logic (but Were Ashamed to Ask)*. Chicago: Univ. of Chicago Pr., 1981.

5F. Rhetoric, Communications, and Debate

Eemeren, Frans H. van, and Rob Grootendorst. *Argumentation, Communication, and Fallacies*. Hillsdale, N.J.: Lawrence Erlbaum 1992.

Eemeren, Frans H. van, Rob Grootendorst, J. Anthony Blair, and Charles A. Willard, eds. *Argumentation: Perspectives and Approaches, Proceedings of the Conference on Argumentation, 1986*. Dordrecht: Foris, 1987.

Makau, Josina M., and Debian L. Marty. *Cooperative Argumentation: A Model for Deliberative Community*. Prospect Heights, Ill.: Waveland, 2001.

Miller, Gerald R., and Thomas R. Nilsen, eds. *Perspectives on Argumentation*. Chicago: Scott Foresman, 1966.

Perelman, Chaïm. *The Realm of Rhetoric*. Trans. William Kluback. Notre Dame, Ind.: Univ. of Notre Dame Pr., 1982.

Perelman, Chaïm, and L. Olbrechts-Tyteca. *The New Rhetoric: A Treatise on Argumentation*. Trans. John Wilkinson and Purcell Weaver. Notre Dame, Ind.: Univ. of Notre Dame Pr., 1969.

Tindale, Christopher William. *Acts of Arguing: A Rhetorical Model of Argument*. Albany, N.Y.: SUNY, 1999.

5G. Pre-College Teaching

Ennis, Robert Hugh. *Logic in Teaching*. Englewood Cliffs, N.J.: Prentice-Hall, 1969.

Kneller, George Frederick. *Logic and Language of Education*. New York: Wiley, 1966.

Lipman, Matthew. *Harry Stottlemeier's Discovery*. Caldwell, N.J.: Universal Diversified Services, 1974. [a fifth grade logic textbook]

Neubert, Gloria A., and James B. Binko. *Inductive Reasoning in the Secondary Classroom*. Washington, D.C.: National Education Association, 1992.

Pritchard, Michael. "Philosophy for Children." *Stanford Encyclopedia of Philosophy*. http://plato.stanford.edu/entries/children [This says much about the teaching of logic and philosophy in the fifth grade.]

Smith, Bunnie Othanel, and Milton O. Meux. *A Study of the Logic of Teaching*. Urbana: Univ. of Illinois Pr., 1970.

Stanley, Maurice. *Introduction to Logical Analysis: Principles of Clear Thinking for Gifted High School and Junior College Students*. Manassas, Va.: Gifted Education, 1989.

6. Individual Figures and Topics

6A. Bertrand Russell

Linsky, Bernard. *Russell's Metaphysical Logic*. Stanford, Calif.: Center for the Study of Language and Information, 1999.

Neale, Stephen. *Descriptions*. Cambridge, Mass.: MIT, 1990. [He defends Russell's approach.]

Russell, Bertrand. *Introduction to Mathematical Philosophy*. London: George Allen & Unwin, 1960. [first published 1919]

———. *Logic and Knowledge: Essays 1901–1950*. Ed. Robert Charles Marsh. New York: Macmillan, 1956. [Selections include "On Denoting," "Mathematical Logic as Based on the Theory of Types," "The Philosophy of Logical Atomism," and "Logical Atomism."]

———. "Mathematical Logic as Based on the Theory of Types." *American Journal of Mathematics* 3 (1908): 222–62.

———. "Mysticism and Logic." In *Mysticism and Logic, and Other Essays*, 1–32. New York: Longmans Green, 1918.

———. "On Denoting." *Mind* 14, no. 56 (October 1905): 479–93.

———. "The Philosophy of Logical Atomism." *Monist* 28, no. 4 (October 1918): 495–527; 29, no. 1–3 (January-April-July 1919): 32–63, 190–222, and 345–80.

———. *The Principles of Mathematics*. 2nd ed. London: George Allen & Unwin,

1956. [This was published in 1903 as "vol. 1"; instead of doing further volumes, Russell worked with Alfred North Whitehead on *Principia Mathematica*.]

————. *Russell's Logical Atomism*. Ed. David Pears. London: Fontana, 1972. [This has Russell's 1918 "The Philosophy of Logical Atomism" and 1924 "Logical Atomism."]

Russell, Bertrand, and Alfred North Whitehead. *Principia Mathematica*. 3 vols. 2nd ed. Cambridge: Cambridge Univ. Pr., 1935. [The first edition came out in 1910–1913.]

Schilpp, Paul Arthur, ed. *The Philosophy of Bertrand Russell*. 3rd ed. New York: Tudor, 1951. [Note Kurt Gödel's selection on "Russell's Mathematical Logic."]

Strawson, Peter F. "On Referring." *Mind* 49, no. 235 (July 1950): 320–44. [This criticizes Russell's theory of descriptions.]

6B. Kurt Gödel

Boolos, George. "Gödel's Second Incompleteness Theorem Explained in Words of One Syllable." *Mind* 103, no. 409 (January 1994): 1–3.

Gensler, Harry J. *Gödel's Theorem Simplified*. Lanham, Md.: Univ. Pr. of America, 1984.

Gödel, Kurt. *Collected Works*. 3 vols. Ed. Solomon Feferman. Oxford: Clarendon, 1986–1995.

————. "On Formally Undecidable Propositions of *Principia Mathematica* and Related Systems I." In *The Undecidable: Basic Papers on Undecidable Propositions, Unsolvable Problems and Computable Functions*, ed. Martin Davis, 5–38. Hewlett, N.Y.: Raven, 1965. http://home.ddc.net/ygg/etext/godel/godel3.htm

Hofstadter, Douglas R. *Gödel, Escher, Bach: An Eternal Golden Braid*. New York: Basic Books, 1979. [Delightfully playful and nontechnical, this book relates Gödel's theorem to the world of painting, music, computers, and clever dialogues between the Tortoise and the Crab.]

Lucas, John R. "Minds, Machines and Gödel." *Philosophy* 36, no. 137 (April-July 1961): 112–27.

Nagel, Ernest, and James R. Newman. *Gödel's Proof*. New York: New York Univ. Pr., 1958.

Shanker, S. G., ed. *Gödel's Theorem in Focus*. London: Croom Helm, 1987.

Tieszen, Richard. "Gödel's Path from the Incompleteness Theorems (1931) to Phenomenology (1961)." *Bulletin of Symbolic Logic* 4, no. 2 (June 1998): 181–203. [about Gödel's late interest in Husserl]

6C.   Willard Van Orman Quine

See also Quine's entries under 3 (Textbooks).

Grice, H. Paul, and Peter F. Strawson, "In Defense of a Dogma." *Philosophical Review* 65, no. 2 (April 1956): 141–58 [They respond to Quine's attack on the analytic/synthetic distinction in "Two Dogmas of Empiricism."]

Hookway, Christopher. *Quine: Language, Experience and Reality.* Stanford, Calif.: Stanford Univ. Pr., 1988.

Marcus, Ruth Barcan. "A Functional Calculus of First Order Based on Strict Implication." *Journal of Symbolic Logic* 11, no. 1 (March 1946): 1–16. [Quine responds in "The Problem of Interpreting Modal Logic."]

———. "Modalities and Intensional Languages." *Synthese* 13, no. 4 (December 1961): 303–22. [She defends quantified modal logic against Quine, who responds in "Reply to Professor Marcus."]

Orenstein, Alex. *W. V. Quine.* Princeton, N.J.: Princeton Univ. Pr., 2002.

Orenstein, Alex, and Petr Kotatko, eds. *Knowledge, Language and Logic: Questions for Quine.* Dordrecht: Kluwer Academic, 2000.

Quine, Willard Van Orman. *From a Logical Point of View: 9 Logico-Philosophical Essays.* 2nd ed. Cambridge, Mass.: Harvard Univ. Pr., 1961. [This has some of Quine's major essays, including "On What There Is" and "Two Dogmas of Empiricism."]

———. "On What There Is." *Review of Metaphysics* 2, no. 12 (September 1949): 21–38. [an influential essay on ontology]

———. *Ontological Relativity, and Other Essays.* New York: Columbia Univ. Pr., 1969.

———. *Philosophy of Logic.* 2nd ed. Cambridge, Mass.: Harvard Univ. Pr., 1986. [a good summary of Quine's views]

———. "The Problem of Interpreting Modal Logic." *Journal of Symbolic Logic* 12, no. 2 (June 1947): 43–48. [He responds to Marcus's "A Functional Calculus of First Order Based on Strict Implication."]

———. "Reply to Professor Marcus." *Synthese* 13, no. 4 (December 1961): 323–30. [He responds to Marcus's "Modalities and Intensional Languages."]

———. *Set Theory and Its Logic.* Cambridge, Mass.: Harvard Univ. Pr., 1963.

———. "Two Dogmas of Empiricism." *Philosophical Review* 60, no. 1 January 1951): 20–43. [He criticizes the analytic/synthetic distinction and the reduction of material objects to sense experience; Grice and Strawson later defend the former in "In Defense of a Dogma."]

———. *The Ways of Paradox, and Other Essays.* New York: Random, 1966.

———. "What Price Bivalence." *Journal of Philosophy* 78, no. 2 (February 1981): 90–95. [He weighs the pragmatic pros and cons in accepting bivalence.]

———. *Word and Object.* Cambridge, Mass.: MIT, 1960.

Quine, Willard Van Orman, and Nelson Goodman. "Steps toward a Constructive

Nominalism." *Journal of Symbolic Logic* 11, no. 4 (December 1947): 105–22. [Quine later recanted from his nominalism and came to accept sets.]

Quine, Willard Van Orman, and Rudolf Carnap. *Dear Carnap, Dear Van: The Quine-Carnap Correspondence and Related Work.* Ed. Richard Creath. Berkeley: Univ. of California, 1990.

Sarkar, Sahotra, ed. *Decline and Obsolescence of Logical Empiricism: Carnap vs. Quine and the Critics.* New York: Garland, 1996.

Schilpp, Paul Arthur, and Lewis Edwin Hahn, eds. *The Philosophy of W. V. Quine.* La Salle, Ill.: Open Court, 1986.

6D. Modal Logic

See also entries by Marcus and Quine under 6C (Quine).

Armstrong, David Malet. *A Combinatorial Theory of Possibility.* Cambridge: Cambridge Univ. Pr., 1989.

Bennett, Jonathan. "Entailment." *Philosophical Review* 78, no. 2 (April 1969): 197–236.

Benthem, Johan van. *A Manual of Intensional Logic.* Stanford, Calif.: Center for the Study of Language and Information, 1988. [This discusses modal, epistemic, temporal, and conditional logics.]

Bigelow, John, and Robert Pargetter. *Science and Necessity.* Cambridge: Cambridge Univ. Pr., 1990.

Burks, Arthur. "The Logic of Causal Propositions." *Mind* 60, no. 239 (July 1951): 363–82.

Carnap, Rudolf. *Meaning and Necessity: A Study in Semantics and Modal Logic.* Chicago: Univ. of Chicago Pr., 1947.

Chellas, Brian F. *Modal Logic: An Introduction.* Cambridge: Cambridge Univ. Pr., 1980.

Chihara, Charles S. *The Worlds of Possibility: Modal Realism and the Semantics of Modal Logic.* Oxford: Clarendon, 1998.

Chisholm, Roderick M. "Identity through Possible Worlds: Some Questions." *Noûs* 1, no. 1 (March 1967): 1–8.

Davies, Martin. *Meaning, Quantification, Necessity: Themes in Philosophical Logic.* London: Routledge & Kegan Paul, 1981.

Feys, Robert. *Modal Logics.* Ed. Joseph Dopp. Louvain: E. Nauwelaerts, 1965.

Forbes, Graeme. *Languages of Possibility.* Oxford: Blackwell, 1989.

———. *The Metaphysics of Modality.* Oxford: Clarendon, 1985.

Girle, Rod. *Modal Logics and Philosophy.* Montreal: McGill-Queen's Univ. Pr., 2000.

Hintikka, Jaakko. *Models for Modalities.* Dordrecht: D. Reidel, 1969.

Hughes, George Edward, and M. J. Cresswell. *A New Introduction to Modal Logic.* London: Routledge, 1996. [This book is intended to replace their older classic,

An Introduction to Modal Logic (London: Methuen, 1968).]

Jubien, Michael. *Ontology, Modality, and the Fallacy of Reference*. Cambridge: Cambridge Univ. Pr., 1993.

Knuuttila, Simo. *Modalities in Medieval Philosophy*. London: Routledge, 1993.

Konyndyk, Kenneth. *Introductory Modal Logic*. Notre Dame, Ind.: Univ. of Notre Dame Pr., 1986.

Kripke, Saul A. "A Completeness Theorem in Modal Logic." *Journal of Symbolic Logic* 24, no. 1 (March 1959): 1–14.

———. *Naming and Necessity*. Cambridge, Mass.: Harvard Univ. Pr., 1980.

———. "Semantical Considerations on Modal Logic." *Acta Philosophica Fennica* 16 (1963): 83–94.

Lagerlund, Henrik. *Modal Syllogistics in the Middle Ages*. Boston: Brill, 2000.

Lappin, Shalom. "Moral Judgments and Identity across Possible Worlds." *Ratio* 20, no. 1 (June 1978): 69–74.

Lemmon, Edward John. "Is There Only One Correct System of Modal Logic?" *Aristotelian Society Supplement* 33 (July 1959): 23–40.

Lewis, Clarence Irving, and Cooper Harold Langford. *Symbolic Logic*. New York: Dover, 1932. [This book, which introduced various systems of modal logic, started the contemporary interest in modal logic.]

Lewis, David K. *On the Plurality of Worlds*. Oxford: Blackwell, 1986.

Linsky, Leonard. *Oblique Contexts*. Chicago: Univ. of Chicago Pr., 1983.

———, ed. *Reference and Modality*. London: Oxford Univ. Pr., 1971.

Loux, Michael J., ed. *The Possible and the Actual: Readings in the Metaphysics of Modality*. Ithaca, N.Y.: Cornell, 1979.

Lycan, William G. *Modality and Meaning*. Dordrecht: Kluwer Academic, 1994.

Marcus, Ruth Barcan. "Extensionality." *Mind* 69, no. 273 (January 1960): 55–62. [on quantified modal logic]

———. *Modalities: Philosophical Essays*. New York: Oxford Univ. Pr., 1993.

McCall, Storrs. *Aristotle's Modal Syllogisms*. Amsterdam: North-Holland, 1963.

Meyer, Robert K. "Entailment." *Journal of Philosophy* 68, no. 21 (4 November 1971): 808–18.

Patterson, Richard. *Aristotle's Modal Logic: Essence and Entailment in the Organon*. Cambridge: Cambridge Univ. Pr., 1995.

Plantinga, Alvin. *The Nature of Necessity*. Oxford: Clarendon, 1974. [a classic defense of modality and Aristotelian essentialism]

Rosenberg, Jay F. *Beyond Formalism: Naming and Necessity for Human Beings*. Philadelphia: Temple Univ. Pr., 1994.

Schwartz, Stephen P., ed. *Naming, Necessity, and Natural Kinds*. Ithaca, N.Y.: Cornell, 1977.

Sinnott-Armstrong, Walter, Diana Raffman, and Nicholas Asher, eds. *Modality, Morality, and Belief: Essays in Honor of Ruth Barcan Marcus*. Cambridge: Cambridge Univ. Pr., 1995.

Smullyan, Arthur Francis. "Modality and Description." *Journal of Symbolic Logic*

13, no. 1 (March 1948): 31–37.

Trundle, Robert C. *Medieval Modal Logic and Science*. Lanham, Md.: Univ. Pr. of America, 1999.

White, Alan R. *Modal Thinking*. Ithaca, N.Y.: Cornell Univ. Pr, 1975. [This is about the ordinary language use of modal concepts, including "must," "may," "can," and "ought."]

Wright, Georg Henrik von. *An Essay in Modal Logic*. Amsterdam: North-Holland, 1951.

6E.   Ethics and Deontic/Imperative Logic

Alchourrón, Carlos E. "Logic of Norms and Logic of Normative Propositions." *Logique et Analyse* 12, no. 47 (September 1969): 242–69.

Anderson, Alan Ross. "The Formal Analysis of Normative Systems." In *The Logic of Decision and Action*, ed. Nicholas Rescher, 147–213. Pittsburgh, Pa.: Univ. of Pittsburgh Pr., 1966.

———. "The Logic of Norms." *Logique et Analyse* 1, no. 2 (April 1958): 84–91.

Åqvist, Lennart. "Choice-Offering and Alternative-Presenting Disjunctive Commands." *Analysis* 25, no. 5 (April 1965): 182–84.

Audi, Robert. *Practical Reasoning*. London: Routledge, 1989.

Beardsley, Elizabeth Lane. "Imperative Sentences in Relation to Indicatives." *Philosophical Review* 53, no. 2 (March 1944): 175–85.

Bennett, Jonathan. "Review of 12 Articles on Imperative Logic." *Journal of Symbolic Logic* 35, no. 2 (June 1970): 314–18. [Many other issues of this journal have similar very useful review articles.]

Castañeda, Hector-Neri. "Actions, imperatives, and obligations." *Proceedings of the Aristotelian Society* 68 (1967–1968): 25–48.

———. "Ethics and Logic: Stevensonian Emotivism Revisited." *Journal of Philosophy* 64, no. 20 (October 1967): 671–83.

———. "Imperative Reasonings." *Philosophy and Phenomenological Research* 21, no. 1 (September 1960): 21–49.

———. "On the Semantics of the Ought-to-Do." *Synthese* 21, no. 4 (December 1970): 449–68.

———. "Outline of a Theory on the General Logical Structure of the Language of Action." *Theoria* 26, no. 3 (1960): 151–82.

Chisholm, Roderick M. "Contrary-to-Duty Imperatives and Deontic Logic." *Analysis* 24, no. 2 (December 1963): 33–36.

———. "The Ethics of Requirement." *American Philosophical Quarterly* 1, no. 2 (April 1964): 147–53.

———. "Supererogation and Offense: A Conceptual Scheme for Ethics." *Ratio* 5 (old series), no. 1 (June 1963): 1–14.

Clarke, Jr., D. S. "The Logical Form of Imperatives." *Philosophia* 5, no. 4 (October 1975): 417–27.

———. "Mood Consistency in Mixed Inferences." *Analysis* 30, no. 3 (January 1970): 100–103.

Cresswell, M. J. "Some Further Semantics for Deontic Logic." *Logique et Analyse* 10, no. 38 (June 1967): 179–91.

Dahl, Norman O. "'"Ought" Implies "Can"' and Deontic Logic." *Philosophia* 4, no. 4 (October 1974): 485–511.

Feldman, Fred. *Doing the Best We Can: An Essay in Informal Deontic Logic.* Dordrecht: D. Reidel, 1986.

Fisher, Mark. "A Logical Theory of Commanding." *Logique et Analyse* 4, no. 15–16 (October 1961): 154–69.

———. "Strong and Weak Negation of Imperatives." *Theoria* 28, no. 2 (1962): 196–200.

———. "A System of Deontic-Alethic Modal Logic." *Mind* 71, no. 282 (May 1962): 231–36.

Forrester, James W. *Being Good and Being Logical: Philosophical Groundwork for a New Deontic Logic.* Armonk, N.Y.: M.E. Sharpe, 1996.

Geach, Peter T. "Imperative and Deontic Logic." *Analysis* 18, no. 3 (January 1958): 49–56.

———. "Imperative Inference." *Analysis Supplement* 23 (1963): 37–42.

Gensler, Harry J. "Acting Commits One to Ethical Beliefs." *Analysis* 43, no. 1 (January 1983): 40–43.

———. "Ethical Consistency Principles." *Philosophical Quarterly* 35, no. 139 (April 1985): 156–70.

———. *Formal Ethics.* London: Routledge, 1996.

———. "How Incomplete is Prescriptivism?" *Mind* 93, no. 369 (January 1984): 103–107.

———. "The Prescriptivism Incompleteness Theorem." *Mind* 85, no. 340 (October 1976): 589–96.

Gombay, Andre. "What *Is* Imperative Inference?" *Analysis* 27, no. 5 (April 1967): 145–52.

Greenspan, P. S. "Conditional Oughts and Hypothetical Imperatives." *Journal of Philosophy* 72, no. 10 (22 May 1975): 259–66.

Hare, Richard Mervyn. *The Language of Morals.* Oxford: Clarendon, 1952. [Chapters 1–4 are on imperative logic, pp. 1–78.]

———. *Practical Inferences.* Berkeley: Univ. of California, 1972.

Hilpinen, Risto, ed. *Deontic Logic: Introductory and Systematic Readings.* Dordrecht: D. Reidel, 1971.

———. *New Studies in Deontic Logic: Norms, Actions, and the Foundations of Ethics.* Dordrecht: D. Reidel, 1981.

Hofstadter, Albert, and J. C. C. McKinsey. "On the Logic of Imperatives." *Philosophy of Science* 6 (1939): 446–57.

Kamp, Hans. "Free Choice Permission." *Proceedings of the Aristotelian Society* 74 (1973–1974): 57–74.

Kenny, A. J. "Practical Inference." *Analysis* 26, no. 3 (January 1966): 65–75.

Körner, Stephan, ed. *Practical Reason: Papers and Discussions*. New Haven, Conn.: Yale, 1974.

Lappin, Shalom. "Moral Judgments and Identity across Possible Worlds." *Ratio* 20, no. 1 (June 1978): 69–74.

Lewis, Clarence Irving. *The Ground and Nature of the Right*. New York: Columbia Univ. Pr., 1955.

MacKay, Alfred F. "Inferential Validity and Imperative Inference Rules." *Analysis* 29, no. 5 (April 1969): 145–56.

————. "The Principle of Mood Consistency." *Analysis* 31, no. 3 (January 1971): 91–96.

Prior, Arthur N. *Papers in Logic and Ethics*. Ed. P. T. Geach and A. J. Kenny. Amherst: Univ. of Massachusetts Pr., 1976.

Rescher, Nicholas. "An Axiom System for Deontic Logic." *Philosophical Studies* 9, no. 1–2 and 4 (January-February and June 1958): 24–30 and 64.

————. *The Logic of Commands*. London: Routledge & Kegan Paul, 1966.

————, ed. *The Logic of Decision and Action*. Pittsburgh, Pa.: Univ. of Pittsburgh Pr., 1966.

Roberts, Colin. "Zellner and Imperative Inferences." *Mind* 84, no. 333 (January 1975): 111–13.

Ross, Alf. "Imperatives and Logic." *Philosophy of Science* 11 (1944): 30–46.

Schueler, G. F. "*Modus Ponens* and Moral Realism." *Ethics* 98, no. 3 (April 1988): 492–500.

Sosa, Ernest. "The Logic of Imperatives." *Theoria* 32, no. 3 (1966): 224–35.

————. "On Practical Inference and the Logic of Imperatives." *Theoria* 32, no. 3 (1966): 211–23.

————. "On Practical Inference with an Excursus on Theoretical Inference." *Logique et Analyse* 13, no. 49–50 (March-June 1970): 213–30.

————. "The Semantics of Imperatives." *American Philosophical Quarterly* 4, no. 1 (January 1967): 57–64.

Turnbull, Robert G. "Imperatives, Logic, and Moral Obligation." *Philosophy of Science* 27 (1960): 374–90.

Williams, Bernard A. O. "Consistency and Realism." *Aristotelian Society Supplement* 45 (1966): 1–22.

————. "Imperative Inference." *Analysis Supplement* 23 (1963): 30–36.

Wright, Georg Henrik von. "Deontic Logic." *Mind* 60, no. 237 (January 1951): 1–15. [This classic article kicked off interest in deontic logic.]

————. *An Essay in Deontic Logic and the General Theory of Action*. Amsterdam: North-Holland, 1968.

————. *The Logic of Preference*. Edinburgh, Scotland: Edinburgh Univ. Pr., 1963. [an axiological logic of "good" and "better"]

————. *Norm and Action: A Logical Enquiry*. New York: Humanities, 1963.

————. *Practical Reason*. Ithaca, N.Y.: Cornell, 1983.

Zellner, Harold. "A Note on R. M. Hare and the Paradox of the Good Samaritan." *Mind* 82, no. 326 (April 1973): 281–82.

6F. Logic of Belief and Knowledge

Boër, Steven E., and William G. Lycan. *Knowing Who.* Cambridge, Mass.: MIT, 1986.

Boh, Ivan. *Epistemic Logic in the Later Middle Ages.* London: Routledge, 1993.

Chisholm, Roderick M. "Epistemic Statements and the Ethics of Belief." *Philosophy and Phenomenological Research* 16, no. 4 (June 1956): 447–60.

———. "The Logic of Knowing." *The Journal of Philosophy* 60, no. 25 (5 December 1963): 773–95. [He sketches Hintikka's *Knowledge and Belief* and raises questions about identity over possible worlds.]

Ellis, Brian David. *Rational Belief Systems.* Totowa, N.J.: Rowman & Littlefield, 1979.

Fisher, Mark. "The Epistemology of Belief and the Epistemology of Degrees of Belief." *American Philosophical Quarterly* 29, no. 2 (April 1992): 111–24.

———. "Justified Inconsistent Beliefs." *American Philosophical Quarterly* 16, no. 4 (October 1979): 247–57.

———. "Remarks on a Logical Theory of Belief Statements." *Philosophical Quarterly* 14, no. 55 (April 1964): 165–69.

Hintikka, Jaakko. "Individuals, Possible Worlds, and Epistemic Logic." *Noûs* 1, no. 1 (March 1967): 33–62.

———. *Knowledge and Belief: An Introduction to the Logic of the Two Notions.* Ithaca, N.Y.: Cornell, 1962.

Hintikka, Jaakko, and Merrill B. Hintikka, eds. *The Logic of Epistemology and the Epistemology of Logic: Selected Essays.* Dordrecht: Kluwer Academic, 1989.

Kaplan, Mark. "A Bayesian Theory of Rational Acceptance." *Journal of Philosophy* 78, no. 6 (June 1981): 305–30.

Kielkopf, Charles F. "A Note on Hintikka's Logic of Belief as an Ethics of Belief." *Philosophical Studies* 23, no. 1–2 (February 1972): 135–37.

Schick, Frederick. "Consistency." *Philosophical Review* 75, no. 4 (October 1966): 467–95.

Schlesinger, George N. *The Range of Epistemic Logic.* Aberdeen, Scotland: Aberdeen Univ. Pr.: 1985.

6G. Temporal Logic

Foundations of Temporal Logic: The WWW-site for Prior-Studies. http://www.kommunikation.aau.dk/prior/index2.htm

McArthur, Robert P. *Tense Logic.* Dordrecht: D. Reidel, 1976.

Mellor, D. H. *Real Time.* 2nd ed. London: Routledge, 1998.

Øhrstrøm, Peter, and Per F. V. Hasle. *Temporal Logic: From Ancient Ideas to Artificial Intelligence.* Dordrecht: Kluwer Academic, 1995.

Prior, Arthur N. *Papers on Time and Tense.* Oxford: Clarendon, 1968.

————. *Past, Present and Future.* Oxford: Clarendon, 1967.

————. *Time and Modality.* Oxford: Clarendon, 1957.

Rescher, Nicholas, and Alasdair Urquhart. *Temporal Logic.* New York: Springer-Verlag, 1971.

6H. Mathematically Oriented Logic

Ackermann, Wilhelm. *Solvable Cases of the Decision Problem.* Amsterdam: North-Holland, 1954.

Benacerraf, Paul, and Hilary Putnam, eds. *Philosophy of Mathematics: Selected Readings.* 2nd ed. Cambridge: Cambridge Univ. Pr., 1983.

Boolos, George S., and Richard C. Jeffrey. *Computability and Logic.* 3rd ed. Cambridge: Cambridge Univ. Pr., 1989.

Calinger, Ronald, ed. *Classics of Mathematics.* Englewood Cliffs, N.J.: Prentice-Hall, 1995. [Pp. 635–753 have classic selections from Georg Cantor, Gottlob Frege, Giuseppe Peano, Bertrand Russell (including the Russell-Frege correspondence on Russell's paradox), Henri Poincaré, David Hilbert, Ernst Zermelo, Alfred North Whitehead, L. E. J. Brouwer, and Kurt Gödel.]

Church, Alonzo. *Introduction to Mathematical Logic.* Princeton, N.J.: Princeton Univ. Pr., 1956.

————. "A Note on the Entscheidungsproblem." *Journal of Symbolic Logic* 1, no. 1 (March 1936): 40–41. [Church's theorem shows that quantificational logic has no decision procedure for validity.]

Curry, Haskell Brooks. *Outlines of a Formalist Philosophy of Mathematics.* Amsterdam: North-Holland, 1951.

Davis, Martin, ed. *The Undecidable: Basic Papers on Undecidable Propositions, Unsolvable Problems and Computable Functions.* Hewlett, N.Y.: Raven, 1965. [Pp. 5–38 have Kurt Gödel's classic, "On Formally Undecidable Propositions of *Principia Mathematica* and Related Systems I"; the book also has related technical papers by Gödel, Church, Turing, Rosser, Kleene, and Post.]

Detlefsen, Michael. *Hilbert's Program: An Essay on Mathematical Instrumentalism.* Dordrecht: D. Reidel, 1986.

Feferman, Solomon. *The Number Systems: Foundations of Algebra and Analysis.* New York: Chelsea, 1989.

Fraenkel, Abraham A. *Set Theory and Logic.* Reading, Mass.: Addison-Wesley, 1966.

Fraenkel, Abraham A., Yehoshua Bar-Hillel, and Azriel Levey. *Foundations of Set Theory.* 2nd ed. Amsterdam: North-Holland, 1973.

Hamilton, A. G. *Logic for Mathematicians.* Cambridge: Cambridge Univ. Pr., 1978.

Heijenoort, Jean van. *From Frege to Gödel: A Source Book in Mathematical Logic, 1879–1931.* Cambridge, Mass.: Harvard Univ. Pr., 1967.

Henkin, Leon. "The Completeness of the First-Order Functional Calculus." *Journal of Symbolic Logic* 14, no. 3 (September 1949): 159–66.

Hintikka, Jaakko, ed. *The Philosophy of Mathematics.* Oxford: Oxford Univ. Pr., 1969.

Hunter, Geoffrey. *Metalogic: An Introduction to the Metatheory of Standard First Order Logic.* London: Macmillan, 1971.

Kleene, Stephen Cole. *Introduction to Metamathematics.* New York: Van Nostrand, 1952.

Kneebone, G. T. *Mathematical Logic and the Foundations of Mathematics.* London: Van Nostrand, 1963.

Lakatos, Imre. *Proofs and Refutations: The Logic of Mathematical Discovery.* Ed. John Worrall and Elie Zahar. Cambridge: Cambridge Univ. Pr., 1976.

Lehman, Hugh. *Introduction to the Philosophy of Mathematics.* Totowa, N.J.: Rowman & Littlefield, 1979.

Post, Emil L. "Introduction to a General Theory of Elementary Propositions." *American Journal of Mathematics* 43, no. 3 (July 1921): 163–85.

Putnam, Hilary. "Mathematics without Foundations." *Journal of Philosophy* 64, no. 1 (19 January 1967): 5–22.

Ramsey, Frank Plumpton, *The Foundations of Mathematics and Other Logical Essays.* Ed. R. B. Braithwaite. London: Routledge, 1931.

Resnik, Michael D. *Frege and the Philosophy of Mathematics.* Ithaca, N.Y.: Cornell, 1980.

Rosser, John Barkley. *Logic for Mathematicians.* New York: McGraw-Hill, 1953.

Stoll, Robert Roth. *Set Theory and Logic.* San Francisco, Calif.: W. H. Freeman, 1963.

Tarski, Alfred. *Undecidable Theories.* Amsterdam: North-Holland, 1953.

Turing, Alan M. "Computing Machinery and Intelligence." *Mind* 59, no. 236 (October 1950): 433–60.

Wang, Hao. "The Axiomatization of Arithmetic." *Journal of Symbolic Logic* 22, no. 2 (June 1957): 145–58.

———. *A Survey of Mathematical Logic.* Peking: Science, 1963. [These papers were later reprinted as *Logic, Computers, and Sets* (New York: Chelsea, 1970).]

Zuckerman, Martin M. *Sets and Transfinite Numbers.* New York: Macmillan, 1974.

6I. Traditional Logic and Syllogisms

Bosanquet, Bernard. *The Essentials of Logic.* London: Macmillan, 1910.

Bradley, Francis Herbert. *The Principles of Logic.* London: Oxford Univ. Pr., 1922.

Dopp, Joseph. *Formal Logic.* Trans. J. Roland E. Ramirez and Robert D. Sweeney. New York: J. F. Wagner, 1960.

Gensler, Harry J. "A Simplified Decision Procedure for Categorical Syllogisms." *Notre Dame Journal of Formal Logic* 14, no. 4 (October 1973): 457–66. http://projecteuclid.org/Dienst/UI/1.0/Summarize/euclid.ndjfl/1093891100

———. "Why the Star Test for Syllogisms Works." http://www.jcu.edu/philosophy/gensler/star.htm

Johnson, William Ernest. *Logic*. 3 vols. Cambridge: Cambridge Univ. Pr., 1921–1924.

Joseph, Horace W. B. *An Introduction to Logic*. Oxford: Clarendon, 1916.

Joseph, Miriam. *The Trivium: The Liberal Arts of Logic, Grammar, and Rhetoric*. Ed. Marguerite McGlinn. Philadelphia: Paul Dry Books, 2002.

Joyce, George Hayward. *Principles of Logic*. London: Longmans Green, 1929.

Maritain, Jacques. *An Introduction to Logic*. Trans. Imelda Choquette. New York: Sheed & Ward, 1937. [also published as *Formal Logic*]

Mercier, Désiré. *Elements of Logic*. New York: Manhattanville, 1912.

Patzig, Günther. *Aristotle's Theory of the Syllogism: A Logicophilological Study of Book A of the Prior Analytics*. Trans. Jonathan Barnes. Dordrecht: D. Reidel, 1969.

Sellars, Roy Wood. *The Essentials of Logic*. Boston: Houghton Mifflin, 1917.

Shen, Eugene. "The Ladd-Franklin Formula in Logic: The Antilogism." *Mind* 36, no. 141 (January 1927): 54–60.

Sullivan, Daniel J. *Fundamentals of Logic*. New York: McGraw-Hill, 1963.

Veatch, Henry Babcock. *Intentional Logic*. New Haven, Conn.: Yale, 1952.

6J. Deviant Logics and Conditionals

See also dialectical logic in 6P (Continental Philosophy and Logic).

Ackermann, Robert John. *An Introduction to Many-Valued Logics*. London: Routledge & Kegan Paul, 1967.

Anderson, Alan Ross, Nuel D. Belnap Jr., and J. Michael Dunn. *Entailment: The Logic of Relevance and Necessity*. 2 vols. Princeton, N.J.: Princeton Univ. Pr., 1975 and 1992.

Appiah, Anthony. *Assertion and Conditionals*. Cambridge: Cambridge Univ. Pr., 1985.

Auwera, Johan van der. *Language and Logic: A Speculative and Condition-Theoretic Study*. Amsterdam: J. Benjamins, 1985.

Bandemer, Hans, and Siegfried Gottwald. *Fuzzy Sets, Fuzzy Logic, Fuzzy Methods with Applications*. New York: J. Wiley, 1995.

Brouwer, Luitzen Egbertus Jan. *Brouwer's Cambridge Lectures on Intuitionism*. Ed. D. van Dalen. Cambridge: Cambridge Univ. Pr., 1981.

Buckley, James J., and Esfandiar Eslami. *An Introduction to Fuzzy Logic and Fuzzy Sets*. Heidelberg, Germany: Physica-Verlag, 2002.

Carnielli, Walter A., Marcelo E. Coniglio, and Itala M. Loffredo D'Ottaviano.

Paraconsistency: The Logical Way to the Inconsistent. New York: Marcel Dekker, 2002.

Chellas, Brian F. "Basic Conditional Logic." *Journal of Philosophical Logic* 4, no. 1 (February 1974): 133–53.

Cooper, William S. "The Propositional Logic of Ordinary Discourse." *Inquiry* 11, no. 3 (Autumn 1968): 295–320.

da Costa, Newton C. A. "On the Theory of Inconsistent Formal Systems." *Notre Dame Journal of Formal Logic* 15, no. 4 (October 1974): 497–510.

Davis, Wayne A. *Implicature: Intention, Convention, and Principle in the Failure of Gricean Theory*. Cambridge: Cambridge Univ. Pr., 1998.

Došen, Kosta. "The First Axiomatization of Relevant Logic." *Journal of Philosophical Logic* 21, no. 4 (November 1992): 339–56.

Dubois, Didier, Henri Prade, and Erich Peter Klement, eds. *Fuzzy Sets, Logics and Reasoning about Knowledge*. Dordrecht: Kluwer Academic, 1999.

Dummett, Michael A. E. *Elements of Intuitionism*. 2nd ed. Oxford: Clarendon, 2000.

Edgington, Dorothy. "Do Conditionals Have Truth-Conditions?" *Critica* 18 (1986): 3–30.

Epstein, Richard L. *The Semantic Foundations of Logic: Propositional Logics*. New York: Oxford Univ. Pr., 1995.

Faris, J. A. "Interderivability of '⊃' and 'If.'" In *Logic and Philosophy*, ed. Gary Iseminger, 203–10. New York: Appleton-Century-Crofts, 1968.

Goguen, J. A. "The Logic of Inexact Concepts." *Synthese* 19, no. 2 (1968–1969): 325–73.

Grice, H. Paul. *Studies in the Way of Words*. Cambridge, Mass.: Harvard Univ. Pr., 1989.

Griffin, Nicholas. *Relative Identity*. Oxford: Clarendon, 1977.

Haack, Susan. *Deviant Logic, Fuzzy Logic: Beyond the Formalism*. Chicago: Univ. of Chicago Pr., 1996.

———. *Deviant Logic: Some Philosophical Issues*. Cambridge: Cambridge Univ. Pr., 1975.

———. *Philosophy of Logics*. Cambridge: Cambridge Univ. Pr., 1978.

Heyting, Arend. *Intuitionism: An Introduction*. Amsterdam: North-Holland, 1956.

Jackson, Frank. *Conditionals*. Oxford: Blackwell, 1987.

———, ed. *Conditionals*. Oxford: Oxford Univ. Pr., 1991. [See especially Jackson's "On Assertion and Indicative Conditionals," pp. 111–35, which defends the material-implication analysis of indicative conditionals.]

Jaśkowski, Stanisław. "Propositional Calculus for Contradictory Deductive Systems." *Studia Logica* 24 (1969): 143–57.

Karp, Carol Ruth. *Languages with Expressions of Infinite Length*. Amsterdam: North-Holland, 1964.

Keisler, H. Jerome. *Model Theory for Infinitary Logic: Logic with Countable Conjunctions and Finite Quantifiers*. Amsterdam: North-Holland, 1971.

Kosko, Bart. *Fuzzy Thinking: The New Science of Fuzzy Logic.* New York: Hyperion, 1993. [A Zen popularization of fuzzy logic, seen as Buddha over Aristotle: "Everything is a matter of degree."]

———. *Neural Networks and Fuzzy Systems: A Dynamical Systems Approach to Machine Intelligence.* Englewood Cliffs, N.J.: Prentice-Hall, 1992.

Kvart, Igal. *A Theory of Counterfactuals.* Indianapolis, Ind.: Hackett, 1986.

Lambert, Karel, ed. *Philosophical Applications of Free Logic.* New York: Oxford Univ. Pr., 1991.

Lambert, Karel, and Bas C. van Fraassen. *Derivation and Counterexample: An Introduction to Philosophical Logic.* Encino, Calif.: Dickenson, 1972. [much on free logic]

Lewis, David K. *Counterfactuals.* Cambridge, Mass.: Harvard Univ. Pr., 1973.

Malinowski, Grzegorz. *Many-Valued Logics.* Oxford: Clarendon, 1993.

McGee, Vann. 1985. "A Counterexample to Modus Ponens." *Journal of Philosophy* 82, no. 9 (September 1985): 462–71.

McLaughlin, Robert N. *On the Logic of Ordinary Conditionals.* Albany, N.Y.: SUNY, 1990.

Nute, Donald. *Topics in Conditional Logic.* Dordrecht: D. Reidel, 1980.

———, ed. Special Issue on Conditionals. *Journal of Philosophical Logic* 10, no. 2 (May 1981).

Pollock, John L. *Subjunctive Reasoning.* Dordrecht: D. Reidel, 1976.

Priest, Graham. *Beyond the Limits of Thought.* Cambridge: Cambridge Univ. Pr., 1995.

———. *In Contradiction.* Dordrecht: Martinus Nijhoff, 1986.

———. *An Introduction to Non-Classical Logic.* Cambridge: Cambridge Univ. Pr., 2001.

———. "What's So Bad about Contradictions?" *Journal of Philosophy* 95, no. 8 (August 1998): 410–26.

———, ed. Special Issue on Impossible Worlds. *Notre Dame Journal of Formal Logic* 38, no. 4 (Fall 1997).

Priest, Graham, J. C. Beall, and Bradley Armour-Garb, eds. *The Law of Non-Contradiction.* Oxford: Clarendon, 2004.

Priest, Graham, Richard Routley [Sylvan], and Jean Norman, eds. *Paraconsistent Logic: Essays on the Inconsistent.* Munich, Germany: Philosophia, 1989.

Read, Stephen. *Relevant Logic.* Oxford: Blackwell, 1988.

Rescher, Nicholas. *Hypothetical Reasoning.* Amsterdam: North-Holland, 1964.

———. *Many-Valued Logic.* New York: McGraw-Hill, 1969.

Rescher, Nicholas, and Robert Brandom. *The Logic of Inconsistency.* Totowa, N.J.: Rowman & Littlefield, 1979.

Rine, David C., ed. *Computer Science and Multiple-Valued Logic: Theory and Applications.* Amsterdam: North-Holland, 1984.

Rosser, John Barkley, and Atwell R. Turquette. *Many-Valued Logics.* Amsterdam: North-Holland, 1952.

Routley [Sylvan], Richard, and Val Routley [Plumwood]. "The Semantics of First-Order Entailment." *Noûs* 6, no. 4 (November 1972): 335–87.

Sanford, David H. *If P, then Q: Conditionals and the Foundations of Reasoning.* London: Routledge, 1989.

Segerberg, Krister. "Notes on Conditional Logic." *Studia Logica* 48, no. 2 (June 1989): 157–68.

Smiley, Timothy, and Graham Priest. "Can Contradictions Be True?" *Aristotelian Society Supplement* 67 (1993): 17–54. [a debate]

Stalnaker, Robert. "A Theory of Conditionals." In *Studies in Logical Theory*, ed. Nicholas Rescher, 98–112. American Philosophical Quarterly Monograph Series, no. 2. Oxford: Blackwell, 1968.

Suber, Peter. *A Bibliography of Non-Standard Logics.* http://www.earlham.edu/~peters/courses/logsys/nonstbib.htm

Sylvan [Routley], Richard, and Jean Norman, eds. *Directions in Relevant Logic.* Dordrecht: Kluwer Academic, 1989.

Tanaka, Koji. "Three Schools of Paraconsistency." *Australasian Journal of Logic* 1 (1 July 2003): 28–42.

Tennant, Neil. *Anti-Realism and Logic: Truth as Eternal.* Oxford: Clarendon, 1987.

Waismann, Friedrich. *How I See Philosophy.* Ed. R. Harré. London: Macmillan, 1968.

Weingartner, Paul, ed. *Alternative Logics: Do Sciences Need Them?* Berlin, Germany: Springer, 2003.

Woods, John. *Paradox and Paraconsistency: Conflict Resolution in the Abstract Sciences.* Cambridge: Cambridge Univ. Pr., 2003.

Woods, Michael. *Conditionals.* Ed. David Wiggins. Oxford: Clarendon, 1997.

Zadeh, Lotfi Asker. *Fuzzy Logic.* Stanford, Calif.: Center for the Study of Language and Information, 1988.

———. *An Introduction to Fuzzy Logic.* Dordrecht: Klewer, 1992.

Zinoviev, Aleksandr. *Philosophical Problems of Many-Valued Logic.* Dordrecht: D. Reidel, 1963.

6K.   Inductive Logic, Probability, and Science

Achinstein, Peter, ed. *The Concept of Evidence.* Oxford: Oxford Univ. Pr., 1983.

Ayer, Alfred Jules. *Probability and Evidence.* London: Macmillan, 1972.

Barker, Stephen Francis. *Induction and Hypothesis: A Study of the Logic of Confirmation.* Ithaca, N.Y.: Cornell, 1957.

Benenson, Frederick C. *Probability, Objectivity and Evidence.* London: Routledge & Kegan Paul, 1984.

Brewka, Gerhard. *Nonmonotonic Reasoning: Logical Foundations of Commonsense.* Cambridge: Cambridge Univ. Pr., 1991.

Brewka, Gerhard, Jürgen Dix, and Kurt Konolige. *Nonmonotonic Reasoning: An*

Overview. Stanford, Calif.: Center for the Study of Language and Information, 1997.

Byrne, Edmund F. *Probability and Opinion: A Study in the Medieval Presuppositions of Post-Medieval Theories of Probability.* The Hague, Netherlands: Martinus Nijhoff, 1968.

Carnap, R. *Logical Foundations of Probability.* 2nd ed. Chicago: Univ. of Chicago Pr., 1962.

―――. "On Inductive Logic." *Philosophy of Science* 12 (1945): 72–97.

Cohen, L. Jonathan. *An Introduction to the Philosophy of Induction and Probability.* Oxford: Clarendon, 1989.

―――. *The Probable and the Provable.* Oxford: Clarendon, 1977.

De Finetti, Bruno. *Theory of Probability.* 2 vols. Trans. Antonio Machi and Adrian Smith. West Sussex, England: John Wiley and Sons, 1974.

Elgin, Catherine Z. *Nelson Goodman's New Riddle of Induction.* New York: Garland, 1997.

Foster, Marguerite H., and Michael L. Martin, eds. *Probability, Confirmation, and Simplicity: Readings in the Philosophy of Inductive Logic.* New York: Odyssey, 1966.

Friedman, Kenneth S. *Predictive Simplicity: Induction Exhum'd.* Oxford: Pergamon, 1990.

Ginsberg, Matthew L. *Readings in Nonmonotonic Reasoning.* Los Altos, Calif.: Morgan Kaufman, 1987.

Goodman, Nelson. *Fact, Fiction, and Forecast.* 4th ed. Cambridge, Mass.: Harvard Univ. Pr., 1983.

Haack, Susan. "The Justification of Deduction." *Mind* 85, no. 337 (January 1976): 112–19.

Hacking, Ian. *The Emergence of Probability: A Philosophical Study of Early Ideas about Probability, Induction and Statistical Inference.* London: Cambridge Univ. Pr., 1975.

―――. *An Introduction to Probability and Inductive Logic.* Cambridge: Cambridge Univ. Pr., 2001.

Harman, Gilbert. "The Inference to the Best Explanation." *Philosophical Review* 74, no. 1 or 409 (January 1965): 88–95.

Hempel, Carl Gustav. *Philosophy of Natural Science.* Englewood Cliffs, N.J.: Prentice-Hall, 1966.

Hintikka, Jaakko, and Patrick Suppes, eds. *Aspects of Inductive Logic.* Amsterdam: North-Holland, 1966.

Hullett, James, and Robert Schwartz. "Grue: Some Remarks." *Journal of Philosophy* 64, no. 9 (11 May 1967): 259–71.

Jeffrey, Richard C. *The Logic of Decision.* New York: McGraw-Hill, 1965.

Josephson, John R., and Susan G. Josephson, eds. *Abductive Inference: Computation, Philosophy, Technology.* Cambridge: Cambridge Univ. Pr., 1994.

Keynes, John Maynard. *The Treatise on Probability.* London: Macmillan, 1929.

Kneale, William Calvert. *Probability and Induction*. Oxford: Clarendon, 1949.

Kuhn, Thomas S. *The Structure of Scientific Revolutions*. 2nd ed. Chicago: Univ. of Chicago Pr., 1970.

Kyburg, Henry Ely, and Howard E. Smokler, eds. *Studies in Subjective Probability*. New York: Wiley, 1964.

Lavrač, Nada, and Sašo Džeroski. *Inductive Logic Programming*. New York: Ellis Horwood, 1994.

Levi, Isaac. *For the Sake of the Argument: Ramsey Test Conditionals, Inductive Inference, and Nonmonotonic Reasoning*. Cambridge: Cambridge Univ. Pr., 1996.

———. *Gambling with Truth; An Essay on Induction and the Aims of Science*. New York: Knopf, 1967.

Mellor, D. H. *The Matter of Chance*. Cambridge: Cambridge Univ. Pr., 1971.

Nagel, Ernest. *The Structure of Science: Problems in the Logic of Scientific Explanation*. New York: Harcourt, Brace and World, 1961.

Popper, Karl Raimund. *Conjectures and Refutations: The Growth of Scientific Knowledge*. New York: Basic Books, 1962.

———. *The Logic of Scientific Discovery*. New York: Basic Books, 1959.

Reichenbach, Hans. *Nomological Statements and Admissible Operations*. Amsterdam: North-Holland, 1954.

Rescher, Nicholas. *Induction: An Essay on the Justification of Inductive Reasoning*. Pittsburgh, Pa.: Univ. of Pittsburgh Pr., 1980.

Skyrms, Brian. *Choice and Chance: An Introduction to Inductive Logic*. 4th ed. Belmont, Calif.: Wadsworth, 2000.

Swain, Marshall, ed. *Induction, Acceptance, and Rational Belief*. Dordrecht: D. Reidel, 1970.

Swinburne, Richard. *The Justification of Induction*. London: Oxford Univ. Pr., 1974.

Walton, Douglas N. *Argumentation Schemes for Presumptive Reasoning*. Mahwah, N.J.: L. Erlbaum, 1996.

Weatherford, Roy. *Philosophical Foundations of Probability Theory*. London: Routledge & Kegan Paul, 1982.

Wisdom, John Oulton. *Foundations of Inference in Natural Science*. London: Methuen, 1952.

Wright, Georg Henrik von. *The Logical Problem of Induction*. 2nd ed. Westport, Conn.: Greenwood, 1957.

———. *A Treatise on Induction and Probability*. Paterson, N.J.: Littlefield Adams, 1960.

6L. Fallacies, Informal Logic, and Critical Thinking

Blair, J. Anthony, and Ralph H. Johnson, eds. *Informal Logic: The First International Symposium*. Inverness, Calif.: Edgepress, 1980.

Copi, Irving M. *Informal Logic*. New York: Macmillan, 1986.

Curtis, Gary N. *Fallacy Files*. http://www.fallacyfiles.org

Downes, Stephen. *Stephen's Guide to the Logical Fallacies*. http://datanation.com/fallacies/index.htm

Engel, S. Morris. *With Good Reason: An Introduction to Informal Fallacies*. New York: St. Martin's, 1976.

Ess, Charles. *Informal Fallacies*. http://www.drury.edu/ess/Logic/Informal/Overview.html

Fearnside, W. Ward, and William B. Holther. *Fallacy: The Counterfeit of Argument*. Englewood Cliffs, N.J.: Prentice-Hall, 1959.

Hansen, Hans V., and Robert C. Pinto, eds. *Fallacies: Classical and Contemporary Readings*. Univ. Park: Pennsylvania State Univ. Pr., 1995.

Levi, Don S. *In Defense of Informal Logic*. Dordrecht: Kluwer Academic, 2000.

Lynch, Michael. *Critical Thinking*. http://www.olemiss.edu/courses/logic/handindx.html

Miller, Lauren, and Michael Connelly. *Critical Thinking across the Curriculum*. http://www.kcmetro.cc.mo.us/longview/ctac/corenotes.htm

Pirie, Madsen. *The Book of the Fallacy*. London: Routledge & Kegan Paul, 1985.

Toulmin, Stephen Edelston. *The Uses of Argument*. Cambridge: Cambridge Univ. Pr., 1958.

Trelogan, T. K. *Arguments and Their Evaluation*. http://www.univnorthco.edu/philosophy/arg.html

Walton, Douglas N. *Argument Structure: A Pragmatic Theory*. Toronto: Univ. of Toronto Pr., 1996.

———. *Informal Logic: A Handbook for Critical Argumentation*. Cambridge: Cambridge Univ. Pr., 1989.

Whyte, Jamie. *Crimes against Logic: Exposing the Bogus Arguments of Politicians, Priests, Journalists, and Other Serial Offenders*. New York: McGraw-Hill, 2004.

6M.   Truth and Paradoxes

Barwise, Jon and John Etchemendy. *The Liar: An Essay on Truth and Circularity*. New York: Oxford Univ. Pr., 1987.

Blackburn, Simon, and Keith Simmons. *Truth*. Oxford: Oxford Univ. Pr., 1999.

Cargile, James. *Paradoxes: A Study in Form and Predication*. Cambridge: Cambridge Univ. Pr., 1979.

Copi, Irving M. *The Theory of Logical Types*. London: Routledge & Kegan Paul, 1971.

Davidson, Donald. *Inquiries into Truth and Interpretation*. 2nd ed. Oxford: Clarendon, 2001.

Dummett, Michael A. E. *Truth and Other Enigmas*. Cambridge, Mass.: Harvard Univ. Pr., 1978.

Erickson, Glenn, W., and John A. Fossa. *Dictionary of Paradox*. Lanham, Md.: Univ. Pr. of America, 1998.

Fine, Kit. "Vagueness, Truth and Logic." *Synthese* 30, no. 3/4 (April/May 1975): 265–300.

Grover, Dorothy. *A Prosentential Theory of Truth*. Princeton, N.J.: Princeton Univ. Pr., 1992.

Hyde, Dominic. "From Heaps and Gaps to Heaps of Gluts." *Mind* 106, no. 424 (October 1997): 641–60.

Kirkham, Richard L. *Theories of Truth: A Critical Introduction*. Cambridge, Mass.: MIT, 1992.

Kripke, Saul. "Outline of a Theory of Truth." *Journal of Philosophy* 72, no. 19 (6 November 1975): 690–716. [also in the Martin anthology]

Martin, Robert L. *The Paradox of the Liar*. New Haven, Conn.: Yale, 1970.

———, ed. *Recent Essays on Truth and the Liar Paradox*. Oxford: Clarendon, 1984.

McGee, Vann. *Truth, Vagueness, and Paradox*. Indianapolis, Ind.: Hackett, 1990.

Ottossøn, Jenny. *Paradoxes*. http://www.algonet.se/~jen-tale/para_gen.html

Sainsbury, Richard Mark. *Paradoxes*. Cambridge: Cambridge Univ. Pr., 1995.

Simmons, Keith. *Universality and the Liar: An Essay on Truth and the Diagonal Argument*. Cambridge: Cambridge Univ. Pr., 1993.

Soames, Scott. *Understanding Truth*. New York: Oxford Univ. Pr., 1999.

Tarski, Alfred. "The Semantic Conception of Truth and the Foundations of Semantics." *Philosophy and Phenomenological Research* 4 (1944): 341–76.

6N. Ontology and Abstract Entities

Armstrong, David Malet. *Universals and Scientific Realism*. 2 vols. Cambridge: Cambridge Univ. Pr., 1978.

———. *Universals: An Opinionated Introduction*. Boulder, Colo.: Westview, 1989.

Bigelow, John. *The Reality of Numbers: A Physicalist's Philosophy of Mathematics*. Oxford: Clarendon, 1988.

Bigelow, John, and Robert Pargetter. *Science and Necessity*. Cambridge: Cambridge Univ. Pr., 1990.

Blackburn, Simon. *Essays in Quasi-Realism*. Oxford: Clarendon, 1993.

———. *Spreading the Word: Groundings in the Philosophy of Language*. Oxford: Clarendon, 1984.

Field, Hartry H. *Realism, Mathematics and Modality*. Oxford: Blackwell, 1989.

———. *Science without Numbers: A Defence of Nominalism*. Princeton, N.J.: Princeton Univ. Pr., 1980.

Goodman, Nelson, and Willard Van Orman Quine. "Steps toward a Constructive Nominalism." *Journal of Symbolic Logic* 11, no. 4 (December 1947): 105–22.

Hale, Bob. *Abstract Objects*. Oxford: Blackwell, 1987.

Hellman, Geoffrey. *Mathematics without Numbers*. Oxford: Clarendon, 1989.
Jubien, Michael. *Ontology, Modality, and the Fallacy of Reference*. Cambridge: Cambridge Univ. Pr., 1993.
Lewis, David K. "Truth in Fiction." *American Philosophical Quarterly* 15, no. 1 (January 1978): 37–46.
Loux, Michael J., ed. *Universals and Particulars: Readings in Ontology*. 2nd ed. Notre Dame, Ind.: Univ. of Notre Dame Pr., 1976.
Maddy, Penelope. *Realism in Mathematics*. Oxford: Clarendon, 1990.
McGinn, Colin. *Logical Properties: Identity, Existence, Predication, Necessity, Truth*. Oxford: Clarendon, 2000.
Moreland, James Porter. *Universals*. Montreal: McGill-Queen's Univ. Pr., 2001.
Munitz, Milton K., ed. *Logic and Ontology*. New York: New York Univ. Pr., 1973.
Simons, Peter M. *Parts: A Study in Ontology*. Oxford: Clarendon, 1987.
Teichmann, Roger. *Abstract Entities*. New York: St. Martin's, 1992.
Wright, Crispin. *Realism, Meaning, and Truth*. 2nd ed. Oxford: Blackwell, 1993.

6O. Philosophy of Language

Austin, John Langshaw. *How to Do Things with Words*. Ed. J. O. Urmson and Marina Sbisà. Cambridge, Mass.: Harvard Univ. Pr., 1975.
Black, Max. *Margins of Precision: Essays in Logic and Language*. Ithaca, N.Y.: Cornell, 1970.
Carnap, Rudolf. *The Logical Syntax of Language*. London: Routledge & Kegan Paul, 1937.
Dowty, David R., Robert E. Wall, and Stanley Peters. *Introduction to Montague Semantics*. Dordrecht: D. Reidel, 1981.
Evans, Gareth, and John McDowell, eds. *Truth and Meaning: Essays in Semantics*. Oxford: Clarendon, 1976.
Gaskin, Richard, ed. *Grammar in Early Twentieth-Century Philosophy*. London: Routledge, 2001.
Keefe, Rosanna, and Peter Smith. *Vagueness: A Reader*. Cambridge, Mass.: MIT, 1996.
Linsky, Leonard. *Names and Descriptions*. Chicago: Univ. of Chicago Pr., 1977.
Lycan, William G. *Logical Form in Natural Language*. Cambridge, Mass.: MIT, 1984.
Machina, K. "Truth, Belief and Vagueness." *Journal of Philosophical Logic* 5, no. 1 (February 1976): 47–78.
Martin, Richard Milton. *Semiotics and Linguistic Structure: A Primer of Philosophic Logic*. Albany, N.Y.: SUNY, 1978.
Martinich Aloysius P., ed. *The Philosophy of Language*. 4th ed. Oxford: Oxford Univ. Pr., 2000.
Moore, A. W., ed. *Meaning and Reference*. Oxford: Oxford Univ. Pr., 1993.
Moser, Paul K., ed. *A Priori Knowledge*. Oxford: Oxford Univ. Pr., 1987.

O'Hear, Anthony, ed. *Logic, Thought and Language*. Cambridge: Cambridge Univ. Pr., 2002.

Platts, Mark de Bretton. *Ways of Meaning: An Introduction to a Philosophy of Language*. London: Routledge & Kegan Paul, 1979.

Proust, Joëlle. *Questions of Formal Logic and the Analytic Proposition from Kant to Carnap*. Ed. Anastasios Albert Brenner. Minneapolis: Univ. of Minnesota Pr., 1989.

Searle, John R. *Speech Acts: An Essay in the Philosophy of Language*. London: Cambridge Univ. Pr., 1969.

Searle, John R., and Daniel Vanderveken. *Foundations of Illocutionary Logic*. Cambridge: Cambridge Univ. Pr., 1985.

Waismann, Friedrich. *How I See Philosophy*. Ed. R. Harré. London: Macmillan, 1968.

Williamson, Timothy. *Vagueness*. London: Routledge, 1994.

6P.   Continental Philosophy and Logic

See also Hegel, Marx, and Husserl in 2G (Nineteenth-Century Logic).

Fay, Thomas A. *Heidegger: The Critique of Logic*. The Hague, Netherlands: Nijhoff, 1977.

Gaskin, Richard, ed. *Grammar in Early Twentieth-Century Philosophy*. London: Routledge, 2001.

Harris, Errol E. *Formal, Transcendental, and Dialectical Thinking: Logic and Reality*. Albany, N.Y.: SUNY, 1987.

Heidegger, Martin. *The Metaphysical Foundations of Logic*. Trans. Michael Heim. Bloomington: Indiana Univ. Pr., 1984.

Mader, Mary Beth. "Suffering Contradiction: Kofman on Nietzsche's Critique of Logic." In *Enigmas: Essays on Sarah Kofman*, ed. Penelope Deutscher and Kelly Oliver, 87–96. Ithaca, N.Y.: Cornell, 1999.

Marquit, Erwin, Philip Moran, and Willis H. Truitt, eds. *Dialectical Contradictions: Contemporary Marxist Discussions*. Minneapolis, Minn.: Marxist Educational, 1982. [a defense of Marxism on the reality of contradictions]

Marxists.org Internet Archive. http://www.marxists.org. Click Search at the bottom and then search for "dialectical logic" or just "logic." See especially "Contradiction and Harmony," A. Spirkin, http://www.marxists.org/reference/archive/spirkin/works/dialectical-materialism/ch02-s11.html, and "Pre-Logic, Formal Logic, Dialectical Logic," Mustafa Cemal, http://www.marxists.org/reference/archive/hegel/txt/system2.htm.

Patočka, Jan. *An Introduction to Husserl's Phenomenology*. Trans. Erazim Kohák. Ed. James Dodd. Chicago: Open Court, 1996. [much on Husserl's logic]

Seebohm, Thomas M., Dagfinn Føllesdal, and Jitendra Nath Mohanty, eds. *Phenomenology and the Formal Sciences*. Dordrecht: Kluwer Academic, 1991.

Solomon, Robert C., ed. *Phenomenology and Existentialism*. New York: Harper & Row, 1972. [much on logic]

Tieszen, Richard. "Gödel's Path from the Incompleteness Theorems (1931) to Phenomenology (1961)." *Bulletin of Symbolic Logic* 4, no. 2 (June 1998): 181–203. [about Gödel's late interest in Husserl]

Tragesser, Robert S. *Phenomenology and Logic*. Ithaca, N.Y.: Cornell, 1977.

White, David A. *Logic and Ontology in Heidegger*. Columbus: Ohio State Univ. Pr., 1985.

6Q. Gender and Logic

Falmagne, Rachel Joffe, and Marjorie Hass, eds. *Representing Reason: Feminist Theory and Formal Logic*. Lanham, Md.: Rowman & Littlefield, 2002.

Hass, Marjorie. "Feminist Readings of Aristotelian Logic." In *Feminist Interpretations of Aristotle*, ed. Cynthia A. Freeland, 19–40. Univ. Park: Pennsylvania State Univ. Pr., 1998.

Irigaray, Luce. "Is the Subject of Science Sexed?" In *Feminism and Science*, ed. Nancy Tuana, 58–68. Bloomington: Indiana Univ. Pr., 1989.

Nye, Andrea. *Words of Power: A Feminist Reading of the History of Logic*. New York: Routledge, 1990. [Nye sees logic as male and unintelligible.]

Plumwood [Routley], Val. "The Politics of Reason: Toward a Feminist Logic." *Australasian Journal of Philosophy* 71, no. 4 (December 1993): 436–62.

6R. African Thought and Intercultural Issues

Apostel, Leo. "African Logic." In *African Philosophy: Myth or Reality?* chap. 6, 181–213. Gent, Belgium: Scientific, 1981.

Gyekye, Kwame. "Philosophy, Logic and the Akan Language." In *Readings in African Philosophy: An Akan Collection*, ed. Safro Kwame, 69–83. Lanham, Md.: Univ. Pr. of America, 1995.

Sogolo, Godwin S. "Logic and Rationality." In *The African Philosophy Reader*, ed. P. H. Coetzee and A. P. J. Roux, 217–33. London: Routledge, 1998.

Wiredu, Kwasi. *Cultural Universals and Particulars: An African Perspective*. Bloomington: Indiana Univ. Pr., 1996. [See especially chapter 7, "Formulating Modern Thought in African Languages," pp. 81–104, and chapter 3, "Are There Cultural Universals?" pp. 21–33.]

6S. Logic Puzzles

Angiolino, Andrea. *Mind-Sharpening Logic Games*. Pittstown, N.J.: Main Street, 2003.

Gardner, Martin. *My Best Mathematical and Logic Puzzles*. New York: Dover,

1994.

Killoran, David M. *LSAT Logic Games Bible*. Hilton Head Island, S.C.: Power-score, 2004.

Smullyan, Raymond M. *To Mock a Mockingbird and Other Logic Puzzles*. New York: Knopf, 1985.

———. *What is the Name of this Book?* Englewood Cliffs, N.J.: Prentice-Hall, 1978.

7. Other

Agazzi, Evandro, ed. *Modern Logic: A Survey*. Dordrecht: D. Reidel, 1981.

Ayer, Alfred Jules. *Language, Truth, and Logic*. New York: Dover, 1946.

Baggini, Julian, and Peter S. Fosl. *The Philosopher's Toolkit: A Compendium of Philosophical Concepts and Methods*. Malden, Mass.: Blackwell, 2003.

Barwise, Jon, and Gerard Allwein, eds. *Logical Reasoning with Diagrams*. Oxford: Oxford Univ. Pr., 1996.

Barwise, Jon, and John Perry. *Situations and Attitudes*. Cambridge, Mass.: MIT, 1983.

Belnap, Nuel. D. "Tonk, Plonk and Plink." *Analysis* 22, no. 6 (June 1962): 130–34.

Belnap, Nuel D., and Thomas B. Steel Jr. *The Logic of Questions and Answers*. New Haven, Conn.: Yale, 1976.

Blanché, Robert. *Axiomatics*. Trans. G. B. Keene. London: Routledge & Kegan Paul, 1962.

Boolos, George. *Logic, Logic, and Logic*. Ed. Richard Jeffrey. Cambridge, Mass.: Harvard Univ. Pr., 1998.

Burks, Arthur Walter. *Chance, Cause, Reason: An Inquiry into the Nature of Scientific Evidence*. Chicago: Univ. of Chicago Pr., 1977. [Among other things, this includes his modal logic of causal statements, in chapters 6–7.]

Carnap, Rudolf. *The Logical Structure of the World: Pseudoproblems in Philosophy*. Trans. Rolf A. George. Berkeley: Univ. of California, 1967.

Cherniak, Christopher. *Minimal Rationality*. Cambridge, Mass.: MIT, 1986.

Chihara, Charles S. *Ontology and the Vicious-Circle Principle*. Ithaca, N.Y.: Cornell, 1973.

Cocchiarella, Nino B. *Logical Studies in Early Analytic Philosophy*. Columbus: Ohio State Univ. Pr., 1986.

Copeland, B. J. *Logic and Reality: Essays on the Legacy of Arthur Prior*. Oxford: Clarendon, 1996.

Copi, Irving M., and James A. Gould, eds. *Contemporary Philosophical Logic*. New York: St. Martin's, 1978.

———. *Contemporary Readings in Logical Theory*. New York: Macmillan, 1967.

Cornman, James W., et al. *Studies in Logical Theory*. Oxford: Blackwell, 1968.

Davis, J. W., D. J. Hockney, and W. K. Wilson, eds. *Philosophical Logic.* Dordrecht: D. Reidel, 1969.

Dummett, Michael A. E. *The Logical Basis of Metaphysics.* Cambridge, Mass.: Harvard Univ. Pr., 1991.

Engel, Pascal. *The Norm of Truth: An Introduction to the Philosophy of Logic.* Trans. Miriam Kochan and Pascal Engel, Toronto: Univ. of Toronto Pr., 1991.

Englebretsen, George. *Essays on the Philosophy of Fred Sommers: In Logical Terms.* Lewiston, N.Y.: E. Mellen, 1990.

———. *Line Diagrams for Logic: Drawing Conclusions.* Lewiston, N.Y.: E. Mellen, 1998.

———. *Logical Negation.* Assen, Netherlands: Van Gorcum, 1981.

Fetzer, James H., ed. *Principles of Philosophical Reasoning.* Totowa, N.J.: Rowman & Allanheld, 1984.

Fraassen, Bas C. van. *Formal Semantics and Logic.* New York: Macmillan, 1971.

Gabbay, Dov M., and Wansing, Heinrich, eds. *What Is Negation?* Dordrecht: Kluwer Academic, 1999.

Geach, Peter Thomas. *Logic Matters.* Berkeley: Univ. of California, 1972.

———. *Reason and Argument.* Berkeley: Univ. of California, 1976.

Gentzen, Gerhard. *The Collected Papers of Gerhard Gentzen.* Ed. M. E. Szabo. Amsterdam: North-Holland, 1969.

Grayling, A. C. *An Introduction to Philosophical Logic.* Sussex, England: Harvester, 1982.

Haack, Susan. *Manifesto of a Passionate Moderate: Unfashionable Essays.* Chicago: Univ. of Chicago Pr., 1999.

Harman, Gilbert. *Change in View: Principles of Reasoning.* Cambridge, Mass.: MIT, 1986.

Heath, Peter L. "Nothing." In *Encyclopedia of Philosophy,* 8 vols., ed. Paul Edwards, 8:524–25. New York: Macmillan and the Free Press, 1967.

Heijenoort, Jean van. *Selected Essays.* Naples, Italy: Bibliopolis, 1985.

Hempel, Carl G. "Problems and Changes in the Empiricist Criterion of Meaning." *Revue Internationale de Philosophie* 4, no. 11 (January 1950): 41–63.

Hughes, R. I. G., ed. *A Philosophical Companion to First-Order Logic.* Indianapolis, Ind.: Hackett, 1993.

Jackson, Frank, and Graham Priest, eds. *Lewisian Themes: The Philosophy of David K. Lewis.* Oxford: Clarendon, 2004.

Kaplan, David. "On the Logic of Demonstratives." *Journal of Philosophical Logic* 8, no. 1 (February 1979): 81–98.

Körner, Stephan, ed. *Philosophy of Logic: Papers and Discussions from the 1974 Bristol Conference on Critical Philosophy.* Berkeley: Univ. of California, 1976.

Lambert, Karel, ed. *The Logical Way of Doing Things.* New Haven, Conn.: Yale, 1969.

Lambert, Karel, and Bas C. van Fraassen. *Derivation and Counterexample: An Introduction to Philosophical Logic.* Encino, Calif.: Dickenson, 1972.

Leblanc, Hugues, ed. *Truth, Syntax and Modality*. Amsterdam: North-Holland, 1973.

Leśniewski, Stanisław. *Collected Works*. 2 vols. Ed. Stanisław J. Surma, Jon T. Srzednicki, and D. I. Barnett. Dordrecht: Kluwer Academic, 1992.

Lewis, David K. *Convention: A Philosophical Study*. Cambridge, Mass.: Harvard Univ. Pr., 1969.

———. *Papers in Philosophical Logic*. Cambridge: Cambridge Univ. Pr., 1998.

———. *Philosophical Papers*. 2 vols. Oxford: Oxford Univ. Pr., 1983 and 1986.

Łukasiewicz, Jan. *Selected Works*. Trans. Olgierd Wojtasiewicz. Ed. L. Borkowski. Amsterdam: North-Holland, 1970.

Mackie, John Leslie. *Truth, Probability and Paradox: Studies in Philosophical Logic*. Oxford: Clarendon, 1973.

Martin, Richard Milton. *Mind, Modality, Meaning, and Method*. Albany, N.Y.: SUNY, 1983.

Martin, Robert M. *There are Two Errors in the the Title of This Book: A Sourcebook of Philosophical Puzzles, Problems, and Paradoxes*. Revised ed. Peterborough, Ont.: Broadview, 2002.

McCall, Storrs, ed. *Polish Logic: 1920–1939 Papers by Ajdukiewicz and Others*. Trans. B. Gruchman et al. Oxford: Clarendon, 1967.

Montague, Richard. *Formal Philosophy: Selected Papers of Richard Montague*. Ed. Richmond H. Thomason. New Haven, Conn.: Yale, 1974.

Moore, Robert C. *Logic and Representation*. Stanford, Calif.: Center for the Study of Language and Information, 1995.

Orenstein, Alex. *Existence and the Particular Quantifier*. Philadelphia: Temple Univ. Pr., 1978.

Pollock, John L. *Technical Methods in Philosophy*. Boulder, Colo.: Westview, 1990.

Pospesel, Howard, and David Marans. *Arguments: Deductive Logic Exercises*. 2nd ed. Englewood Cliffs, N.J.: Prentice-Hall, 1978. [A collection of real-life arguments that can be analyzed using the tools of modern logic.]

Putnam, Hilary. *Philosophy of Logic*. New York: Harper & Row, 1971.

Ramsey, Frank Plumpton. *Foundations: Essays in Philosophy, Logic, Mathematics, and Economics*. Ed. D. H. Mellor. Atlantic Highlands, N.J.: Humanities, 1978.

Read, Stephen. *Thinking about Logic: An Introduction to the Philosophy of Logic*. Oxford: Oxford Univ. Pr., 1995.

Rescher, Nicholas. *Topics in Philosophical Logic*. Dordrecht: D. Reidel, 1969. [See especially his "Map of Logic," pp. 6–9.]

Richter, Duncan. *Historical Dictionary of Wittgenstein's Philosophy*. Lanham, Md.: Scarecrow Pr., 2004.

Shanker, Stuart G., ed. *Philosophy of Science, Logic, and Mathematics in the Twentieth Century*. London: Routledge, 1996. [This has a good chronology.]

Shapiro, Stewart. *Foundations without Foundationalism: A Case for Second-Order*

Logic. Oxford: Clarendon, 1991.

Shoesmith, D. J., and T. J. Smiley. *Multiple-Conclusion Logic.* Cambridge: Cambridge Univ. Pr., 1978.

Simons, Peter M. *Philosophy and Logic in Central Europe from Bolzano to Tarski: Selected Essays.* Dordrecht: Kluwer Academic, 1992.

Smullyan, Raymond M. *First-Order Logic.* New York: Springer-Verlag, 1968.

Sorensen, Roy A. *Blindspots.* Oxford: Clarendon, 1988.

———. *Thought Experiments.* Oxford: Oxford Univ. Pr., 1992.

Stalnaker, Robert. *Inquiry.* Cambridge, Mass.: MIT, 1984.

Strawson, Peter F. *Introduction to Logical Theory.* London: Methuen, 1952.

———. *Logico-Linguistic Papers.* London: Methuen, 1971.

Strawson, Peter F., ed. *Philosophical Logic.* London: Oxford Univ. Pr., 1967.

Tarski, Alfred. *Logic, Semantics, Metamathematics: Papers from 1923 to 1938.* 2nd ed. Trans. J. H. Woodger. Ed. John Corcoran. Indianapolis, Ind.: Hackett, 1983. [This has Tarski's famous paper, "The Semantic Conception of Truth and the Foundations of Semantics," first published in English in *Philosophy and Phenomenological Research* 4 (1944): 341–76.]

Tennant, Neil. *Natural Logic.* Edinburgh, Scotland: Edinburgh Univ. Pr., 1978.

Wang, Hao. *Beyond Analytic Philosophy: Doing Justice to What We Know.* Cambridge, Mass.: MIT, 1986.

Weston, Anthony. *A Rulebook for Arguments.* Indianapolis, Ind.: Hackett, 1992.

Wittgenstein, Ludwig. *Philosophical Investigations.* Trans. G. E. M. Anscombe. New York: Macmillan, 1953.

———. *Tractatus Logico-Philosophicus.* Trans. D. F. Pears and B. F. McGuinness. London: Routledge & Kegan Paul, 1961.

Wolfram, Sybil. *Philosophical Logic: An Introduction.* London: Routledge, 1989.

Wright, Georg Henrik von. *Philosophical Logic.* Ithaca, N.Y.: Cornell, 1983.

———. *Truth, Knowledge, and Modality.* Oxford: Blackwell, 1984.

———, ed. *Logic and Philosophy.* The Hague, Netherlands: M. Nijhoff, 1980.